Contributions to Environmental Sciences & Innovative Business Technology

Contributions to Environmental Sciences & Innovative Business Technology (CESIBT) is an interdisciplinary series of peer-reviewed books dedicated to addressing emerging research trends relevant to the interplay between Environmental Sciences, Innovation, and Business Technology in their broadest sense. This series constitutes a comprehensive up-to-date interdisciplinary reference that develops integrated concepts for sustainability and discusses the emerging trends and practices that will define the future of these disciplines.

This series publishes the latest developments and research in the various areas of Environmental Sciences, Innovation, and Business Technology, combined with scientific quality and timeliness. It encompasses the theoretical, practical, and methodological aspects of all branches of these scientific disciplines embedded in the fields of Environmental Sciences, Innovation, and Business Technology.

The series also draws on the best research papers from EuroMid Academy of Business and Technology (EMABT) and other international conferences to foster the creation and development of sustainable solutions for local and international organizations worldwide. By including interdisciplinary contributions, this series introduces innovative tools that can best support and shape both the economical and sustainability agenda for the welfare of all countries, through better use of data, a more effective organization, and global, local, and individual work. The series can also present new case studies in real-world settings offering solid examples of recent innovations and business technology with special consideration for resolving environmental issues in different regions of the world.

The series can be beneficial to researchers, instructors, practitioners, consultants, and industrial experts, in addition to governments from around the world. Published in collaboration with EMABT, the Springer CESIBT series will bring together the latest research that addresses key challenges and issues in the domain of Environmental Sciences & Innovative Business Technology for sustainable development.

Maria José Sousa ·
João Rodrigues dos Santos ·
Miguel Sousa · Subhendu Kumar Pani
Editors

Economic Sustainability Potentiated by Artificial Intelligence

Springer

Editors
Maria José Sousa
University of Lisbon
Lisbon, Portugal

Miguel Sousa
University of Essex
Colchester, UK

João Rodrigues dos Santos
Center for Transdisciplinary Development Studies
Europeia University
Lisbon, Portugal

Subhendu Kumar Pani
Krupajal Engineering College
Bhubaneswar, Odisha, India

ISSN 2731-8303 ISSN 2731-8311 (electronic)
Contributions to Environmental Sciences & Innovative Business Technology
ISBN 978-981-96-8088-7 ISBN 978-981-96-8089-4 (eBook)
https://doi.org/10.1007/978-981-96-8089-4

This Springer imprint is published by the registered company Springer Nature Singapore Pte Ltd.
The registered company address is: 152 Beach Road, #21-01/04 Gateway East, Singapore 189721, Singapore

Foreword

The transformative potential of artificial intelligence (AI) in promoting economic sustainability is both unprecedented and critically important in today's rapidly evolving global landscape. From climate change to widening socioeconomic disparities, we face complex challenges that demand innovative solutions. AI, with its capacity to analyze vast amounts of data and generate actionable insights, stands out as a powerful enabler—optimizing processes, enhancing decision-making, and driving sustainable growth across diverse industries. Applications of AI extend from advanced data analytics to real-time monitoring of supply chains, predictive maintenance of critical infrastructures, and novel approaches in renewable energy management.

This book, *Economic Sustainability Potentiated by Artificial Intelligence*, offers a timely and comprehensive exploration of how AI can underpin the transition to more sustainable economic models—objectives highlighted not only in the United Nations' 2030 Agenda for Sustainable Development but also in the European Green Deal policies. It illustrates how intelligent systems facilitate decarbonization, improve resource efficiency, foster inclusive development, and support a broad range of specific contexts—from social economy and ethical concerns in aviation to agriculture and wine tourism. Moreover, the book extends to exploring AI-driven digital commerce for sustainable urban development, immersive learning in virtual reality for soft skills, engineering education in a sustainable context, and the transformative role of AI-based participatory budgeting. The book also explores how digital leadership reshapes human resource management in public administration and how IT project management can be effectively implemented in Portuguese public administration, highlighting the role of governance in guiding AI deployment.

Yet, the convergence of AI and sustainability also presents challenges. Ethical governance, data privacy and security, and the carbon footprint of AI systems themselves are increasingly pressing concerns. Integrating AI into policy frameworks requires careful debate—balancing technological capabilities with the preservation of human values, social equity, and environmental responsibility. In this context, Europe's evolving regulatory landscape, including the AI Act, underscores the need for clear guidelines and robust oversight to ensure AI's responsible development.

More than a theoretical exploration, this book provides actionable insights and real-world case studies, bridging the gap between academic research, industry practice, and policymaking. It illuminates how businesses are leveraging AI to create circular economic models, how governments are incorporating AI-driven evidence into public policy, and how local communities are benefiting from more efficient and data-informed resource management. By bringing together diverse perspectives, it empowers readers to harness AI's potential in ways that support ethical principles and catalyze economic resilience.

At this fundamental moment in human history, *Economic Sustainability Potentiated by Artificial Intelligence* serves as both a guide and a call to action. While AI can undoubtedly amplify our capabilities, it is our collective vision, rigorous governance, and commitment to equitable progress that will determine how sustainable our chosen pathway becomes. I am confident that the forward-looking ideas and strategies set forth in this book will inspire innovation, encourage collaboration across sectors and disciplines, and, ultimately, help shape a more sustainable global economy.

Gabriel Osório de Barros
Economist and Deputy
Director-General, PLANAPP—Centre
for Planning, Policy and Foresight
in Public Administration
Lisbon, Portugal

About This Book

The book *Economic Sustainability Potentiated by Artificial Intelligence* makes an in-depth analysis of the impact of artificial intelligence (AI) on the contemporary economic landscape. It highlights how AI revolutionizes industrial processes, and also how it plays a crucial role in promoting economic sustainability. Several sectors in which AI fits, from business strategies to efficient resource management, are explored. Theoretical and field research reveal how the synergy between technology and the economy can boost sustainable development.

Throughout the pages of this book, readers will be guided through an enlightening narrative on how the responsible integration of AI can maximize efficiency and promote ethical business practices. This book not only explores the current state of the intersection between AI and economics, but also casts a visionary eye to the future, inspiring business leaders, academics, and decision-makers to embrace artificial intelligence as a driving force for lasting economic sustainability. Moreover, the book intends to bridge the gap between academia and the industry, presenting innovative frameworks and practical solutions to promote sustainable economy based on artificial intelligence and digitalization.

Contents

About the Editors

Maria José Sousa has a Ph.D. in Management, with Aggregation in Management and Public Management, is a lecturer and researcher at ISCTE/University Institute of Lisbon and was a pro-rector between March 2022 and January 2024. She is a member of the Advisory Board of Banco Português de Fomento for Public Policies and a member of the scientific committee of the Ph.D. "in Law and Economics" (Cycle XXXIX), Università degli Studi del Molise. She is the coordinator of several European innovation projects (funded by programs such as Erasmus+, Horizon Europe and ICMPD) and is also an external expert of COST Association—European Cooperation in Science and Technology.

João Rodrigues dos Santos is a university professor at Instituto Superior de Educação e Ciências (ISEC Lisboa) and at IADE/Univeridade Europeia. He has a Ph.D. in Economics and Management (Universidad Europea de Madrid), a Master's degree in International Security and Civil Protection Management, and a degree in Geography and Territorial Planning (Universidade Nova de Lisboa). As an integrated researcher, he is a member of the Center for Transdisciplinary Development Studies at the University of Trás-os-Montes and Alto Douro. At ISEC Lisboa, and he directs the bachelor's and master's degrees in Management and Public Administration. He is the author of two books in two of his areas of interest (Economics and Territorial Management) and frequently publishes chapters and articles of an academic, technical, and scientific nature. João Rodrigues dos Santos currently collaborates with CNN Portugal, analyzing and commenting on national and international economic news.

Miguel Sousa is a master student at University of Northumbria, UK, and a bachelor's in business and management from University of Essex. His interests are in economics/business studies, digital marketing, and information systems. He is a co-author of several articles about digitalization published in journals such as *Neural Computer and Applications*, *Transnational Corporations Review*, and *Journal of Medical Systems.*

Subhendu Kumar Pani received the Ph.D. degree from Utkal University, Odisha, India, in 2013. He is currently working as the principal with Krupajal Computer Academy (KCA), Bhubaneswar. He has published 51 International Journal articles (25 Scopus index). His professional activities include roles as the book series editor (CRC Press, Apple Academic Press, and Wiley-Scrivener), an associate editor, an editorial board member, and a reviewer of various international journals. He is an associate with number of conference societies. He has more than 150 international publications, five authored books, 15 edited and upcoming books, and 20 book chapters into his account. His research interests include data mining, big data analysis, Web data analytics, fuzzy decision-making, and computational intelligence. He is a fellow in SSARS Canada and a life member in IE, ISTE, ISCA, OBA.OMS, SMIACSIT, SMUACEE, and CSI. He was a recipient of five researcher awards.

Economic Sustainability Potentiated by Artificial Intelligence

1

Maria José Sousa, João Rodrigues dos Santos, Miguel Sousa, and Subhendu Kumar Pani

Abstract

This review examines the interplay of artificial intelligence (AI) and related technologies in fostering socioeconomic resilience, sustainability, and innovation. It highlights how AI-driven systems optimize resources, enhance decision-making, and promote inclusivity in governance and policymaking. From enabling participatory budgeting to addressing digital divides, AI demonstrates potential in transforming public administration and urban management. The role of AI in advancing the circular economy and reducing environmental impacts is emphasized, alongside its applications in automation for energy efficiency and waste reduction. Ethical considerations, such as fairness and transparency, remain crucial in ensuring responsible AI deployment. The review also explores smart cities as a convergence of these advancements, showcasing AI's capacity to improve urban living conditions. While these technologies promise economic growth and environmental benefits, challenges such

M. J. Sousa (✉)
University of Lisbon, Lisbon, Portugal
e-mail: maria.jose.sousa@iscte-iul.pt

J. R. dos Santos
Universidade Europeia, Lisbon, Portugal
e-mail: joao-rodrigues.santos@universidadeeuropeia.pt

Centro de Estudos Transdisciplinares para o Desenvolvimento da Universidade de Trás-os-Montes, Vila Real, Portugal

M. Sousa
University of Essex, Colchester, UK

S. K. Pani
Krupajal Computer Academy, Bhailo, India

M. J. Sousa et al. (eds.), *Economic Sustainability Potentiated by Artificial Intelligence*, Contributions to Environmental Sciences & Innovative Business Technology,
https://doi.org/10.1007/978-981-96-8089-4_1

as job displacement and high energy consumption underscore the need for balanced approaches. This synthesis provides a foundation for understanding AI's transformative role across sectors.

Keywords

Artificial intelligence (AI) • Sustainable economy • Digitalisation • Ethics • Policy • Business

1.1 Introduction

Economic sustainability has emerged as a pivotal concept in the twenty-first century, emphasizing the need to balance growth with long-term environmental, social, and financial stability. As global challenges such as climate change, resource depletion, and social inequality intensify, societies increasingly seek innovative solutions to build resilient economies. Artificial intelligence (AI) has proven to be one of the most transformative technologies of our time, offering unparalleled potential to address these challenges. By optimizing resource use, enabling better decision-making, and fostering inclusive economic practices, AI serves as a critical enabler of sustainable economic development. This introduction explores how AI can potentiate economic sustainability, laying the foundation for a deeper discussion of its applications, benefits, and challenges.

Economic sustainability refers to the ability of an economy to support a defined level of economic production indefinitely while maintaining environmental integrity and ensuring social equity. It involves the efficient use of resources, reducing waste, and fostering innovation to meet present needs without compromising the ability of future generations to meet their own. The principles of economic sustainability align closely with the United Nations' Sustainable Development Goals (SDGs), which advocate for inclusive and sustainable economic growth, full and productive employment, and responsible consumption and production practices (United Nations 2015).

Achieving economic sustainability requires integrating environmental, social, and economic considerations into decision-making processes. This integration presents significant challenges, particularly in a world characterized by rapid urbanization, technological advancements, and globalization. However, advancements in AI offer promising solutions to these challenges, enabling governments, businesses, and communities to adopt more sustainable practices.

AI encompasses a range of technologies, including machine learning, natural language processing, and computer vision, that enable machines to perform tasks traditionally requiring human intelligence. These technologies have already demonstrated their capacity to revolutionize industries by automating processes, improving efficiency, and enabling data-driven insights. In the context of economic sustainability, AI's transformative power lies in its ability to analyze vast amounts of data, identify patterns, and make predictions that inform sustainable decision-making.

One of AI's most significant contributions to economic sustainability is its potential to optimize resource allocation. For instance, AI algorithms can analyze energy consumption patterns in real time, allowing utilities to distribute power more efficiently and reduce waste. Similarly, AI-powered predictive analytics can enhance supply chain management, minimizing overproduction and reducing the environmental footprint of industries (Kumar et al. 2020).

AI also plays a crucial role in advancing environmental sustainability, a cornerstone of economic resilience. By monitoring and analyzing environmental data, AI can help detect and mitigate the effects of climate change. For example, AI-driven models are used to predict weather patterns, monitor deforestation, and optimize the deployment of renewable energy systems (Rolnick et al. 2019). These applications not only protect natural resources but also create economic opportunities in green technology sectors.

Economic sustainability is not solely about environmental considerations; it also requires addressing social inequities and ensuring broad-based economic participation. AI can contribute to this goal by democratizing access to resources and opportunities. For instance, AI-driven financial technologies (fintech) are expanding access to banking services for underserved populations. By using alternative data sources and machine learning algorithms, these systems can assess creditworthiness without traditional financial histories, enabling more inclusive lending practices (Ghosh et al. 2021).

Furthermore, AI can revolutionize education and workforce development, equipping individuals with the skills needed to thrive in a sustainable economy. AI-powered personalized learning platforms can tailor educational content to individual needs, helping workers adapt to rapidly changing job markets. By fostering a more skilled and adaptable workforce, AI supports economic resilience and equity (Luckin et al. 2018).

While AI holds immense promise for economic sustainability, its adoption is not without challenges. One significant concern is the environmental impact of AI itself. Training large AI models requires substantial computational power, which can lead to high energy consumption and carbon emissions. Addressing this issue requires developing energy-efficient AI systems and leveraging renewable energy sources for computational tasks (Strubell et al. 2020).

Ethical considerations also play a critical role in ensuring that AI contributes to sustainable and inclusive economic growth. Bias in AI algorithms, for example, can exacerbate social inequalities if not properly addressed. Ensuring transparency, fairness, and accountability in AI systems is essential to maximizing their positive impact while minimizing potential harm (Binns 2018).

Data privacy is another critical concern. AI systems rely on vast amounts of data, raising questions about how personal and sensitive information is collected, stored, and used. Robust data governance frameworks are necessary to protect individuals' privacy while enabling the responsible use of AI for economic sustainability (Floridi et al. 2018).

Governments, businesses, and civil society must work together to harness AI's potential for economic sustainability. Policy frameworks that promote innovation while addressing ethical and environmental concerns are essential. For instance, governments can incentivize the development of AI technologies that contribute to sustainability goals through grants, tax credits, and public–private partnerships.

Additionally, international cooperation is crucial for addressing global challenges such as climate change and economic inequality. Collaborative initiatives can help standardize AI practices, share best practices, and ensure that the benefits of AI are distributed equitably across nations and communities (OECD 2019).

1.2 Methodology

The methodology for this study involved a comprehensive analysis of academic articles and the application of VOSviewer for network visualization. A systematic literature review was conducted by identifying relevant publications from prominent databases such as Scopus. Search terms included combinations of keywords "artificial intelligence," "sustainability," and "economy." Inclusion criteria focused on peer-reviewed articles, publications from 2010 onwards, and those addressing the interplay of AI with sustainable economic growth. Articles were screened for relevance based on their abstracts and then thoroughly analyzed to identify recurring themes and insights.

To visualize the interconnectedness of the topics, VOSviewer software was employed. This tool facilitated the creation of bibliometric maps highlighting co-occurrence patterns of keywords. The visualization process helped identify central themes, emerging trends, and research gaps within the field. Then a qualitative analysis of the literature helped to understand the current state of art and potential directions for future research.

1.3 Structure of the Articles Analysis Using Network Approach

The analysis of the network allows to identify key trends in the field, and in this research, the VOSviewer network highlights both the theoretical underpinnings and practical implications of AI, sustainability, and socioeconomic growth, as showed in Fig. 1.1.

The network visualization provides insights into the interconnectedness of key research themes related to AI, sustainability, and economic growth. These themes connect various clusters (represented in colors), reflecting the interplay between practical implementations and theoretical considerations in the field.

One cluster emphasizes governance and policy concerns, with terms like "public sector," "policy consideration," and "fintech" showcasing the relevance of AI in public administration and financial technologies. Another cluster highlights

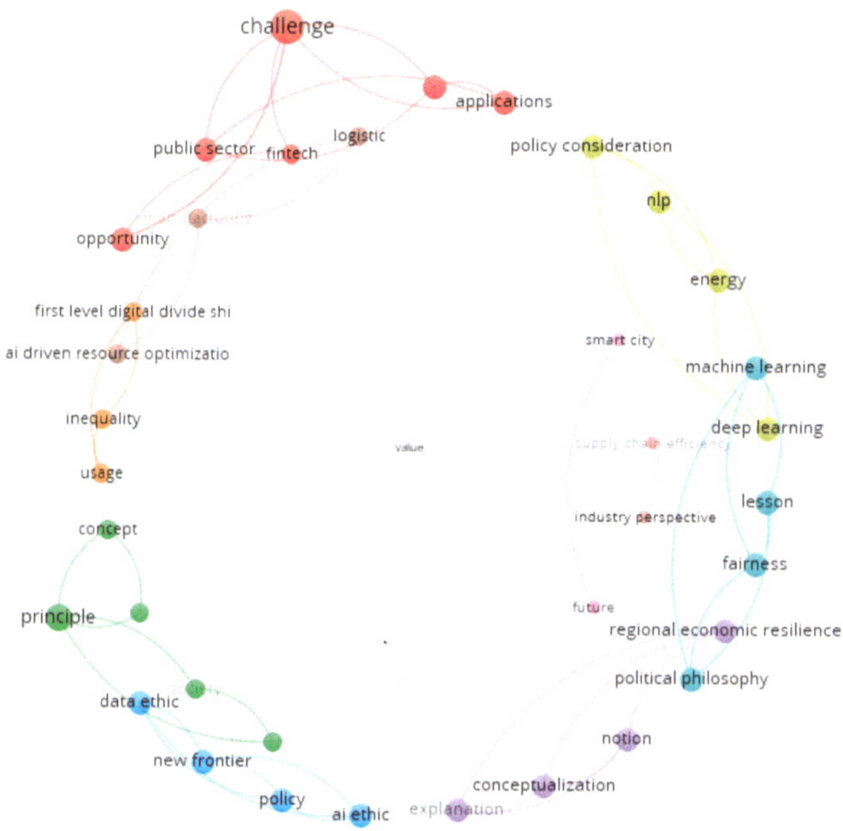

Fig. 1.1 VOSviewer network about AI and sustainable economy

resource optimization, inequality, and the digital divide, underlining the ethical and socioeconomic challenges AI poses.

The green cluster focuses on foundational principles, such as “data ethics” and “concepts,” suggesting an ongoing dialogue on the ethical frameworks guiding AI adoption. Meanwhile, the blue and yellow clusters explore technical aspects like “machine learning,” “deep learning,” and “energy,” demonstrating the interplay between technological advancements and their implications for efficiency and fairness.

Finally, the purple cluster reflects broader philosophical underpinnings, with terms like “conceptualization” and “political philosophy” pointing to the theoretical grounding of AI’s societal impact. This network highlights the complexity of integrating AI into sustainable economic systems, revealing both opportunities and challenges for the society and organizations.

In the next section, a qualitative analysis of the articles and the concepts are made in order to understand the topic being studied.

1.4 Analysis of the Concepts and Research Attributes

Economic resilience reflects an economy's ability to withstand, adapt to, and recover from external shocks. Scholars like Martin and Sunley (2015) emphasize that resilience involves both resisting disruptions and recovering swiftly. Recent research highlights the transformative role of technological advancements, such as artificial intelligence (AI) and digital tools, in enhancing resilience. These innovations improve forecasting, optimize resource allocation, and foster adaptive capacities within economies, as noted by Hallegatte et al. (2020). At the heart of these advancements are intelligent systems, which encompass AI-driven technologies that simulate human intelligence for problem-solving and decision-making. According to Russell and Norvig (2021), such systems enable real-time data analysis and automation, fostering efficiency and innovation across industries. For instance, machine learning algorithms can identify patterns, predict outcomes, and provide actionable insights, revolutionizing sectors like manufacturing, health care, and logistics.

Resource optimization, a crucial component of sustainable development, is another area where AI demonstrates immense potential. By leveraging predictive analytics and dynamic allocation strategies, AI technologies maximize efficiency in resource use. Chien et al. (2022) demonstrate how these technologies optimize supply chain processes, reduce waste, and minimize costs, which is particularly relevant in industries facing significant resource constraints. This capability is increasingly influencing policymaking, as AI-driven tools offer data analytics and machine learning capabilities to improve decision-making processes. Janssen et al. (2021) argue that AI enhances transparency, efficiency, and inclusivity in policymaking by incorporating diverse data sources and stakeholder perspectives. For example, tools like predictive modeling and sentiment analysis allow policymakers to anticipate societal trends and adapt policies accordingly, thereby improving governance.

Participatory budgeting, a democratic process that empowers citizens to influence public fund allocation, has also benefited from AI integration. Cabannes (2019) notes that AI enhances the scalability and efficiency of participatory budgeting frameworks by analyzing citizen inputs and prioritizing projects aligned with public preferences. This democratization of decision-making aligns closely with the principles of digital inclusion, which aims to ensure equitable access to digital technologies and the Internet. Studies by van Deursen and van Dijk (2019) highlight how digital inclusion fosters economic opportunities, education, and civic engagement. AI plays a dual role in this context: while it bridges gaps through personalized learning platforms and telemedicine, it also necessitates ethical considerations to avoid exacerbating digital divides.

The growing adoption of data-driven decision-making further underscores AI's transformative role across various domains. Davenport and Harris (2017) assert that data-driven organizations achieve superior operational efficiency and innovation by leveraging advanced analytics tools powered by AI. These tools provide real-time insights, helping leaders make informed decisions in governance, health care, and business. Governments worldwide are increasingly adopting AI in governance to streamline processes, enhance service delivery, and foster transparency. Wirtz et al. (2020) illustrate how AI tools, such as natural language processing systems, improve citizen engagement and policy communication, transforming public administration.

AI also plays a pivotal role in advancing the circular economy, an economic model focused on resource reuse, recycling, and waste reduction. According to the Ellen MacArthur Foundation (2020), AI facilitates efficient resource management and product lifecycle analysis, aiding the transition toward sustainable practices. These technologies predict material flows, optimize recycling processes, and design sustainable products, contributing to environmental sustainability and economic efficiency. Automation, closely linked to AI, intersects with sustainability by reducing environmental impact. Automated systems powered by AI enable precision in industries such as agriculture, minimizing waste and energy consumption. The Global e-Sustainability Initiative (GeSI, 2021) reports that automation technologies have the potential to significantly cut global greenhouse gas emissions by optimizing industrial processes and promoting renewable energy adoption.

AI's contributions to socioeconomic growth are equally noteworthy, enhancing productivity, creating new markets, and improving public services. However, as Brynjolfsson and McAfee (2014) caution, the economic expansion enabled by AI also presents challenges such as job displacement and inequality. Policymakers must therefore adopt balanced strategies to ensure inclusive growth while harnessing AI's potential. The ethical implications of AI further complicate its deployment. Floridi and Cowls (2019) emphasize the importance of ethical frameworks to guide AI development and deployment, addressing concerns such as bias, transparency, accountability, and privacy. Ensuring fairness and minimizing harm are central to the responsible use of AI in societal and organizational contexts.

The environmental impact of AI, particularly its substantial energy demands, adds another layer of complexity. Strubell et al. (2019) highlight the significant carbon footprint associated with AI research and deployment. Despite these challenges, AI also offers innovative solutions to environmental issues, such as monitoring ecosystems, optimizing renewable energy systems, and reducing waste through predictive analytics. This dual role underscores the importance of sustainable innovation, which integrates environmental, social, and economic considerations into technological advancements. Porter and Kramer (2011) discuss the concept of shared value, where businesses create economic value while addressing societal challenges. AI is a key enabler of such innovation, offering tools to develop green technologies and foster eco-friendly practices.

Finally, smart cities illustrate the integration of AI in urban development, using technology to improve living conditions through efficient resource management and enhanced public services. AI-driven systems underpin smart city initiatives, enabling traffic management, energy optimization, and waste reduction. Batty et al. (2012) observe that smart cities address urban challenges while promoting sustainability and improving quality of life. The interplay of these elements—from governance to environmental sustainability—demonstrates the transformative potential of AI and its capacity to drive progress across multiple domains.

1.5 Conclusion

AI's ability to analyze complex systems, optimize resource use, and foster innovation positions it as a powerful tool for advancing economic sustainability. By addressing environmental challenges, promoting inclusive growth, and enabling data-driven decision-making, AI can help societies transition toward resilient and equitable economies. However, realizing this potential requires careful consideration of ethical, environmental, and social implications. Through collaborative efforts and forward-thinking policies, stakeholders can ensure that AI serves as a force for sustainable development, benefiting both current and future generations.

References

Batty M, Axhausen KW, Giannotti F et al (2012) Smart cities of the future. Eur Phys J Special Topics 214(1):481–518

Binns R (2018) Fairness in machine learning: lessons from political philosophy. In: Proceedings of the 2018 conference on fairness, accountability, and transparency, pp 149–159. https://doi.org/10.1145/3287560.3287583

Brynjolfsson E, McAfee A (2014) The Second Machine Age: Work, Progress, and Prosperity in a Time of Brilliant Technologies. W. W. Norton & Company

Cabannes Y (2019) Participatory budgeting: Concept and principles. Public Admin Develop 39(2):73–84

Chien SY et al (2022) AI-driven resource optimization in manufacturing and logistics. J Ind Eng 68(3):115–130

Davenport TH, Harris JG (2017) Competing on analytics: the new science of winning. Harvard Business Review Press.

Ellen MacArthur Foundation (2020) AI and the circular economy. Retrieved from https://www.ellenmacarthurfoundation.org

Floridi L, Cowls J (2019) A unified framework of five principles for AI in society. Harvard Data Sci Rev 1(1)

Floridi L, Taddeo M, Turilli M (2018) Data ethics: the new frontier of AI ethics and policy. Philos Technol 31(2):123–136. https://doi.org/10.1007/s13347-017-0263-5

GeSI (2021) Enabling sustainability through automation. Retrieved from https://gesi.org

Ghosh S, Garg P, Kapoor A (2021) AI in fintech: opportunities and challenges. J Finan Innov 6(1):35–49. https://doi.org/10.1007/s40821-021-00159-w

Hallegatte S, Rentschler J, Walsh B (2020) Building back better: economic resilience post-crisis. World Bank Publications

Janssen M et al (2021) AI for policymaking: opportunities and challenges. Gov Inf Q 38(1):101579

Kumar P, Prakash S, Yadav A (2020) AI and supply chain efficiency: an industry perspective. J Oper Manage Res 13(3):212–225. https://doi.org/10.1007/s12063-020-00168-6

Luckin R, Holmes W, Griffiths M, Forcier LB (2018) Intelligence unleashed: an argument for AI in education. Pearson Education. https://www.pearson.com

Martin R, Sunley P (2015) On the notion of regional economic resilience: Conceptualization and explanation. J Econ Geogr 15(1):1–42

OECD (2019) Artificial intelligence in society. Organ Econ Co-Oper Develop. https://doi.org/10.1787/eedfee77-en
Porter ME, Kramer MR (2011) Creating shared value. Harv Bus Rev 89(1/2):62–77
Rolnick D, Donti PL, Kaack LH, Kochanski K, Lacoste A, Sankaran K, Bengio Y (2019) Tackling climate change with machine learning. arXiv preprint arXiv:1906.05433
Russell S, Norvig P (2021) Artificial intelligence: a modern approach. Pearson.
Strubell E, Ganesh A, McCallum A (2019) Energy and policy considerations for deep learning in NLP. arXiv preprint arXiv:1906.02243
Strubell E, Ganesh A, McCallum A (2020). Energy and policy considerations for deep learning in NLP. In: Proceedings of the 58th annual meeting of the Association for Computational Linguistics (pp. 3645–3650). https://doi.org/10.18653/v1/2020.acl-main.364
United Nations (2015) Transforming our world: the 2030 agenda for sustainable development. https://www.un.org/sustainabledevelopment
van Deursen AJAM, van Dijk JAGM (2019) The first-level digital divide shifts to inequalities of usage. New Media Soc 21(3):497–517
Wirtz BW, Weyerer JC, Geyer C (2020) Artificial intelligence and the public sector: Applications and challenges. Int J Public Adm 43(8):650–661

AI in the Sustainability of Social Economy

2

Helena Belchior Rocha, Elsa Justino, and Serdar Yener

Abstract

Social economy represents a vital sector that prioritizes social and environmental objectives over profit maximization. This sector includes a diverse organization such as cooperatives, mutual benefit societies, associations, and foundations. These entities operate based on principles of democratic governance, reinvestment of profits for social purposes, and autonomy from public authorities. In recent years, the integration of artificial intelligence (AI) into various sectors has shown immense potential to drive efficiency, innovation, and sustainability. The social economy is no exception. AI can provide powerful tools to enhance the effectiveness and impact of social economy initiatives, helping these organizations to better achieve their missions. This chapter explores the intersection of social economy and AI, examining how AI can address the challenges faced by social economy organizations and contribute to sustainable development.

Keywords

Sustainability • Social economy • Artificial intelligence • Innovation

H. B. Rocha (✉) · E. Justino
Political Science and Public Policies Department, Iscte University Institute of Lisbon, Lisbon, Portugal
e-mail: helena_rocha@iscte-iul.pt

E. Justino
e-mail: elsa_justino@iscte-iul.pt

S. Yener
Boyabat Faculty of Economics and Administrative Sciences, Department of International Trade and Business, Sinop University, Sinop, Turkey
e-mail: serdar_yener@hotmail.com

M. J. Sousa et al. (eds.), *Economic Sustainability Potentiated by Artificial Intelligence*, Contributions to Environmental Sciences & Innovative Business Technology,
https://doi.org/10.1007/978-981-96-8089-4_2

2.1 Framework of Social Economy

Social economy encompasses a wide array of organizations that are driven by social objectives rather than profit. These organizations include:

Cooperatives, that are member-owned businesses that operate for the benefit of their members, who use their services or products; organizations that provide insurance, health care, and other services to their members, called mutual benefit societies.

Groups formed for a common purpose, often related to social, cultural, or environmental goals (Associations) and Foundations, the nonprofit entities that support charitable activities through grants and other funding mechanisms (OECD 2022).

All these institutions are managed by the principles of social economy which are: democratic governance, decision-making processes that involve the participation of members or stakeholders; reinvestment of profits, meaning that profits are reinvested into the organization or used for social purposes rather than distributed to shareholders; autonomy with independence from public authorities, although they may collaborate with governments and other entities; role in society, playing an essential role in promoting social inclusion, reducing inequalities, and fostering sustainable development. Often, these organizations address gaps left by the public and private sectors, providing services and support to vulnerable populations (OECD 2022).

Despite their significant contributions, social economy organizations face several challenges. Social economy organizations often face significant financial constraints, primarily due to limited access to funding and investment. Traditional financial institutions may view these organizations as high-risk investments, leading to difficulties in securing loans and venture capital. Additionally, social enterprises often rely on grants and donations, which can be unpredictable and insufficient to cover operational costs (EC 2021).

Inconsistent legal frameworks and a lack of supportive policies pose substantial challenges for social economy organizations. Regulatory environments vary widely across regions, creating uncertainty and hindering the growth of these entities (OECD 2021). For example, complex licensing requirements and restrictive zoning laws can impede the establishment and expansion of social enterprises. Moreover, the absence of standardized regulations can lead to difficulties in verifying and maintaining the credibility of social economy initiatives (OECD 2021).

Social economy organizations often struggle with operational challenges such as weak asset bases, lack of skilled labor, and maintaining democratic governance (EC 2021). These organizations typically operate with limited resources, making it difficult to invest in infrastructure and technology. Additionally, attracting and retaining skilled employees can be challenging due to competitive labor markets and the often-lower salaries offered by social enterprises (OECD 2022). Also maintaining democratic governance also requires continuous engagement and participation from members, which can be resource intensive.

Therefore, leveraging AI in social economy organizations can overcome some of these challenges, improving their efficiency and impact. The following points will delve into the methodology used and the applications benefit of AI in the social economy and provide practical recommendations for integrating AI into these organizations.

2.2 Methodology

The methodology employed in this chapter is based in a comprehensive literature review, which serves as the foundation for analyzing the intersection between artificial intelligence (AI) and the social economy. This approach allows for a systematic exploration of existing academic studies, policy reports, and industry publications to identify current trends, challenges, and potential applications of AI within social economy organizations.

Literature review is an essential methodological instrument in social sciences and policy research, as it facilitates the aggregation of knowledge from diverse sources, enabling researchers to synthesize findings and generate new insights (Snyder 2019). Through systematically reviewing existing publications, this chapter ensures an understanding of the state of AI integration into the social economy, uncovering patterns, gaps, and innovative practices. This method also helps to validate theoretical assumptions and informs policy recommendations by drawing from empirical evidence and examples.

The research process began with the identification of relevant literature, including peer-reviewed journal articles, reports from international organizations (such as the OECD and the European Commission), and white papers from technology firms and NGOs. Keywords such as "AI in social economy," "sustainable innovation," "AI for social impact," and "digital transformation in social enterprises" were used to source material from academic databases, including Scopus, Web of Science, and Google Scholar.

Selected literature was critically evaluated based on relevance, methodological rigor, and contribution to the discourse on sustainability and digital transformation in social organizations. The inclusion of examples and policy documents allowed for a multidimensional perspective, reflecting both theoretical frameworks and practical applications. Thematic coding techniques were employed to identify recurring themes such as AI-driven financial inclusion, operational efficiency, and policy frameworks supporting social economy innovation (Booth et al. 2016).

This literature-based approach provides a holistic view of the opportunities AI offers to social economy organizations while addressing barriers such as regulatory challenges, limited technological infrastructure, and ethical concerns. It also highlights the transformative role of AI in enhancing financial management, optimizing resource allocation, and fostering inclusive growth (WEF 2024; Omotayo et al. 2024).

In line with examples in literature review methodology, this study adopts a transparent and replicable process, ensuring the reliability of findings and the potential for future research expansion. The insights derived from this methodological framework contribute to a broader understanding of how AI can drive sustainability and resilience within the social economy, reinforcing its role in promoting social inclusion and reducing inequalities.

2.3 AI to Help Overcome Challenges in the Social Economy

AI can enhance financial inclusion and access to funding for social economy organizations.

> Addressing these challenges requires innovative solutions that can surmount traditional barriers to financial inclusion. One promising avenue is the integration of artificial intelligence (AI) and data analytics into the financial ecosystem of developing economies (Omotayo et al. 2024, p. 289).

With the leverage of AI-driven financial technologies (FinTech), these organizations can:

Improve Credit Assessment: AI algorithms can analyze vast amounts of data to assess creditworthiness more accurately, enabling social enterprises to secure loans and investments that might otherwise be inaccessible (Omotayo et al. 2024; Djeundje et al. 2021; Dwivedi. et al. 2021).

Automate Fundraising: AI tools can streamline the fundraising process by identifying potential donors and investors, predicting donation patterns, and optimizing fundraising campaigns (Cherniavska et al. 2023).

Enhance Financial Management: AI-powered financial management systems can help social economy organizations manage their finances more efficiently, reducing costs and improving financial stability (Sampene et. al. 2022).

Regulatory Barriers: AI can help and simplify regulatory environments, making it easier for social economy organizations to comply with legal requirements (Kumar & Chaturvedi 2021; Jones 2020).

Regulatory Compliance: AI systems can monitor regulatory changes in real-time and ensure that organizations remain compliant with evolving laws and regulations. This reduces the administrative burden and minimizes the risk of non-compliance (Deloitte 2021).

Policy Analysis: AI can analyze large datasets to identify trends and insights that inform policy advocacy efforts, helping social economy organizations push for more supportive regulatory frameworks (Parmeet 2024; Comunale and Manera 2024).

Automated Reporting: AI can automate the generation of compliance reports, ensuring accuracy and reducing the time and resources needed for regulatory reporting (Parmeet 2024; Comunale and Manera 2024).

Also, for operational challenges AI can address various operational challenges faced by social economy organizations, enhancing their efficiency and effectiveness.

Resource Optimization: AI can optimize resource allocation by analyzing data on resource usage and identifying areas for improvement. This helps organizations make better use of their limited resources (Mohammed 2023; Waltersmann et al. 2021).

Skill Development: AI-powered training platforms can provide personalized learning experiences for employees, helping them acquire the skills needed to operate effectively in the social economy (Morandini et al. 2023).

Decision Support: AI can support decision-making processes by providing data-driven insights and predictive analytics, enabling organizations to make informed decisions and plan strategically (Laat et al. 2020).

> In an era defined by rapid technological advance, the transformative potential of artificial intelligence (AI) stands as both a beacon of promise and a challenge. It has become increasingly evident that AI is not merely a tool for innovation but, when deployed responsibly and ethically, can be a force for social innovation – improving equity, shaping societies and redefining the global landscape (WEF 2024, p. 3).

In this sense, integrating AI into their operations, social economy organizations can overcome these challenges and enhance their capacity to achieve their social and environmental objectives. This not only improves their sustainability but also amplifies their impact on society.

2.4 How Can Organizations Implement These AI Solutions?

Implementing AI solutions in social economy organizations requires a structured approach to ensure successful integration and maximize impact. Some of the steps, for example, start by identifying the specific challenges or opportunities that AI can address within your organization. Set clear, measurable goals to guide the implementation process (WEF 2024).

AI systems rely heavily on high-quality data. Evaluate the data you currently have and identify any gaps. Ensure that your data is accurate, relevant, and accessible. This might involve cleaning existing data, integrating data from different sources, and establishing data governance practices (WEF 2024).

Another important thing is to build a cross-functional team. Assemble a team that includes members from various areas such as IT, operations, finance, and program management. This team should work collaboratively to oversee the AI implementation process, ensuring that all perspectives are considered and that the AI solutions align with organizational goals, and then begin with pilot projects that are likely to deliver quick wins and demonstrate the value of AI. Also, it is important to choose AI tools and platforms that are scalable and flexible enough to grow with your organization. Look for solutions that can be easily integrated with your existing systems and that offer the flexibility to adapt to changing needs and technologies (Trabelsi 2024).

Implement strong data governance policies to protect data privacy and security. This means establishing clear protocols for data access, usage, and storage. Ensuring compliance with relevant regulations and ethical standards is crucial for maintaining trust and integrity (WEF 2024).

Not less important is to encourage a culture of continuous learning and adaptation. For that it is essential to provide training and development opportunities to help staff understand and effectively use AI tools, promoting an environment where experimentation and innovation are valued (Trabelsi 2024).

Define key performance indicators (KPIs) to measure the success of your AI initiatives. Regularly monitor and evaluate these metrics to assess the impact of AI on your organization and make data-driven decisions for future improvements.

> Looking ahead, the continued collaboration between social innovators, policymakers, technologists and communities will be vital in shaping an ecosystem where AI acts as a catalyst for social good (WEF 2024).

However, implementing AI solutions in social economy organizations can present several challenges. Some of the obstacles are:

Insufficient or Low-Quality Data: AI systems require large volumes of high-quality data to function effectively. Social economic organizations may struggle with data availability, and/or limited access to relevant data can hinder AI training and performance. The poor-quality data can lead to biased or inaccurate AI outputs. To mitigate this, organizations should invest in data collection and management practices to ensure data is accurate, comprehensive, and representative and collaborate with other organizations to share data and resources (Alhosani and Alhashmi 2024).

Outdated Infrastructure: Many social economy organizations operate with limited technological infrastructure, which can be a barrier to AI implementation. AI requires significant computational resources that may not be available with outdated equipment and because existing systems may not be compatible with new AI technologies. In this case it is advised to upgrade infrastructure to support AI technologies. This might involve investing in cloud-based solutions that offer scalable computing power and storage (Alhosani and Alhashmi 2024).

Integrating AI into existing workflows and systems can be complex and time-consuming, so it is important to ensure new AI tools work seamlessly with current systems and staff need to be trained to use new AI tools effectively. In order to do it organizations need a plan for a phased integration approach, starting with pilot projects and providing comprehensive training programs for staff to build AI literacy and competence (Alhosani and Alhashmi 2024).

Resistance to changing established workflows and processes is another obstacle, and in this case it is suggested to communicate the benefits of AI clearly to all stakeholders. Emphasize how AI can augment human work and involve all in the implementation process (Alhosani and Alhashmi 2024).

Then, there are the ethical and bias concerns, and AI systems can inadvertently perpetuate biases present in the training data, leading to ethical concerns which means ensuring AI is used responsibly and implement robust data governance and ethical guidelines. Regularly audit AI systems for bias and take corrective actions as needed as well as promote transparency in AI decision-making processes (Hanna et al. 2024; Ferrara 2024).

Moreover, implementing AI can be expensive, and social economy organizations often operate with limited budgets; besides the initial costs, there are ongoing costs of continuous maintenance and updates. In this case seeking funding and grants specifically aimed at technology adoption in the social sector and exploring partnerships with tech companies and academic institutions to share resources and expertise can be a solution (Belchior-Rocha et al. 2018).

The report of the World Economic Forum (WEF 2024),

> …finds three primary impact areas where AI is making significant contributions: healthcare, with 25% of innovators using AI to advance access to health; environmental sustainability, with 20% of social innovators applying AI to tackle climate solutions; and economic empowerment, notably prevalent in lower-income countries where 80% of all initiatives aimed at enhancing livelihoods are based. But AI is also revolutionizing practices in other areas such as agriculture through predictive analytics and precision farming, addressing climate resilience and boosting productivity (WEF 2024, p. 4).

In this sense, anticipating these challenges and planning accordingly, social economy organizations can successfully implement AI solutions and harness their potential to drive sustainability and social impact.

2.5 AI Applications in Social Economy

AI emerged as a transformative tool that enhances the efficiency, reach, and impact of social economy organizations. Some of the areas where AI applications have driven significant improvements are as follows:

Financial inclusion and access to funding through microfinance and credit assessment, AI algorithms assess creditworthiness by analyzing alternative data sources, enabling social organizations to secure loans that might otherwise be inaccessible (Omowole et al. 2024). In fundraising optimization, AI tools predict donor behavior, identify potential investors, and personalize fundraising campaigns, maximizing resource mobilization (Cherniavska et al. 2023).

Resource optimization and operational efficiency, with supply chain optimization, social organizations use AI to optimize logistics, reducing waste and enhancing supply chain transparency. For example, AI-powered platforms analyze data to predict demand and adjust production cycles. For resource allocation, AI-driven analytics identify inefficiencies and suggest areas for resource reallocation, ensuring that limited resources are used effectively.

Social economy organizations leverage AI to offer personalized learning experiences, upskilling employees and volunteers to meet evolving demands (Morandini et al. 2023), and AI tools provide interactive, adaptive training for marginalized communities, boosting employability and social inclusion through digital literacy programs.

Enhancing social services, AI supports telemedicine initiatives, delivering healthcare solutions to underserved communities. For instance, AI-driven diagnostics and health monitoring apps improve early disease detection and treatment. Chatbots and virtual counselors powered by AI provide mental health support, making services more accessible and reducing stigma (Nwankwo et al. 2024).

In environmental sustainability and climate action, social organizations can apply AI for predictive analytics in agriculture, improving yield forecasts and promoting sustainable farming practices (WEF 2024). In recycling processes, AI algorithms optimize and track waste patterns, contributing to circular economic initiatives.

The World Economic Forum's AI for Impact Initiative highlights AI applications that address global challenges. One initiative focuses on applying AI in low-income communities to promote economic empowerment and access to essential services, demonstrating how technology can drive social innovation on a large scale (WEF 2024).

AI may also help measure the impact of activities done, increase the sustainability and efficiency of fundraising efforts, and more readily detect fraud, all of which can increase an organization's efficiency. They can enhance their procedures and more effectively convey them to their stakeholders by using impact analysis (Faruq et al. 2024). This problem lies at the core of AI's application and advancement. Large-scale study of stakeholder sentiment is already made possible by these techniques. For instance, using data gathered from social networks would enable the identification of patterns about the effects of services, programs, or fundraising initiatives as well as how the general public views them.

Following and learning from successful examples, social economy organizations can effectively implement AI solutions to overcome challenges and achieve their sustainability.

2.6 Conclusion

Organizations have already seized this opportunity to develop tools that use AI to improve their effectiveness, but all this is not possible without political support for the social economy.

The European Commission has recognized the importance of the social economy and launched the Social Economy Action Plan to support its development. This plan aims to create a conducive environment for social economy organizations by improving access to finance, enhancing visibility, and fostering innovation. It also seeks to streamline regulatory frameworks across member states, making it easier for social organizations to operate and scale their activities.

Political support has an imperative role in addressing the financial and regulatory challenges faced by social economy organizations. Governments can implement policies that provide financial incentives, such as tax breaks and grants, to encourage investment in the social economy (Capraro et al. 2024). Additionally, creating clear and supportive regulatory frameworks can reduce barriers to entry and promote the growth of social organizations and to foster an inclusive economic environment, and political support can help social economy organizations contribute more effectively to sustainable development and social cohesion. Political support remains essential in ensuring equitable access to AI technologies, increasing an inclusive digital transformation that aligns with social economic principles.

Finally, integrating AI solutions, social economy organizations can enhance their operational capacity, drive sustainability, and expand their impact across diverse sectors, helping to achieve a sustainable transformation in our society.

References

Alhosani K, Alhashmi SM (2024) Opportunities, challenges, and benefits of AI innovation in government services: a review. Discov Artif Intell 4:18. https://doi.org/10.1007/s44163-024-00111-w

Belchior-Rocha H, Almeida F, Abreu R (2018) IoT sustainability in higher education: an exploratory study. In Simões D, Barbosa B, Filipe S (eds) Smart marketing with the internet of things. IGI Global, Hershey, pp 185–197

Booth A, Sutton A, Papaioannou D (2016) Systematic approaches to a successful literature review. Sage

Capraro V, Lentsch A, Acemoglu D, Akgun S, Akhmedova A, Bilancini E, Bonnefon JF, Brañas-Garza P, Butera L, Douglas KM, Everett JAC, Gigerenzer G, Greenhow C, Hashimoto DA, Holt-Lunstad J, Jetten J, Johnson S, Kunz WH, Longoni C, Lunn P, Natale S, Paluch S, Rahwan I, Selwyn N, Singh V, Suri S, Sutcliffe J, Tomlinson J, van der Linden S, Van Lange PAM, Wall F, Van Bavel FJ, Viale R (2024) The impact of generative artificial intelligence on socioeconomic inequalities and policy making. PNAS Nexus 3(6):191. https://doi.org/10.1093/pnasnexus/pgae191

Cherniavska O, Belov A, Shmygol N, Järvis M, Cherniavska O, Tsalko T (2023) Artificial intelligence tools for university fundraising 5.0: a comprehensive analysis. In: IEEE 5th International conference on modern electrical and energy system (MEES), Kremenchuk, Ukraine, pp 1–6. https://doi.org/10.1109/MEES61502.2023.10402529.

Comunale M, Manera A (2024) the economic impacts and the regulation of AI: a review of the academic literature and policy actions. IMF Working Paper No. 2024/65

Deloitte (2021) AI and the financial sector: regulatory compliance and risk management. Retrieved from Deloitte website.

Djeundje VB, Crook J, Calabrese R, Hamid M (2021) Enhancing credit scoring with alternative data. Expert Syst Appl 163:113766

Dwivedi YK, Hughes L, Ismagilova E, Aarts G, Coombs C, Crick T et al (2021) Artificial Intelligence (AI): multidisciplinary perspectives on emerging challenges, opportunities, and agenda for research, practice and policy. Int J Inform Manage 57:101994

EC (European Commission) (2021) COM 778 final. Building an economy that works for people: an action plan for the social economy. In: Communication from the Commission to the European Parliament, The council, The European Economic and Social Committee and the Committee of the Regions

Faruq O, Haque S, Sufian M, Hossain MA, Talukder T, Shayed U (2024) AI-driven strategies for enhancing non-profit organizational impact. https://doi.org/10.62127/aijmr.2024.v02i05.1088.
Ferrara E (2024) Fairness and bias in artificial intelligence: a brief survey of sources, impacts, and mitigation strategies. Science 6(1):3. https://doi.org/10.3390/sci6010003
Hanna M, Pantanowitz L, Jackson B, Palmer O, Visweswaran S, Pantanowitz J, Deebajah M, Rashidi H (2024) Ethical and bias considerations. In: Artificial intelligence (ai)/machine learning, modern pathology, 100686, ISSN 0893-3952. https://doi.org/10.1016/j.modpat.2024.100686
Jones R (2020) The role of artificial intelligence in regulatory compliance. J Compliance Regul 12(2):87–104
Kumar S, Chaturvedi D (2021) AI in regulatory compliance: automating the future. J Finan Transform 53:29–33
Laat M, Joksimovic S, Ifenthaler D (2020) Artificial intelligence, real-time feedback and workplace learning analytics to support in situ complex problem-solving: a commentary. Int J Inform Learn Technol 37:267–277. https://doi.org/10.1108/IJILT-03-2020-0026
Mohammed A (2023) Artificial intelligence and resource optimization: a study of Fintech start-ups. Resour Policy 80:103250. ISSN 0301-4207. https://doi.org/10.1016/j.resourpol.2022.103250.
Morandini S, Fraboni F, De Angelis M, Puzzo G, Giusino D, Pietrantoni L (2023) The impact of artificial intelligence on workers' skills: upskilling and reskilling in organisations. Inform Sci 26:39–68. https://doi.org/10.28945/5078
Nwankwo E, Emeihe E, Ajegbile M, Olaboye J, Maha C (2024) Telemedicine and AI to improve healthcare access in rural settings. Int J Life Sci Res Arch 07:59–077. https://doi.org/10.53771/ijlsra.2024.7.1.0061
OECD (2022) Designing legal frameworks for social enterprises: practical guidance for policy makers. Local Economic and Employment Development (LEED), OECD Publishing, Paris. https://doi.org/10.1787/172b60b2-en
OECD/KDI (2021). Case Studies on the Regulatory Challenges Raised by Innovation and the Regulatory Responses. OECD Publishing, Paris. https://doi.org/10.1787/8fa190b5-en
Omotayo BA, Wilhelmina AA, Adeola OA, Olubusola O, Chinwe CO, Onyeka CO (2024) Leveraging AI and data analytics for enhancing financial inclusion in developing economies. Finan Account Res J 6(3):288–303. P-ISSN: 2708-633X, E-ISSN: 2708-6348. https://doi.org/10.51594/farj.v6i3.856
Omowole B, Oghenekome U, Chukwunweike M, Somto W. (2024) Integrating fintech and innovation in microfinance: Transforming credit accessibility for small businesses. Int J Frontline Res Rev 3:090–100. https://doi.org/10.56355/ijfrr.2024.3.1.0032
Parmeet SC (2024) The impact of AI on economic forecasting and policy-making: opportunities and challenges for future economic stability and growth. York University. https://doi.org/10.13140/RG.2.2.24945.70249
Sampene, AK, Agyeman FO, Robert B, Wiredu J (2022) Artificial intelligence as a path way to Africa's transformations. Artif Intell 9(1)
Snyder H (2019) Literature review as a research methodology: an overview and guidelines. J Bus Res 104:333–339. https://doi.org/10.1016/j.jbusres.2019.07.039
Trabelsi MA (2024) The impact of artificial intelligence on economic development. J Electr Business Digital Econ 3(2):142–155. https://doi.org/10.1108/JEBDE-10-2023-0022
Waltersmann L, Kiemel S, Stuhlsatz J, Sauer A, Miehe R (2021) Artificial intelligence applications for increasing resource efficiency in manufacturing companies—a comprehensive review. Sustainability 13(12):6689. https://doi.org/10.3390/su13126689
WEF—World Economic Forum (2024) AI for impact: the role of artificial intelligence in social innovation. In: Collaboration with EY and Microsoft—WHITE PAPER. https://www.weforum.org/publications/ai-for-impact-artificial-intelligence-in-social-innovation/

3 Ethical Concerns of Aviation Professionals Towards AI-Based Systems

An Exploratory Study Understanding Motives Behind Non-acceptance of AI

Ines Berlenga, Maria José Chambel, and Luis Curral

Abstract

The present research sits within a broader study developed under a Ph.D. Programme in Psychology at Faculty of Psychology in Lisbon University (Faculdade de Psicologia da Universidade de Lisboa) and EASA (European Union Aviation Safety Agency). As AI-based systems are entering into the sector, it was important to listen to aviation professionals, assessing their level of acceptance of AI, applied to eight different hypothetical aviation scenarios. More than that, taking into consideration the ones that expressed their opinion as not acceptable (171 professionals), it was important to unveil the motives behind this negative opinion (2395 contents). Conclusions showed that aviation professionals have ethical concerns about the AI-based system itself (30%), about the consequent human negative impact when using such systems (28%), about how their data is used by the technology (11%), and about AI-based systems putting aviation safety at risk (6%). Results show a detailed motive basis for non-acceptance of AI, to which aviation organisations, research centres, and aviation professionals are invited to do further research, to ensure that AI-based

I. Berlenga (✉) · M. J. Chambel · L. Curral
Faculty of Psychology in Lisbon University (Faculdade de Psicologia da Universidade de Lisboa—FPUL), Lisbon, Portugal
e-mail: berlenga@edu.ulisboa.pt; ines.berlenga@easa.europa.eu

M. J. Chambel
e-mail: mjchambel@psicologia.ulisboa.pt

L. Curral
e-mail: lcurral@psicologia.ulisboa.pt

I. Berlenga
EASA (European Union Aviation Safety Agency), Cologne, Germany

M. J. Sousa et al. (eds.), *Economic Sustainability Potentiated by Artificial Intelligence*, Contributions to Environmental Sciences & Innovative Business Technology,
https://doi.org/10.1007/978-981-96-8089-4_3

systems applied to aviation besides being mandatory safe can also be ethically based.

Keywords

Ethics • Artificial intelligence • AI-based systems • Aviation • Aviation professionals • Non-acceptance • Perception

3.1 Aim of the Study

Artificial intelligence (AI) is not the future, it is now. It developed rapidly and its already impacting in our daily life; however, when applied to sectors with safety high impacts, like aviation, AI-based systems should be even better studied, and risks should be well known—different levels of risks will lead to different level of rules.

The present research aimed to understand the main ethical motives behind non-acceptance of AI-based systems applied to aviation. The analysis of the contents aimed to highlight priorities to be taken into consideration when developing, certifying, using, and generically taking decisions concerning AI-based system in the aviation context.

3.2 Ethical Concerns

The European Union (EU) is based on fundamental rights, democracy, and the rule of law. Based on Article 2 of the Treaty on European Union and the Charter of EU fundamental rights, Europe values: dignity, freedoms, equality, solidarity, citizen's rights, and justice. The charter is legally binding, entered into force with the Treaty of Lisbon (December 2009), and it is considered "in the light of changes in society, social progress and scientific and technological developments".

More recently and important for AI topics, the European Union defined the third generation of rights into the Charter of Fundamental Rights which now contains: "data protection, guarantees on bioethics and transparent administration" (EP, 2012).

In the same logic, Yeung et al. (2020, p. 78) believe that "the international human rights framework provides the most promising set of standards for ensuring that AI systems are ethical in their design, development, and deployment."

In the present chapter, we took the Charter of the fundamental rights of the European Union (including the third-generation ones) as providing proper norms in order to secure a strong basis for an ethical AI-based system. On this basis key ethical concepts were chosen as demonstrators of the rights, and concepts were: equal opportunities, non-discrimination and fairness, data protection, labour protection, right for privacy, right for professional development, right for autonomous decision-making, transparency, and accountability.

Another important factor to consider when approaching ethics in AI is that we consider an AI-based system as a tool or an artefact that supports human performance (Introna, 2024), in that sense the AI-based system is not considered as a moral entity in itself but only as an instrument. As such, these AI-based systems should be ensured as ethical as they will create an impact to the humans that will work with/interact with/being supported by it.

AI-based systems ethics goes in line with the notion of machine ethics in the sense of: "be able to take into account societal values, moral and ethical consideration, weigh the respective priorities of values held by different stakeholders in various multicultural contexts, explain its reasoning, and guarantee transparency" (Müller, 2023).

A fundamental milestone for all AI-related activities was the European Union Artificial Intelligence Act (2024/1689), the first comprehensive AI law, the document is driven by risk management, and in this sense different rules apply to different risk levels, and it has a strong demand on transparency and supports innovation. The AI Act sets also an important reference to ethics concerning AI (EC, 2023):

> In order to ensure a consistent and high level of protection of public interests as regards health, safety and fundamental rights, common rules for high-risk AI systems should be established. Those rules should be consistent with the Charter, non-discriminatory and in line with the Union's international trade commitments. They should also take into account the European Declaration on Digital Rights and Principles for the Digital Decade and the Ethics guidelines for trustworthy AI of the High-Level Expert Group on Artificial Intelligence (AI HLEG).

3.3 The Aviation Sector

When linking aviation to AI, it is common understanding that things should be handled with much care. In the case of aviation, we can characterise this type of organisations as high reliability organisations (HROs), and in aviation the tension between control and flexibility often is shown as a tension between safety and the service that can be provided (Parker, 2014); this also applies when a disruptive technology is proposing to play a crucial role, which is now the case of AI-based systems.

Since 2018 the European Union Aviation Safety Agency (EASA) put together a roadmap to pave the wave of AI safe introduction in aviation sector. "As concerns the aviation sector, AI not only affects the products and services provided by the industry; it also triggers the rise of new business models. This affects most of the domains under the mandate of the Agency, and its core processes (certification, rulemaking, organisation approvals, and standardisation) are impacted. (…) Beyond this, the liability, ethical, social and societal dimension of AI should also be considered." (EASA, 2023).

The roadmap refers as one of the four main actions of an EU strategy for AI usage: "enabling conditions for AI development and uptake in the EU, including establishing a regulatory framework that promotes innovation while ensuring safety and ethics;" (EASA, 2023). In this sense the Agency refers AI trustworthiness as a key driver for an ethical AI that the ethical approach to AI is a major factor to enhance trust, and the respect of European shared values would be an important factor to consider AI trustworthy.

One important reference of the developments on ethics for AI in aviation was the works of the already referred High-Level Expert Group on Artificial Intelligence (AI HLEG) (EC, 2024) and in particular the Assessment List for Trustworthy Artificial Intelligence (ALTAI) (EC, 2020). On the basis of the consequent research, EASA published the AI Concept Paper (Issue 2.0) with an Ethics Based Assessment Chapter applied specifically to the aviation sector and in line with the 7 key ethical requirements for trustworthy AI: Human agency and oversight; technical robustness and safety; privacy and data governance; transparency; diversity; non-discrimination and fairness; societal an environmental well-being; and accountability.

These references go in line with the recent AI Act published by the European Parliament (EP, 2024), and we believe a very strong basis for any research in safety-linked sectors, high risk considered, like the case of aviation:

> (50) As regards AI systems that are safety components of products, or which are themselves products, falling within the scope of certain Union harmonisation legislation listed in an annex to this Regulation, it is appropriate to classify them as high-risk under this Regulation if the product concerned undergoes the conformity assessment procedure with a third-party conformity assessment body pursuant to that relevant Union harmonisation legislation. In particular, such products are (…) aviation.

3.4 Factors for AI Readiness—The Human Dimension

It is well accepted that when detaining the computer capacity and the big data availability AI could turn from an old concept to a current practical reality. In a recent systematic literature review, Ali and Khan (2024) referred a set of factors that influence readiness for AI, their top 10 were: Organisation IT infrastructure; top management support; resource availability; collaborative culture; organisation size; organisation capabilities; data quality; compatibility; financial budget; AI-business potentials.

If we extract from the list of factors the ones that are related to human behaviour, it seems that top management support and a collaborative culture play an important role. If top management has a clear intent or willingness as decision-makers to go for AI, the collaborative culture shows us more about what the professionals would need to psychologically and behaviourally engage with AI. The authors define collaborative culture as "a critical factor in organizational readiness, which involves creating trust, sharing accurate information, and ensuring that the right people receive the correct information in appropriate locations." (Ali and Khan 2024, p.24).

Culture was also one of the top five dimensions for organisational readiness when adopting AI that Jöhnk et al. (2021) mentioned in their study. When listen to professionals they refer to three culture factors: (1) Innovativeness; Employees' fear of AI-induced job loss threatens proactive innovativeness; (2) Collaborative work—AI deployment relies on integrating different perspectives, i.e. domain, data, and IT; (3) Change management—Employees' lack of understanding and fear of AI threaten the acceptance of AI-based systems.

It seems that AI still faces human appropriate culture in order to be ready to be successfully accepted and adopted, and more clearly in the study of Yogesh et al. (2021) the common thread among the definitions of AI is the increasing capability of machines to perform roles and tasks currently performed by humans within the workplace and society in general.

We can then think that humans need to trust, need information sharing, need the right people to be properly informed, and need to solve their perception of being threatened or replaced by AI both professionally and in societal terms to accept AI-based system.

Having into consideration all the above, we formulated an assumption that the reasons supporting the non-acceptance of AI-based systems are related to direct threats to humans (both personal and professional) and information sharing (systems transparency).

3.5 Method

This chapter is part of a broader study on ethics for AI in aviation that brought together EASA and research activities under a PhD Programme in Psychology in Faculty of Psychology in Lisbon University (Faculdade de Psicologia da Universidade de Lisboa).

A survey was launched through two main channels: an email to a network of organisational stakeholders working in AI for aviation and a post in LinkedIn with an instruction that should be responded only by aviation professionals that developed their work directly or indirectly with AI.

The survey was based on 8 scenario cases, designed by a team with different aviation backgrounds. Each scenario would tell a story of a hypothetical situation where an aviation professional would be supported by an AI-based system in a work context. These scenarios were described within a certain aviation domain, either in present days or in the future. The scenarios were drafted having in mind the ethical factors chosen from the theorical foundation and resulted on the following:

(1) Artificial intelligence and physiological data monitoring,
(2) Artificial intelligence and pilot support in "go-around" situations,
(3) Artificial intelligence in maintenance,

(4) Artificial intelligence airport allocation of airlines to a terminal,
(5) Artificial intelligence airline crew members attribution to flights,
(6) Artificial intelligence and speech recognition in voice communications,
(7) Artificial intelligence and the risk of de-skilling, and
(8) New competencies when teaming with an AI-based system.

A rating scale of seven points was the basis for the aviation professionals to position their level of ethical acceptance towards each scenario. For the respondents that decided to rate 3 points or less, meaning the non-acceptance range of the scale, they were asked the reply to the following qualitative questions:

(1) In case it is not acceptable (if you chose from 1 to 3), what would be necessary for you to change your position to acceptable?
(2) In the case described above, do you see any other ethical concerns? If yes, please shortly explain.

A text box gave freedom to the respondents to express their opinion concerning the non-acceptance, the contents were gathered, and a content analysis was extensively done in two phases, the first one to understand the main categories for each scenario and the second phase to go deeper into the specificities of the contents exploring a second level of categorisation factors. The content analysis was made with the support of NVIVO software, and several meetings were held with a team of aviation professionals when there were doubts in the category attribution exercise.

When first reviewing the type of contents of the two different questions, it was observed that there was not a clear distinction between the type of contents that were only related to factors that would turn an opinion to acceptable and the factors that were only related to ethical concerns. Instead, for most of the cases, respondents highlighted the factors why they would not accept the scenario case. In this sense, we took the decision to present the results together as if it was the case of having only one question bringing up the reasons that are behind the non-acceptance of each scenario case.

3.6 Results

Our starting point was a population of 231 aviation professionals, 78% with more than 10 years of experience working in the aviation European market. Within this group, we found 171 professionals mentioning non-acceptance for, at least, one of the scenarios cases. On this basis a total of 2395 contents were analysed and categorised to understand the reasons why AI-based systems were seen as non-ethically acceptable.

The number of aviation professionals that showed non-acceptance for at least one of the eight cases was the following (Fig. 3.1)

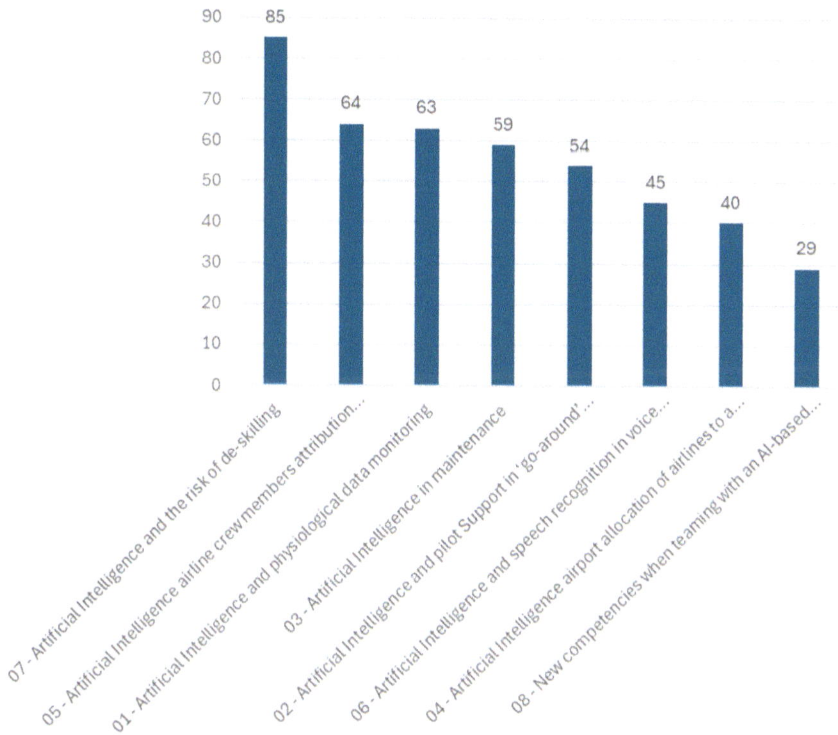

Fig. 3.1 Number of aviation professionals that showed non-acceptance per case

Scenario case 01 *Concerning physiological data monitoring in the cockpit* received the highest number of contents (580) and scenario case 08 concerning *teaming up with an AI-based system* the lower number of contents (156).

The results are distributed in Table 3.1, considering the 16 first-level categories, and referring each scenario case, from the highest frequency to the lowest frequency of responses and the correspondent percentage:

We can observe that 75% of the ethical non-acceptance motives is explained by the first four categories: *AI-based systems*, *human impact*, *data*, and *safety*. Detaining *AI-based systems* and *human impact* 58% of the explanation.

The following 16% is explained by the following categories: *Competences impact*, *training*, *rejection of AI*, *unfair bias*, and *explainability*.

The 9% left are explained by the lowest frequencies representing the categories of: *AI testing*, *fairness*, *regulation and oversight*, *market*, *operations*, *acceptance factors*, and *knowledge*.

The highest category of non-acceptance for all cases is the *AI-based systems* itself, followed by the category concerning *human impact*, apart from case 04.

Table 3.1 Distribution of the first-level categories by total and by case scenario

First-level categories	All cases (% & No. of contents)	Case 01 (% & No. of contents)	Case 02 (% & No. of contents)	Case 03 (% & No. of contents)	Case 04 (% & No. of contents)	Case 05 (% & No. of contents)	Case 06 (% & No. of contents)	Case 07 (% & No. of contents)	Case 08 (% & No. of contents)
AI-based systems	30% 711	20% 115	33% 109	33% 94	47% 99	24% 81	42% 93	19% 53	43% 67
Human impact	28% 662	31% 181	28% 93	37% 105		26% 89	39% 88	18% 50	36% 56
Data	11% 269	22% 129				34% 115	11% 25		
Safety	6% 140	12% 70	10% 33	5% 13			4% 10	2% 6	5% 8
Competences impact	4% 103							38% 103	
Training	3% 78							22% 60	12% 18
Rejection of AI	3% 66	4% 26	6% 19	2% 7			3% 7		4% 7
Unfair bias	3% 63				30% 63				
Explainability	3% 60		6% 21	14% 39					
AI Testing	2% 55		9% 30	9% 25					
Fairness	2% 45					13% 45			

(continued)

Table 3.1 (continued)

First-level categories	All cases (% & No. of contents)	Case 01 (% & No. of contents)	Case 02 (% & No. of contents)	Case 03 (% & No. of contents)	Case 04 (% & No. of contents)	Case 05 (% & No. of contents)	Case 06 (% & No. of contents)	Case 07 (% & No. of contents)	Case 08 (% & No. of contents)
Regulation and oversight	2% 44	4% 21			6% 13	3% 10			
Market	2% 36				17% 36				
Operations	1% 25		8% 25						
Acceptance factors	1% 22	4% 22							
Knowledge	1% 16	3% 16							
Total: % and No.	100% 2395	24% 580	14% 330	12% 283	9% 211	14% 340	9% 223	11% 272	7% 156

The category *data* emerges as the third category for case 01, case 05, and case 06. These three cases are the ones that directly involve human data being collected and processed by an AI-based system.

Safety is the fourth category that appears in all cases except case 04 and case 05 which are the ones that are not linked to a flight operation.

The only case that does not follow the generic tendency is case number 04 (*artificial intelligence airport allocation of airlines to a terminal*), the only one where the human is not interacting directly with the AI-based system. The non-acceptance of this case sits within the system itself, unfair bias, market factors, and the need for regulation.

Some categories are strongly shown in terms of number of contents but are contingent to a specific case, namely: (1) the one concerning the competences impact, just mentioned in case 07 that describe a situation linked to de-skilling; (2) the one about unfair bias only in case 04 where a situation of possible attribution of market benefits is described; (3) the category about market also is only relevant for case 04; (4) the category concerning fairness that are linked only to case 05 describing the attribution of flights to crews; (5) the category on operations that is related to case 02 describing a flight "go-around" manoeuvre; and (6) two categories—*acceptance factors* and *Knowledge* that appear only for case 01 the one concerning physiological data monitoring.

Another observation is that category *rejection of AI* despite having less power of explanation (66 responses overall) is mentioned in most of the cases where a human interacts directly with an AI-based system in a high safety risk situation.

Based on Table 01 and adding the second level of analysis (subcategories), having as criteria of presentation that the subcategories would have at least 10 responses and explain at least 10% of the corresponding first-level category, the results per case were the following:

3.6.1 Artificial Intelligence and Physiological Data Monitoring—Results Presentation and Discussion

Scenario case 01 was concerning a single-pilot operation where the pilot was supported by an AI-based system operated from the ground. The AI-based system would measure physiological data from the pilot for the purpose of cognitive workload assessment.

Case one resulted in a 35% of non-acceptance (52% of acceptance and 13% of no opinion), and this 35% represented a sum of 580 contents.

The case gathered the highest number of contents representing 24% of all contents of the study (580 of 2395).

From the 16 first-level categories in this study, case 01 was present in half of them: *AI-based systems*, *human impact*, *data*, *safety*, *rejection of AI*, *acceptance factors*, *regulation and oversight*, and *knowledge*.

When analysing the subcategories if we apply the criteria (Table 3.2): explaining at least 10% of the category and having at least 10 responses, the results for non-acceptance for case 01 were:

Reasons for ethical non-acceptance of *artificial intelligence and physiological data monitoring* are mainly linked to lack of assurance that the AI-based system performs as expected, to the fear of suffering professional threats, to the feeling of being personally "invaded" by AI, to mistrust that data will be used to a good intent, and to the need of human command and control of the AI-based system.

Table 3.2 Distribution of categories and subcategories more meaningful for case 01

Category	Subcategory	Subcategory responses	% against the category
AI-based systems	Compensating for mistrust in AI	12	10
	Ensuring proper operational domain definition	15	13
	Ensuring system performance	**27**	**23**
	Ensuring human oversight	20	17
Human impact	Psychological impacts	24	13
	Professional impacts	**40**	**22**
	Personal invasion	**46**	**25**
	Mistrust	29	16
Data	Data Protection	39	30
	Data good usage assurance	**57**	**44**
	No data sharing	18	14
Safety	Demonstration of safety	16	23
	Threats aviation safety	23	33
	Human command and control	**29**	**41**
Rejection of AI	Unacceptable situation	13	50

Bold is the "Subcategory" with high frequency of responses within the "Category"

3.6.2 Artificial Intelligence and Pilot Support in "Go-Around" Situations—Results Presentation and Discussion

Scenario case 02 was concerning a two pilot in cockpit supported by an AI-based systems for "go-around" situations. If a threshold was about to be exceeded, the system would provide an acoustic warning and automatically initiate the "go-around" manoeuvre; however, the pilot could overide this decision.

Case 02 resulted in 34% of non-acceptance (50% tendency for acceptance and 16% of no opinion). This 34% represented a sum of 330 contents meaning 14% of the overall 2395 contents.

From the 16 first-level categories in this study, case 02 was represented in 7: *AI-based systems*, *human impact*, *safety*, *rejection of AI*, *explainability*, *AI testing*, and *operations*.

When analysing the subcategories when applying the criteria (Table 3.3): explaining at least 10% of the category and having at least 10 responses, the results for non-acceptance for case 02 were:

Reasons for ethical non-acceptance of *artificial intelligence and pilot support in "go-around" situations* are mainly linked to the lack of human oversight assurance, to accountability impacts in case of failure, to the fact that this situation is a threat to aviation safety, to the need for deep knowledge about the AI-based systems procedures, and to the fact that the operation of landing of an aircraft is more than just the "go-around" manoeuvre criteria.

Table 3.3 Distribution of categories and subcategories more meaningful for case 02

Category	Subcategory	Subcategory responses	% against the category
AI-based systems	Ensuring operational mitigations	19	17
	Ensuring human oversight	**49**	**45**
	Compensating for mistrust in AI	15	14
Human impact	Psychological impacts	13	14
	Human command and control	12	13
	Accountability impacts	**35**	**38**
Safety	**Threats aviation safety**	**18**	**55**
Explainability	**Need for deep knowledge on the AI procedures**	**10**	**48**
Operations	**Landing is more than GA criteria**	**10**	**40**

Bold is the "Subcategory" with high frequency of responses within the "Category"

Table 3.4 Distribution of categories and subcategories more meaningful for case 03

Category	Subcategory	Subcategory responses	% against the category
AI-based systems	Compensating for mistrust in AI	15	16
	Ensuring human oversight	**35**	**37**
	Ensuring system performance	11	12
Human impact	Professional impacts	27	26
	Accountability impacts	**51**	**49**
Explainability	**Need for knowledge of the calculation methods**	**12**	**31**

Bold is the "Subcategory" with high frequency of responses within the "Category"

3.6.3 Artificial Intelligence in Maintenance—Results Presentation and Discussion

Scenario case 03 was concerning an aviation maintenance expert that would access the aircraft based on an AI-based system that evaluate the condition of the aircraft through a colour scheme mechanism, and the expert would have at the end to sign off the airworthiness of the aircraft.

Case 03 resulted in 35% of non-acceptance (49% acceptance and 16% of no opinion). This 35% represented a sum of 283 contents, meaning 12% of the 2395 overall contents.

From the 16 first-level categories, case 03 was represented in 6: *AI-based systems*, *human impact*, *safety*, *rejection of AI*, *explainability*, and *AI testing*.

When analysing the subcategories and applying the criteria (Table 3.4): explaining at least 10% of the category and having at least 10 responses, the results for non-acceptance for case 03 were:

Reasons for ethical non-acceptance of *artificial intelligence in maintenance* are mainly linked to the lack of human oversight, to the accountability impacts in case of error, and to the need of knowledge concerning the calculation methods done by the AI-based system.

3.6.4 Artificial Intelligence Airport Allocation of Airlines to a Terminal—Results Presentation and Discussion

Scenario case 04 was concerning an AI-based system function attributing airport gates to airlines, in order to minimise the walking distance of passengers. The AI-based system was trained with the data set from the domestic airline of that airport and in consequence possibly giving advantage to some airlines despite paying the same airport fees.

Table 3.5 Distribution of categories and subcategories more meaningful for case 04

Category	Subcategory	Subcategory responses	% against the category
AI-based systems	Ensuring continuous safety monitoring	10	10
	Ensuring system performance	**31**	**31**
	Ensuring human oversight	27	27
Unfair bias	**System should ensure fairness**	**20**	**32**
	Bias monitoring and correction in the system	15	24
Market	Bias specific rules installed in the system	10	16
	Fear of corruption	**10**	**28**

Bold is the "Subcategory" with high frequency of responses within the "Category"

Case 04 resulted in 24% of non-acceptance (59% acceptance and 17% of no opinion). This 24% represented a sum of 211 contents which stand for 9% of the 2395 contents of this research.

From the 16 first-level categories in this study, case 04 was represented in 4: *AI-based systems, unfair bias, market,* and *regulation.*

When analysing the subcategories if we apply the criteria (Table 3.5): explaining at least 10% of the category and having at least 10 responses, the results for non-acceptance for case 04 were:

Reasons for ethical non-acceptance of *artificial intelligence airport allocation of airlines to a terminal*, are mainly linked to lack of system performance assurance, doubt that the system ensure fairness treatment and, fear of corruption when managing the data.

3.6.5 Artificial Intelligence Airline Crew Members Attribution to Flights—Results Presentation and Discussion

Scenario case 05 described an airline attributing family-friendly rosters to pilots, done through the means of an AI-based system that worked based on voluntary data given by the pilot about her/his personal social/family environment.

Case 05 resulted in 38% of non-acceptance (45% acceptance and 17% of no opinion). This 38% represented a sum of 340 contents, meaning 14% of the 2395 overall number of contents.

From the 16 first-level categories, case 05 was represented in 5: *AI-based systems, human impact, data, fairness,* and *regulation.*

Table 3.6 Distribution of categories and subcategories more meaningful for case 05

Category	Subcategory	Subcategory responses	% against the category
AI-based systems	Ensuring system transparency	15	19
	Ensuring human oversight	**22**	**27**
Human impact	Personal invasion	29	33
	Professional impacts	**38**	**43**
	Mistrust	11	12
Data	Data good usage assurance	38	33
	No data sharing	15	13
	Data protection	**53**	**46**
Fairness	**Fear of discrimination**	**11**	**24**

Bold is the "Subcategory" with high frequency of responses within the "Category"

When analysing the subcategories when applying the criteria (Table 3.6): explaining at least 10% of the category and having at least 10 responses, the results for non-acceptance for case 05 were:

Reasons for ethical non-acceptance of *artificial intelligence airline crew members attribution to flights* are mainly linked to lack of AI-based system oversight, fear of professional negative impacts, the need for data protection, and the fear of being discriminated.

3.6.6 Artificial Intelligence and Speech Recognition in Voice Communications—Results Presentation and Discussion

Scenario case 06 deals with an AI-based system monitoring voice communication and identifying misunderstanding between pilots and air traffic controllers (ATCO). In case of misunderstanding, the AI-based systems request confirmation from both the pilot and the controller; however, some voice patterns due to individual physiognomy or originated by non-native speakers were more misinterpreted than others, and in consequence some professionals are more subjected to confirmation from the AI-based system.

Case 06 resulted in 29% of non-acceptance (45% acceptance and 26% of no opinion). This 29% represented a sum of 223 contents, which constitute 9% of the sum of the total contents (2395).

From the 16 first-level categories in this research, case 06 was represented in 5: *AI-based systems*, *human impact*, *data*, *safety*, and *rejection of AI*.

Table 3.7 Distribution of categories and subcategories more meaningful for case 06

Category	Subcategory	Subcategory responses	% against the category
AI-based systems	**Ensuring proper operational domain definition**	**23**	25
	Ensuring system performance	21	23
	Ensuring human oversight	11	12
	Ensuring continuous safety monitoring	10	11
Human impact	Professional impacts	15	17
	Fairness	**43**	**49**
Data	**Data misuse risks**	**10**	**40**

Bold is the "Subcategory" with high frequency of responses within the "Category"

When analysing the subcategories if we apply the criteria (Table 3.7): explaining at least 10% of the category and having at least 10 responses, the results for non-acceptance for case 06 were:

Reasons for ethical non-acceptance of *artificial intelligence and speech recognition in voice communications* are mainly linked to the lack of proper operational domain definition, the doubt of fairness assurance, and the risk of data misuse.

3.6.7 Artificial Intelligence and the Risk of De-skilling—Results Presentation and Discussion

Scenario case 07 refers to a futuristic situation where an AI-based system ensures airspace safe separation between aircrafts by detecting conflicts and determine resolutions in a fully automated way. This AI-based system is supervised by a human air traffic controller who can take over in safety-related events which occurs in a very rare occasion for non-significant situations. The AI-based system suffered a severe cyberattack, and the human must substitute the system.

Case 07 resulted in 49% of non-acceptance (33% acceptance and 18% of no opinion). This 49% represented a sum of 272 contents, constituting 11% of the 2395 total number of contents.

From the 16 first-level categories, case 07 was represented in 5: *AI-based systems, human impact, safety, competences impact,* and *training.*

When analysing the subcategories if we apply the criteria (Table 3.8): explaining at least 10% of the category and having at least 10 responses, the results for non-acceptance for case 07 were:

Reasons for ethical non-acceptance of *artificial intelligence and the risk of de-skilling* are mainly linked to the lack of operational mitigation measures, the

Table 3.8 Distribution of categories and subcategories more meaningful for case 07

Category	Subcategory	Subcategory responses	% against the category
AI-based systems	**Ensuring operational mitigations**	**19**	**36**
	Ensuring human oversight	17	32
Human impact	Professional impacts	13	26
	Psychological impacts	14	28
	Accountability impacts	**20**	**40**
Competences impact	New competences	10	10
	Managing competence training	12	12
	Regular manual practise	**48**	**47**
	Attention to de-skilling	16	16
Training	On the job training	14	23
	Recurrent competence training	**26**	**43**
	Ensuring proper quality training	15	25

Bold is the "Subcategory" with high frequency of responses within the "Category"

accountability negative impacts in case of failure, the need for regular job manual practise, and the need for recurrent training.

3.6.8 New Competencies When Teaming with an AI-Based System—Results Presentation and Discussion

Scenario case 08 is dealing with a situation in the future where an air traffic controller will be, in a 3–5-year future, facing a new way of work: teaming with an AI-based system. The AI ATCO will be capable of communicating with the human ATCO, as per current day human team operations. The human ATCO will go through training and competence development to be prepared to team up with AI.

Case 08 resulted in 20% of non-acceptance (58% acceptance and 22% of no opinion). This 20% represented a sum of 156 contents, which constitutes 7% of the overall results (2395).

From the 16 first-level categories in this study, case 08 was represented in 5: *AI-based systems*, *human impact*, *safety*, *training*, and *rejection of AI*.

Table 3.9 Distribution of categories and subcategories more meaningful for case 08

Category	Subcategory	Subcategory responses	% against the category
AI-based systems	Gaining experience	12	18
	Ensuring human oversight	**24**	**36**
Human impact	**Professional impacts**	**25**	**45**

Bold is the "Subcategory" with high frequency of responses within the "Category"

When analysing the subcategories and applying the criteria (Table 3.9): explaining at least 10% of the category and having at least 10 responses, the results for non-acceptance for case 08 are as follows:

Reasons for ethical non-acceptance of *new competencies when teaming with an AI-based system* are mainly linked to the lack of human oversight assurance and to the possible negative professional impacts.

3.6.9 Presentation of the Categories and Subcategories Considering All Cases

How do the categories and subcategories explain the ethical non-acceptance throughout the different cases? Observations on how the categories and subcategories were impacting each of the cases and how this was compared with the overall results are now being presented.

To have more meaningful results, we took the decision not to present all the disperse and scattered low frequency subcategories; however, we ensure to present and explain at least 70% of each category.

Adding to the above, other criteria was taken: presentation only refers to subcategories that represent 10% or more of the correspondent category. If this criteria were not enough to explain 70% of the first-level category, then the following subcategories would be included until the threshold was reached.

The main results are distributed among the categories: *AI-based systems*, *human impact*, *data,* and *safety*.

3.6.10 Category AI-Based Systems—Subcategories Results Presentation and Discussion

The results were as the following (Table 3.10):

The subcategories that fall under *AI-based systems* are concerning ensuring human oversight of the system, system's performance as expected, the correct operational domain definition, ensuring a set of operation mitigations in order to avoid failure, and having to build-up trust towards the system.

Table 3.10 Distribution of the subcategories more meaningful for category AI-based systems along the cases

Subcategory	All cases	Case 01	Case 02	Case 03	Case 04	Case 05	Case 06	Case 07	Case 08
Ensuring human oversight	**29%** **205**	17% 20	**45%** **49**	37% 35	27% 27	27% 22	12% 11	32% 17	36% 24
Ensuring system performance	**16%** **111**	23% 27	6% 6	12% 11	**31%** **31**	7% 6	23% 21	8% 4	7% 5
Ensuring proper operational domain definition	**10%** **71**	13% 15	7% 8	5% 5	8% 8	4% 3	**25%** **23**	4% 2	10% 7
Ensuring operational mitigations	**10%** **68**	3% 4	17% 19	5% 5	5% 5	10% 8	5% 5	**36%** **19**	4% 3
Compensating for mistrust in AI	**9%** **61**	10% 12	**14%** **15**	16% 15	5% 5	10% 8	2% 2	2% 1	4% 3

Bold is the "Subcategory" with high frequency of responses within the "Category"

The presented subcategories explain 73% of the *AI-based system* category and 22% of all contents of the study.

Need for human oversight is the main reason for case 02 on the "go-around" manoeuvre; ensuring system performance has more replies in case 04 the one about the allocation of airlines to terminals; the proper operational domain definition appears more in case 06 concerning the speech recognition situation; and the assurance of operational mitigations is more linked to case 07 that describes a situation of risk of de-skilling.

As a summary, ethical non-acceptance reasons are heavily linked to AI-based systems concerning the risk of not being oversight by humans, not performing as expected, not having the proper operational domain definition, not having mitigation measures as back up for risk of failure and the need for build-up trust towards AI.

3.6.11 Category Human Impact—Subcategories Results Presentation and Discussion

The results were as the following (Table 3.11):

First observation is that *human impact* is not mentioned in case 04—*Airport allocation of airlines to a terminal*, which seems making sense as it describes a situation where a system is impacting airlines and not humans directly.

The subcategories that fall under *human impact* are concerning the professional negative impacts, the accountability in case of failure, feeling personally "invaded", psychological negative impacts, and the need for fairness.

These presented subcategories explain 78% of the *human impact* category and 22% of all contents of the study.

Table 3.11 Distribution of the subcategories more meaningful for category human impact along the cases

Subcategory	All Cases	Case 01	Case 02	Case 03	Case 05	Case 06	Case 07	Case 08
Professional impacts	**25%** **166**	22% 40	9% 8	26% 27	**43%** **38**	17% 15	26% 13	**45%** **25**
Accountability impacts	**19%** **128**	7% 13	38% 35	**49%** **51**	4% 4	3% 3	40% 20	4% 2
Personal invasion	**13%** **85**	25% 46	9% 8		**33%** **29**	2% 2		
Psychological impacts	**11%** **71**	13% 24	14% 13	4% 4		9% 8	**28%** **14**	14% 8
Fairness	**10%** **69**	9% 16	3% 3	2% 2	1% 1	**49%** **43**	2% 1	5% 3

Bold is the "Subcategory" with high frequency of responses within the "Category"

Professional negative impacts is the subcategory that better explain the ethical non-acceptance concerning *human impact*, this is very heavily shown in case 08 that asks about the acceptance of teaming with an AI-based system (45%) and in case 05 where personal information from the pilot and from her/his family was the basis for AI-based system crew allocation definition (43%).

Accountability impacts is the subcategory more present in case 03 the one that would ask the maintenance expert, to sign off a decision on the aircraft airworthiness, done by the AI-based system.

Psychological negative impacts was first ranked in case 07 where a situation of risk of de-skilling was described, and the ATCO could not perform her/his work successfully anymore. The only case where psychological impacts are not mentioned is case 05 that speaks about personal data sharing.

The subcategory *fairness* appears as more frequent in case 06 that describes a situation where the AI-based system could discriminate the ATCO on the basis of her/his voice traits.

3.6.12 Category *Data*—Subcategories Results Presentation and Discussion

The results were as the following (Table 3.12):

The category *data* is strongly present in case 01, 05, and 06, cases directly describing situations where data, in particular personal data, is used by AI. On the other hand, this category does not appear in any of the other cases.

The subcategories that better explain the ethical non-acceptance are the doubt that the data is used only to a good intent and not serving other non-desirable purposes; data protection issues; the decision of not sharing data with the AI; and the risks of data misuse, namely manipulation or corruption.

Table 3.12 Distribution of the subcategories more meaningful for category data along the cases

Subcategory	All Cases	Case 01	Case 05	Case 06
Data good usage assurance	**37%** **100**	**44%** **57**	33% 38	20% 5
Data protection	**37%** **99**	30% 39	**46%** **53**	28% 7
No data sharing	**13%** **36**	**14%** **18**	**13%** **15**	12% 3
Data misuse risks	**10%** **28**	8% 10	7% 8	**40%** **10**

Bold is the "Subcategory" with high frequency of responses within the "Category"

These presented subcategories explain 98% of the *data* category and 11% of all contents of the study.

Data good usage assurance is the subcategory that is more relevant when considering case 01, when dealing with physiological data monitoring, the professionals have doubts that these data will not serve other intentions rather aviation improvement matters.

Data protection is the subcategory more prominent in case 05 as the situation describes the usage of personal data from the person and related family as source for the AI-based system decision.

No data sharing subcategory is more balanced distributed along the cases, with slightly higher relevance in case 01 concerning physiological data monitoring.

The subcategory of *data misuse risk* is more present in case 06 where there seems to be a high risk of discrimination dependent on the characteristics of people's voice.

In sum, *data* is a very ethical relevant factor explaining non-acceptance. Professionals emphasise the risk of data usage not to a good intent or even misuse of their data, they declare that personal data should not be shared with AI-based systems, and data protection should be ensured.

3.6.13 Category *Safety*—Subcategories Results Presentation and Discussion

The results were as the following (Table 3.13):

The category *safety* is not very highly represented but is mentioned in all cases. The main subcategories are about: AI as a threat to aviation safety, that human should be in command and control of the system, that there is a need for demonstration that the situation is safe, and the need for regulation to be defined and oversight to be done.

These presented subcategories explain 98% of the *safety* category and 6% of all contents of the study.

Table 3.13 Distribution of the subcategories more meaningful for category safety along the cases

Subcategory	All cases	Case 01	Case 02	Case 03	Case 06	Case 07	Case 08
Threats aviation safety	**34%** **48**	**33%** **23**	**55%** **18**	8% 1	40% 4	33% 2	
Human command and control	**26%** **36**	**41%** **29**	12% 4	23% 3			
Demonstration of safety	**22%** **31**	**23%** **16**	15% 5	8% 1	30% 3	**50%** **3**	38% 3
Regulation and oversight	**16%** **22**		18% 6	**62%** **8**	30% 3	17% 1	50% 4

Being a *threatto aviation safety* is the subcategory that gathers more responses in case 01 and in case 02, seems that professionals do not accept being physiologically monitored by an AI-based system, and that this situation could lead to an aviation safety concern; also having AI deciding about a "go-around" situation is non-acceptable as it also jeopardises aviation safety.

Human command and control is the subcategory that gathers more contents in category *safety* concerning case 01, in the sense that humans should be in control of themselves and not being monitored by AI.

Subcategory *demonstration of safety* is the one that shows a higher weight in case 01, concerning physiological data monitoring, and also plays a role in case 07 related to the risk of de-skilling.

Concerning the subcategory *regulation and oversight,* it was mentioned in all cases except case 02. It is more relevant in case 03 where a maintenance professional is challenged to sign off a decision made by an AI-based system.

3.6.14 Other Categories—Subcategories Results Presentation and Discussion

As already mentioned, 75% of the ethical reasons for non-acceptance were explained by the four main categories (and consequent subcategories) namely: *AI-based systems*, *human impact*, *data,* and *safety*.

All the other categories represent each of them less than 5% of the overall non-acceptance comments.

Despite being not very representative, as the present research is an exploratory study, we decided to share more information about the least mentioned categories and correspondent subcategories for the purpose of understanding other type of contents mentioned.

It was decided to compile these categories on a single table that represent 16% of the overall study contents. Each subcategory, individually, explains more than 2% and less than 6% of the global results.

Moreover, to harmonise the presentation, we took the same criteria as before, only taking into consideration the subcategories that would explain 70% of the correspondent category.

The results were as the following (Table 3.14):

From the results, we can observe that the category *competences impact* represented mainly by: the need for recurrent manual job practise; the concerning of not getting de-skilled, managing competences by the means of training, and developing new competences. All of these reasons are only applied to case 07 where an ATCO is being put under pressure by being asked to replace an AI-based system in case of failure.

The category *training* seems to be linked to the above, as the consequent subcategories are concerning the need for recurrent training; the assurance that the training has proper quality; and the need for on job training. These are especially meaningful for case 07 but also for case 08 when the professionals are put in a situation where they have to perform their job teaming up with an AI-based system.

The following category is named *rejection of AI*, these contents have to do with a generic opinion against AI, and we encounter these contents mainly in the cases that link human to AI-based systems in an aviation safety situation. Subcategories show the situation as unacceptable that AI should not be applied to aviation and AI is not to be trusted.

Unfair bias is a category that applies only to case 04 concerning the allocation of certain airlines to certain airport terminals, and subcategories expressed that the system should ensure fairness, that bias should be monitored and corrected, and the system itself should contemplate bias specific rules.

Explainability is the last category presented in the upper table, and professionals mentioned that they would need to know about the calculation methods of the system, its procedures, a deep knowledge of the overall system, and also the need for transparency on the way the systems were trained.

As a closure of the results presentation, we would like to mention what type of other categories, despite the fact that they are shown with very low frequencies, were explaining the 9% left of the present study.

These are concerning: (1) *AI testing*, highlighting the need for demonstration, validation, and having a transition period—linked to case 02 and case 03; (2) *Fairness*, given pertinence to the fear of discrimination, ensuring fair decisions and ethical assurance—linked to case 05; (3) *Regulation and oversight*, reinforcing the need for regulation to be developed and oversight to be done by the means of inspection and by an external organisational authority—in particular for case 01, case 04, and case 05; (4) *Market* matters, highlighting the fear of corruption and the potential economic impacts on different stakeholders—linked to case 04; (5) *Operations*, meaning technical operational measures concerning case 02; (6) *Acceptance factors*, mentioning the need of a human role checking, correcting, commanding, and controlling an AI-based system that should only support humans; and (7) Knowledge, stressing the importance of knowledge of the systems governance and reliability.

Table 3.14 Distribution of the subcategories more meaningful for several categories along the cases

Category	Subcategory	All Cases	Case 01	Case 02	Case 03	Case 04	Case 05	Case 06	Case 07	Case 08
Competences impact	**Regular manual practise**	**47%** **48**							47% 48	
	Attention to de-skilling	**16%** **16**							16% 16	
	Managing competence training	**12%** **12**							12% 12	
	New competences	**10%** **10**							10% 10	
	Subcategories	Representing 83% of category & 3% of all content of the study								
Training	**Recurrent competence training**	**33%** **26**							43% 26	
	Ensuring proper quality training	**31%** **24**							25% 15	50% 9
	On the job training	**23%** **18**							23% 14	22% 4
	Subcategories	Representing 87% of category & 3% of all content of the study								
Rejection of AI	**Unacceptable situation**	**32%** **21**	50% 13	37% 7	14% 1					
	Refusal of AI in aviation	**29%** **19**	15% 4	21% 4	14% 1			71% 5		71% 5
	AI mistrust	**15%** **10**	8% 2	16% 3	71% 5					
	Subcategories	Representing 76% of category & 2% of all content of the study								
Unfair bias	**System should ensure fairness**	**32%** **20**				32% 20				
	Bias monitoring and correction in the system	**24%** **15**				24% 15				

(continued)

Table 3.14 (continued)

Category	Subcategory	All Cases	Case 01	Case 02	Case 03	Case 04	Case 05	Case 06	Case 07	Case 08
	Bias specific rules installed in the system	**16%** **10**				16% 10				
	Subcategories	Representing 71% of category & 2% of all content of the study								
Explainability	**Need for knowledge of the calculation methods**	**20%** **12**			31% 12					
	Need for deep knowledge on the AI procedures	**20%** **12**		48% 10	5% 2					
	Deep knowledge of the system	**18%** **11**		10% 2	23% 9					
	Transparency of AI training	**13%** **8**		24% 5	8% 3					
	Subcategories	Representing 72% of category & 2% of all content of the study								

3.7 Conclusions

The opinion of 171 aviation professionals on the motives why they do not find ethically acceptable the application of AI-based system to aviation gives fertile ground to the scientific community to further investigate on ethics for AI, to the authorities to be aware of priority factors impacting the regulation of the sector and to the industry emphasising ethics importance when developing their products.

Higher number of comments and higher level of consequent categorisation resulted for case 01 concerning human physiological data monitored by AI, case 05 concerning the crew allocation to flights done by an AI-based system on the basis of their personal data, and case 02 concerning an aircraft "go-around" manoeuvre initiated by an AI-based system. We can then conclude that aviation professionals show more ethical concerns for the cases either directly linked to a human-AI interaction or linked to an aviation safety situation.

On the other hand, the less ethically negatively evaluated cases were case 04 describing airline allocation to terminals done by the means of AI and case 08 where professionals are being prepared, by the means of competence development, to team up with an AI-based system.

Based on the results, we conclude that aviation professionals have ethical concerns about the technology itself (30%), they are worried about the consequent human negative impact when using AI (28%), they have serious doubts about how their data is being used by the technology (11%), and they somehow think that AI-based systems will jeopardise aviation safety (6%).

We could observe that *AI-based system* category was highly mentioned in all cases (711 in 2395 comments), and we can consider that the main ethical concerns are linked to the need for human oversight of the AI-based system and the assurance that the system performs as expected.

Apart from case 04 that is not linked directly with aviation safety neither with human direct interaction, all the other cases heavily detain *human impact* reasons for being not accepted (662 in 2395). Ethical reasons for non-acceptance of AI are highly linked to consequent human impacts. Aviation professional shows concerns towards the fear of experiencing professional threats, problems linked to responsibility and accountability, feeling of being personally "invaded" by the technology, suffering psychological discomfort, suffering discrimination, and not trusting in AI. We conclude that human impact is an important complex phenomenon that should be carefully further studied and developed if we want to have a successful introduction of AI in an aviation professional working context.

Despite nonbeing a reason for all cases, professionals mentioned as ethically nonacceptable the risk of data misuse, highlighting data protection as critical.

Finally, professionals argued that safety is also an ethical concern, AI-based systems are seen as jeopardising aviation safety, and humans should command, control, and oversight AI.

Based on our theoretical sources, we assumed that the ethical reasons behind non-acceptance were mainly related to threats to humans (on personal context and professional context) and with information sharing, namely having an AI-based system ethically transparent.

From the results just mentioned before, we have a good basis to say that the assumptions were right; however, other factors explain how the aviation professionals find ethically nonacceptable to work with AI in the aviation sector. Factors linked to other characteristics of the technology itself, to data management, and to safety are clearly some of these examples. We invite the scientific and research community to continue exploring ethics in AI, using other high-risk sectors of activity, or other methodologies in order to reinforce this field of knowledge.

Acknowledgements We would like to thank dearly the European Union Aviation Safety Agency (EASA) particularly the AI Programme team lead by Guillaume Soudain and we invite all to read EASA edition: "Ethics for AI in Aviation - Aviation Professionals Survey Results 2024/2025".

References

Ali W, Khan AZ (2024) Factors influencing readiness for artificial intelligence: A systematic literature review. Data Sci Manage. https://doi.org/10.1016/j.dsm.2024.09.005

EASA (2023) EASA artificial intelligence roadmap 2.0: a human-centric approach to AI in aviation. EASA. https://www.easa.europa.eu/

EASA (2024) Artificial intelligence concept paper (Issue 2.0): guidance for level 1 & 2 machine-learning applications. EASA. https://www.easa.europa.eu/

European Commission. (n.d.). Why do we need the Charter? The charter of fundamental rights: what it covers and how it relates to the European Convention on Human Rights. https://ec.europa.eu/

European Commission (2020) assessment list for trustworthy artificial intelligence (ALTAI) for self-assessment. https://digital-strategy.ec.europa.eu/

European Commission (2024) Ethics guidelines for trustworthy AI. High-level expert group on artificial intelligence. https://digital-strategy.ec.europa.eu/

European Parliament (2012) Charter of fundamental rights, C 326/391. https://eur-lex.europa.eu/legal-content/EN/TXT/PDF/?uri=CELEX:32012P0228&from=EN

European Parliament (2023) EU AI Act: first regulation on artificial intelligence. https://www.europarl.europa.eu/

European Parliament. (2024). *Regulation (EU) 2024/1689 of the European Parliament and of the Council of 13 June 2024 laying down harmonised rules on artificial intelligence and amending various regulations and directives (Artificial Intelligence Act).* https://eur-lex.europa.eu/

Introna L (2024) Phenomenological approaches to ethics and information technology. In: Zalta EN, Nodelman U (eds) The Stanford encyclopedia of philosophy (Summer 2024 Edition). https://plato.stanford.edu/archives/sum2024/entries/ethics-it-phenomenology/

Jöhnk J, Weißert M, Wyrtki K (2021) Ready or not, AI comes—an interview study of organizational AI readiness factors. Bus Inf Syst Eng 63(1):5–20. https://doi.org/10.1007/s12599-020-00685-1

Kabashkin I, Misnevs B, Zervina O (2023) Artificial intelligence in aviation: New professionals for new technologies. Appl Sci 13:11660. https://doi.org/10.3390/app132111660

Müller VC (2023) Ethics of artificial intelligence and robotics. In: Zalta EN, Nodelman U (eds) The Stanford encyclopedia of philosophy (Fall 2023 Edition). https://plato.stanford.edu/archives/fall2023/entries/ethics-ai/

Parker S (2014) Beyond motivation: job and work design for development, health, ambidexterity, and more. Annu Rev Psychol 65:661–691. https://doi.org/10.1146/annurev-psych-010213-115208

United Nations Foundation (2023) The universal declaration of human rights is turning 75: Here's what you need to know. https://www.unfoundation.org/

Yeung K, Howes A, Pogrebna G (2020) AI governance by human rights–centered design, deliberation, and oversight. In: Dubber MD, Pasquale F, Das S (eds) The Oxford handbook of ethics of AI. Oxford University Press. https://doi.org/10.1093/oxfordhb/9780190067397.013.23

Yogesh K et al (2021) Artificial intelligence (AI): multidisciplinary perspectives on emerging challenges, opportunities, and agenda for research, practice, and policy. Int J Inf Manage 57:101994. https://doi.org/10.1016/j.ijinfomgt.2019.08.002

4 Artificial Intelligence in Agriculture: Revolutionizing Methods and Practices in Portugal

Maria José Sousa

Abstract

As a result of ongoing technical breakthroughs, artificial intelligence (AI) has become an important topic for researchers and industry experts. The creation of machines that mimic human cognitive functions including learning, reasoning, and self-correction is referred to as artificial intelligence (AI). Its numerous uses in various industries have demonstrated its growing accuracy and effectiveness, and agriculture has likewise embraced AI to boost productivity and revenue. In order to fully comprehend the idea, current research, and projects related to AI in agriculture, a review of the literature was conducted. Additionally, this paper takes a practical approach to the potential of AI in agriculture by addressing the emergence of new techniques and practices, utilizing a case study approach, and examining how experts, academics, and agriculture professionals perceive the application of AI in agriculture. It helps build practical applications by providing insights that are relevant to both academic and real-world contexts.

Keywords

Artificial intelligence • Agriculture • Efficiency • Case studies • Quantitative analysis

M. J. Sousa (✉)
University of Lisbon, Lisbon, Portugal
e-mail: maria.jose.sousa@iscte-iul.pt

M. J. Sousa et al. (eds.), *Economic Sustainability Potentiated by Artificial Intelligence*, Contributions to Environmental Sciences & Innovative Business Technology,
https://doi.org/10.1007/978-981-96-8089-4_4

4.1 Introduction

AI stands as a technological development that has captivated both academic discourse and practical applications across diverse industries. Its evolution has been punctuated by redefined definitions, adapting to technological advancements (Kok et al. 2002). AI represents a branch of computer science dedicated to crafting machines capable of emulating human-like cognitive functions, encompassing learning, reasoning, and self-correction.

The far-reaching implications of AI are manifested across various sectors, with its penetration into industries such as manufacturing, health care, marketing, finance, and tourism (Dwivedi et al. 2021). However, one sector that can witness a transformative impact is agriculture, where AI's applications can potentiate a new era of innovation, efficiency, and sustainability.

In explaining AI's role in agriculture, it becomes imperative to understand its overarching definition and how it pertains to this domain. Defined broadly, AI in agriculture represents a confluence of technology and methodology aiming to revolutionize conventional agricultural practices. Its core objective is to expand, automate, and optimize farming operations, thereby enhancing productivity, mitigating risks, and ensuring sustainable food production.

The significance of agriculture as a global economic basis is very high staggering global value of $3.6 trillion and a consistent contribution of 4% to the global GDP over the past two decades, and the agricultural sector highlights the socioeconomic dimensions of numerous nations (FAO 2022). However, this sector has several challenges, ranging from unpredictable weather patterns, crop-damaging pests, and diseases to the deleterious impacts of climate change and desertification (Komarek et al. 2020). These challenges create some threats to agricultural productivity, exacerbating the already mounting pressures due to an expanding of global population.

At the forefront of these challenges stands the United Nations' 2030 Sustainable Goals, notably the Zero-Hunger program, which underscores the imperative of achieving food security sustainably (United Nations 2015). Against this backdrop, the integration of AI into agricultural practices emerges as an instrumental strategy, offering novel methodologies and technological solutions to address these challenges.

The combination of innovation technology, digitalization, and AI stands tries to mitigate the challenges that agriculture faces (Yela Aranega et al. 2022). This paradigm change indicates the need to reevaluate established agricultural practices and technologies, with a renewed emphasis on maximizing the production of crops, minimizing resource use, and decreasing labor costs.

Moreover, the strategic significance of AI in agriculture extends beyond efficiency gains. It fundamentally reshapes the agricultural business models, opening new possibilities for growth, innovation, and sustainability (Toniolo et al. 2019). These models serve as framework stating how value is created and delivered within the agricultural ecosystem.

However, despite the interest and the potential of AI in reshaping agricultural paradigms, the existing literature on AI's role in agriculture remains a work-in-progress (Cavazza et al. 2023). While there is a proliferation of studies examining the potential synergies between AI and agriculture, the depth of scholarly exploration concerning its implications on agricultural business models remains relatively limited.

The literature on AI in agriculture largely encompasses case studies that evaluate and discuss technological solutions and algorithms, spanning diverse agricultural domains (Panpatte and Ganeshkumar 2021). Yet, the scarcity of success stories within the literature, elucidating tangible benefits reaped by farms leveraging AI technologies, is evident (Sood et al. 2022).

While the literature underscores the necessity of coupling AI with complementary technologies like IoT (Internet of things) and robotics to create a cohesive digital ecosystem in agriculture (Chiles et al. 2021; Dal Mas et al. 2023), it also delineates emerging farming techniques such as vertical farming, aquaculture, insect breeding, and precision agriculture (Davies and Wilson 2020; Saad et al. 2021).

AI also have the potential to address multifaceted sustainability issues, including reducing agricultural pollution, optimizing resource utilization, and fostering environmentally friendly practices (Bogomolov et al. 2021; Rao et al. 2018). However, several gaps persist in the academic literature, as the shortage of comprehensive AI-driven business models in agriculture (Cavazza et al. 2023). Despite the emergence of technological solutions and algorithms, the absence of real-world case studies showing AI's potential implications on the future of agriculture remains a reality.

In response to this call for in-depth analysis rooted in real-world scenarios to contribute for the formulation, implementation, and evaluation of public policies, this research adopts a multi-case study methodology. It will provide an in-depth exploration of AI's application in agriculture and its impacts.

The subsequent sections of this article delineate the current literature, develop the research question, elucidate the chosen methodology, present key findings, and engage in comprehensive discussions regarding recommendations for policy, implications, and limitations of the research.

4.2 Literature Review

4.2.1 Conceptualization of AI in Agriculture

Artificial intelligence has garnered significant attention from both scholars and practitioners in recent years. Defined broadly, AI stands as a branch of computer science aiming to develop machines capable of human-like cognitive processes (Kok et al. 2002). Its applications span various industries and notably agriculture (Dwivedi et al. 2021).

The agricultural sector, with a global value of $3.6 trillion and a consistent contribution of 4% to the global GDP, holds immense significance (FAO 2022). However, it faces formidable challenges—unpredictable risks like weather conditions, crop-destroying pests, and diseases, coupled with the adverse impacts of climate change and desertification (Komarek et al. 2020; FAO 2022). The need for sustainable food security has been highlighted as a crucial goal by the United Nations (United Nations 2015).

In response to these challenges, technology, digitalization, and AI emerge as potential solutions (Yela Aranega et al. 2022). The integration of AI in agriculture aims not only to address these challenges but also to enhance productivity and reduce human effort (Saad et al. 2021). The strategic role of AI in agriculture becomes apparent considering its potential to reshape existing methodologies and technologies for maximizing crop yield and ensuring sustainable food production (FAO 2022).

AI's impact on agricultural business models stands as a focal point in this evolving landscape (Toniolo et al. 2019). Defined as the framework through which organizations create value, business models in agriculture are poised for transformation through AI integration (Bagnoli et al. 2019, 2018; Biloslavo et al. 2018). This transformation not only addresses sector-specific challenges but also aligns with sustainability objectives (Biancone et al. 2022; Blackmore et al. 2015).

Sustainability remains a pivotal theme in this context, with AI-driven solutions showing promise in supporting environmentally friendly practices (Bogomolov et al. 2021; Rao et al. 2018).

4.2.2 AI Technology Applied to Agriculture

Precision agriculture, enabled by AI, empowers farmers to make data-driven decisions by harnessing information from diverse sources such as satellite imagery, drones, sensors, and weather forecasts (Qin et al. 2019; Castellini et al. 2016). Through the utilization of machine learning algorithms, farmers can optimize irrigation, fertilization, and pesticide application, resulting in heightened crop yields, diminished resource usage, and enhanced sustainability.

Crop monitoring and disease detection leverage computer vision and machine learning techniques to analyze images of crops captured by drones or satellites (Mohanty et al. 2016; Castellini et al. 2016). By identifying patterns and anomalies in plant health and growth, AI systems can swiftly detect diseases, pests, nutrient deficiencies, and other issues, enabling farmers to take timely actions to mitigate losses.

Predictive analytics plays a crucial role as AI models analyze historical data on weather patterns, soil conditions, crop yields, and market prices to forecast future trends (Hansen 2002; Tao et al. 2018). These forecasts aid farmers in planning planting schedules, managing inventories, optimizing resource allocation, and minimizing risks associated with climate variability and market fluctuations.

Robotic farming is on the rise with AI-powered robots and autonomous vehicles being developed to execute various tasks on the farm (Li and Wang 2017). These tasks include planting seeds, applying fertilizers and pesticides, weeding, and harvesting crops, thereby operating with precision and consistency, reducing labor costs, and increasing operational efficiency.

Supply chain optimization, facilitated by AI technology, streamlines the entire agricultural supply chain from production to distribution (Tao et al. 2018). Machine learning algorithms analyze data on demand forecasts, transportation routes, storage conditions, and market dynamics to reduce waste and ensure the timely delivery of agricultural products to consumers.

In crop breeding and genomics, AI techniques, including genetic algorithms and deep learning, accelerate the breeding of crops with desirable traits (Zhang et al. 2016). By identifying genetic markers associated with these traits, breeders can efficiently develop improved crop varieties.

Finally, AI-powered farm management systems integrate data from various sources, automate routine tasks, and provide insights tailored to specific farm conditions (Li and Wang 2017). These systems aid farmers in planning, monitoring, and decision-making, optimizing resource utilization and maximizing profitability.

4.2.3 Creating a Model of Analysis

The analysis model provided underscores a comprehensive approach to bolstering agricultural productivity and sustainability by integrating advanced technologies and management practices. Each component plays a pivotal role in this endeavor.

Increased production efficiency is central to optimizing agricultural operations by leveraging advancements in machinery, automation, and resource utilization to enhance productivity, minimize waste, and streamline processes. This entails maximizing the use of available resources such as land, labor, and inputs (Rejesus et al. 2003).

Precision in agricultural management harnesses technology like GPS, sensors, and drones to precisely manage farming variables such as planting, irrigation, fertilization, and pest control. By accurately targeting inputs and actions, farmers can optimize yields while minimizing costs and environmental impact (Gebbers and Böhner 2010).

Harvest optimization strategies and technologies are geared toward maximizing yield and quality. This includes adopting timely harvesting practices, employing post-harvest handling techniques, and utilizing equipment such as harvesters and storage facilities to minimize losses and maintain product quality (Ferguson and Burras 2009).

Intelligent water resource management acknowledges the critical role of water in agriculture and employs technology to monitor, conserve, and efficiently distribute water for irrigation. Techniques like drip irrigation, soil moisture sensors, and water recycling systems optimize water usage while sustaining crop growth (Khan and Qazi 2012).

Improvement in quality focuses on enhancing agricultural products through better farming practices, handling methods, and technologies. By adhering to higher quality standards, farmers can increase market value, reduce waste, and meet consumer demands for superior products (Doerge 1999).

Decision support systems provide farmers with data-driven insights and recommendations for informed decision-making. By analyzing data, modeling scenarios, and employing predictive algorithms, these systems assist in crop selection, planting schedules, resource allocation, pest management, and risk mitigation strategies, thereby enhancing overall farm performance and profitability (Singh et al. 2014) (Fig. 4.1).

From the literature emerged, this model of analysis will guide the empirical research and the answer to the main research question: RQ: What are the main differences in perceptions of experts and academics versus professionals of the agriculture sector regarding AI impacts in agriculture? The following sections will be developed to reply to this question.

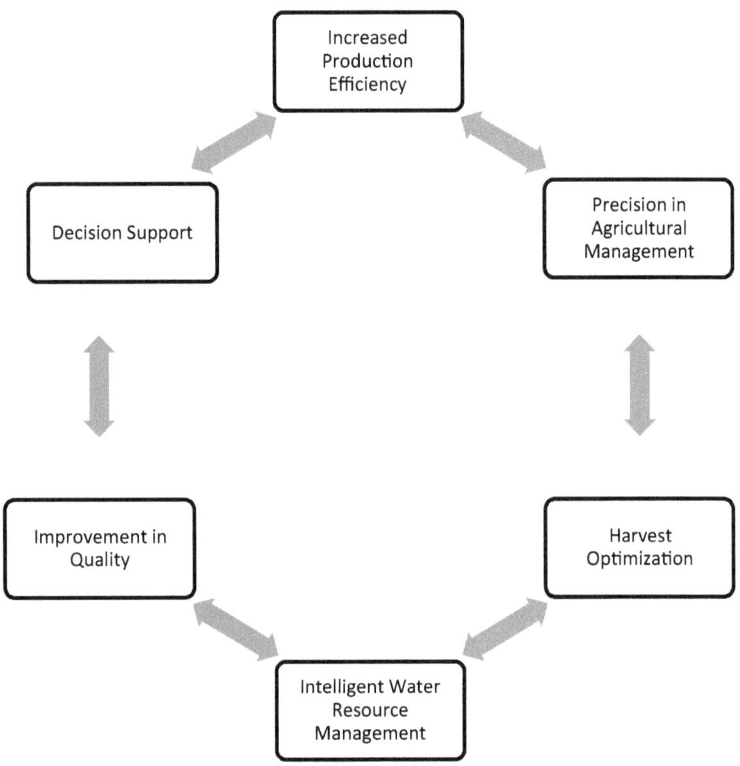

Fig. 4.1 Model of analysis

4.3 EU Public Policies to Promote the Digitalization of Agriculture

European Union (EU) has been actively developing policies and initiatives to promote the use of artificial intelligence in agriculture, aiming to foster innovation, sustainability, and efficiency in the agricultural sector. Some existing EU policies related to AI in agriculture include (Table 4.1).

These policies and initiatives reflect the EU's commitment to fostering AI innovation in agriculture while ensuring responsible and ethical use. They aim to support farmers, promote sustainable practices, and drive digital transformation in the agricultural sector across member states.

4.4 The Context of AI in Agriculture in Portugal

4.4.1 AgTech Companies and Startups in Portugal

Modern agriculture is shaped by technological innovation changing the traditional across Portugal. Several technological startups have emerged redefining agricultural operations with cutting-edge technologies. From software platforms fostering collaboration to environmental sustainability initiatives, these startups are pioneering a revolution in the agriculture sector.

PROGROW stands tall as a software company harnessing the power of state-of-the-art IoT connectivity and data analytics. Its web platform, dedicated to industrial frontline operations, offers a transformative experience, ushering any equipment or workstation into the realm of online efficiency. Similarly, WISECROP, an Online Agricultural Operating System, orchestrates predictive indicators and business management tools, centralizing farm management and amplifying seasonal results. This convergence of technology enables access to vital information on crops, equipment, and irrigation, both on and off the field.

ABOUT AQUA FOOD, an ambitious sustainability project, aspires to produce and commercialize healthy, premium foods with maximal environmental responsibility. Their visionary endeavor metamorphosed into the Bio-Agro Concept startup, epitomizing a futuristic approach to food production. On a parallel trajectory, AGROINSIDER emerges as a guiding light within the agribusiness value chain. Their suite of consulting and technological tools, operating on SaaS solutions, champions the preservation of the invaluable asset—Natural Capital.

Navigating the intricate web of agricultural commerce, AGRI MARKETPLACE emerges as a B2B cloud-based platform facilitating real transactions of agricultural food crops. Within this digital marketplace, fair food trade is fostered through processes that are streamlined, efficient, and transparent, thereby narrowing the gap between farmers and agro-industry buyers. Similarly, WINE WITH SPIRIT, a Winetech company, intertwines innovation with an ancient industry, producing award-winning Portuguese wines that traverse four continents.

Table 4.1 European Union policies to promote the digitalization of agriculture

European AI strategy	The EU's AI strategy focuses on fostering the development and deployment of AI across various sectors, including agriculture. It aims to support research, innovation, and ethical AI use while ensuring compliance with EU values and regulations
Common agricultural policy (CAP) reform	The CAP, a major EU policy for agriculture and rural development, has been undergoing reforms to incorporate digitalization, including AI technologies. The updated CAP seeks to support digital transformation in agriculture, encouraging the adoption of innovative technologies for sustainable and efficient farming practices
Digital Europe Programme	This EU initiative allocates funding to support digital transformation in various sectors, including agriculture. It aims to enhance digital skills, infrastructure, and technologies, fostering the adoption of AI and other digital tools in farming practices
Horizon Europe	The EU's research and innovation framework program, Horizon Europe, includes funding opportunities for AI-related projects in agriculture. It supports research, innovation, and collaboration to develop AI-driven solutions for sustainable agriculture, focusing on areas like precision farming and smart agri-food systems
European data strategy	The EU is working on a comprehensive data strategy to facilitate the sharing and use of agricultural data, which is crucial for AI applications in farming. Efforts to create common data spaces and standards aim to unlock the potential of AI in agriculture by enabling data-driven decision-making
Ethical guidelines for trustworthy AI	The EU has been developing ethical guidelines and regulatory frameworks for AI. These guidelines emphasize the importance of ethical AI design, transparency, accountability, and human oversight, which are relevant considerations in AI applications for agriculture
European Green Deal	The EU's Green Deal sets ambitious sustainability goals, including the reduction of pesticide and fertilizer use. AI technologies in agriculture, such as precision farming, can contribute to achieving these goals by optimizing resource utilization and minimizing environmental impacts

However, beyond technology-driven enterprises, institutions like CREDITO AGRICOLA have evolved from their agricultural roots into universal financial and insurance groups. Their practice of proximity and relationships continues to nurture and support the sector. Complementing this financial landscape, AGRISTARBIO steps in with its organic mineral fertilizer derived from biosolids, fostering sustainability amid the intensifying need for food and feedstock.

ROVENSA specializes in sustainable agricultural solutions, combating the environmental footprint of farming practices. Their commitment to safe, healthy, and nutritious food production resonates across a spectrum of companies like FARMCONTROL, offering IoT cloud software for monitoring agricultural operations, and FLOW TECHNOLOGY, simplifying industry with a manufacturing execution system.

Venturing into specialized domains, AQUAPONICS IBERIA champions sustainable, efficient aquaponics technologies, training and providing consulting services. Meanwhile, AGRIW takes center stage with its Agriculture 4.0 Smart Farming Solution, rooted in artificial intelligence and machine learning, while TERRAPRIMA champions environmental services provided by agro-forestry activities.

Echoing the technological chorus, HARKER lends support to the milk production industry, promoting health, animal welfare, and farm development. Similarly, BIOSANI steps in with biotechnical pheromones for plant protection in organic agriculture, blending expertise with youthful dynamism.

In the realm of quality control, CALSEG assumes a pivotal role, offering inspection services for agri-food products. Simultaneously, IBERO FOREST MASSA pioneers the production of biocarbon, innovating ecological technology through biomass transformation.

Meanwhile, AGRICIENCIA offers information management and viticulture services, and HIDROSOPH consults in irrigation management and environmental services, underlining a collective commitment to sustainability. MAGAGER, entrenched in history since the 1930s, and CUDELL, a platform advocating environmental stewardship, add unique layers to this mosaic of agricultural innovation.

Each organization, whether a technology-driven business or an agent of environmental sustainability, plays a part in the innovation process of AI application to agriculture sector in Portugal.

4.4.2 Portugal Participation in Horizon Europe Projects to Promote Digitalization in Agriculture

Portugal participated in several Horizon Europe projects focused on agriculture, contributing expertise, research, and innovation in various fields related to sustainable farming practices, digitalization, and technological advancements. While the participation in these projects might evolve, here are a few examples of Horizon Europe projects where Portugal has been involved in the agricultural sector (Table 4.2).

Portugal's involvement in these Horizon Europe projects demonstrates its commitment to leveraging technological advancements and innovation for the benefit

Table 4.2 Projects funded by EU to promote the digitalization of agriculture

Project	Portuguese Consortium
SmartAgriHubs: Portugal was part of the SmartAgriHubs project, contributing to the establishment of Digital Innovation Hubs (DIHs) in agriculture. These hubs aimed to facilitate the digital transformation of the agricultural sector by promoting the adoption of innovative technologies, including AI, IoT, and robotics, among farmers and agribusinesses	CONSULAI, TEKEVER, EDIA (Empresa de Desenvolvimento e Infra-Estruturas do Alqueva), UNPARALLEL Innovation, Instituto Nacional de Investigação Agrária e Veterinária, Centro Operativo e de Tecnologia de Regadio, Associação de Beneficiários da Obra da Vigia, e FreedomGrow
IoF2020 (Internet of Food and Farm): Portugal participated in the IoF2020 project, which focused on integrating Internet of things (IoT) technologies in agriculture. This project aimed to enhance farming practices, sustainability, and efficiency by leveraging IoT solutions, which might include AI-driven analytics and decision support systems	UNPARALLEL INNOVATION LDA
AGINFRA+: Portugal contributed to the AGINFRA + project, which aimed to develop an open data infrastructure for the agricultural community. This project sought to provide tools and services, including AI-driven analytics and data management solutions, to support agricultural research and innovation	INESC-ID (Instituto de Engenharia de Sistemas e Computadores, Investigação e Desenvolvimento), LNEC (Laboratório Nacional de Engenharia Civil)
FutureAgriculture: Portugal might have participated in projects like FutureAgriculture, emphasizing sustainable intensification in agriculture through technology adoption. This project likely explored the application of AI, big data analytics, and precision agriculture to enhance agricultural productivity while minimizing environmental impacts	INIAV, Instituto Superior Técnico (IST)
NEFERTITI: Portugal might have been involved in NEFERTITI, focusing on knowledge exchange and innovation uptake among farmers in Europe. While not primarily centered on AI, this project aimed to integrate innovative technologies, including digital solutions and precision agriculture, to improve farming practices	Agroop

of its agricultural sector. However, participation might have varied across different projects and initiatives within Horizon Europe, and new projects may have emerged since my last update. For the most up-to-date and comprehensive insights into Portugal's participation in Horizon Europe projects within the domains of agriculture and artificial intelligence.

4.5 Methodology

Given the aim of exploring AI applications in agriculture in Portugal, the methodology for conducting this research will follow the next steps:

Research Objective Clarification: Clear research objectives that outline the specific aspects of AI in agriculture, to guide the entire research process.
Case Study Selection: Identify and select two innovative case studies in Portugal that have integrated AI technologies or innovative digital solutions in their operations. This case study selection should align with the research objectives and provide a representative example for analysis.
Data Collection: Utilize a mixed-method approach for data collection, including qualitative and quantitative data. Qualitative methods involve document analysis based on the theoretical model developed in this research to gather insights into AI implementation; quantitative methods will consider a sample of observations by questionnaire (Table 4.4) to analyze perceptions of the impacts of AI application to agriculture.
Data Analysis: For qualitative data content analysis will be used to identify recurring themes or patterns. Quantitative data will be statistically analyzed.
Recommendations: Summarize the findings and provide recommendations based on the study's outcomes, highlighting best practices, potential areas for improvement, and implications for public policies in the agricultural sector.

4.6 Empirical Study—The Impact of AI in the Agribusiness in Portugal

4.6.1 Qualitative Analysis

(A) **Portuguese Olive Oil Case Study**

Artificial intelligence can be a vital resource in the production of olive oil in Portuguese agriculture, bringing with it a range of significant benefits. These technological advances can transform traditional agricultural practices, providing substantial improvements in various key areas of olive tree cultivation and olive oil production. The integration of artificial intelligence in olive oil production in

Portugal not only drives efficiency and quality but also promotes a more sustainable and environmentally conscious approach. These benefits can significantly contribute to the competitiveness and excellence of the country's olive oil sector.

Using the model created based on the literature review, the analysis of the application of AI in the Portuguese olive oil production is as follows:

1. **Increased Production Efficiency**: *Esporão*, a prominent olive oil producer, implements AI-driven systems that analyze historical weather data, soil moisture, and plant health metrics. This data-driven approach allows them to precisely schedule irrigation, fertilization, and other cultivation practices, resulting in optimized resource allocation and increased olive oil yield without compromising quality.
2. **Precision in Agricultural Management:** *Sovena*, a leading olive oil company, integrates AI-based drones equipped with imaging technology to monitor olive groves. These drones identify specific areas affected by pests or diseases, enabling targeted interventions that minimize the use of pesticides and herbicides while maintaining a healthy and sustainable olive orchard.
3. **Harvest Optimization:** *Cobrançosa*, an olive oil cooperative, employs AI algorithms that process data from temperature sensors, humidity levels, and fruit ripeness. This technology aids in predicting the optimal harvest time, ensuring that olives are picked at their peak, contributing to enhanced oil quality and efficiency during harvesting.
4. **Intelligent Water Resource Management:** *Herdade das Servas*, utilizing AI-powered irrigation systems, measures soil moisture levels and climate data. This technology allows for precise water delivery to olive trees, reducing water wastage and ensuring the trees receive adequate hydration for optimal growth while conserving a scarce resource.
5. **Improvement in Quality:** *Oliveira da Serra* employs AI-based sensory analysis tools to assess olive oil quality. By analyzing chemical compositions and sensory attributes, they ensure that only the highest-quality olive oil, meeting strict standards, reaches the market, enhancing consumer trust and satisfaction.
6. **Decision Support:** *The National Association of Olive Growers* utilizes AI-powered platforms that process extensive datasets on market trends, climate patterns, and consumer preferences. This data-driven insight equips farmers with valuable information for informed decision-making on plantation management, production strategies, and market positioning to optimize their olive oil business.

(B) **Portuguese Winery Case Study**

In winery, AI stands as an invaluable asset revolutionizing the production processes. Its implementation in Portuguese has reshaped conventional practices and significantly enhanced various facets of grape cultivation and wine production. It leads to higher efficiency and quality but also leads to a more sustainable and environmentally conscious approach.

Using the model created based on the literature review, the analysis of the application of AI in the Portuguese winery is as follows:

1. **Enhanced Production Efficiency**: *Quinta do Vallado*, a renowned winery in Portugal's Douro Valley, utilizes AI-powered systems to analyze weather data, soil moisture, and vine health. This data-driven approach enables them to optimize vineyard management, resulting in increased grape yield and improved wine quality.
2. **Precision in Agricultural Management**: *Symington Family Estates* employs AI-powered drones equipped with imaging technology to identify specific vineyard areas affected by pests or diseases. This targeted approach allows for precise intervention, reducing the need for chemical treatments and promoting sustainable vineyard practices.
3. **Harvest Optimization**: *Herdade do Rocim*, an Alentejo winery, utilizes AI algorithms to analyze grape maturity and weather conditions. This technology assists in determining the optimal time for grape harvesting, ensuring that the grapes are picked at their peak ripeness for superior wine production.
4. **Intelligent Water Resource Management**: *Adega Mayor*, located in the Alentejo region, implements AI-driven irrigation systems that use sensors to monitor soil moisture levels. This data guides precise irrigation, ensuring that the vines receive adequate water without unnecessary waste, contributing to sustainable water management.
5. **Improvement in Quality**: *Casa Ferreirinha* employs AI-powered quality control systems that analyze chemical compositions and sensory characteristics of wines at different stages of production. This meticulous analysis ensures consistency and high-quality standards in their wine offerings.
6. **Decision Support**: *Wine Intelligence*, a market research company, provides Portuguese winemakers with AI-generated insights based on consumer preferences, market trends, and competitor analysis. This information assists wineries in making informed decisions regarding production volumes, marketing strategies, and product positioning.

Comparative analysis—qualitative approach

To make a comparative analysis of the case studies, Table 4.3 was created following the dimensions of the model of analysis:

4.6.2 Quantitative Analysis

To answer the research question, it was applied a quantitative methodologic approach supported by a questionnaire to identify differences between experts, academics, and agriculture professionals on AI's impact on the agricultural sector. The information was collected via a structured questionnaire that was prepared after the

Table 4.3 Comparative analysis of the case studies

	Wine case study	Olive oil case study
Increased production efficiency	AI optimizes grape cultivation by meticulously managing vineyards. By analyzing diverse datasets encompassing weather patterns, soil conditions, and fertilization, AI streamlines resource allocation, leading to heightened productivity and superior wine quality during harvests	AI is optimizing the olive tree cultivation process, enabling more effective management of plantations. It analyzes data on weather conditions, fertilization, and soil management, allowing precise resource optimization, and resulting in higher productivity and quality in harvests
Precision in agricultural management	AI systems facilitate precise management of vineyard threats such as pests, diseases, and weeds. This precision minimizes reliance on chemical interventions, fostering sustainable practices and mitigating environmental impacts	AI systems allow for a more precise approach to managing pests, diseases, and weeds. This precision reduces reliance on pesticides and herbicides, promoting more sustainable agricultural practices and minimizing environmental impacts
Harvest Optimization:	AI predicts the ideal time for grape harvesting, considering factors like grape ripeness and weather forecasts. This predictive capability ensures efficient harvesting, resulting in elevated wine quality	AI can predict the optimal time for olive harvesting based on a variety of factors, such as fruit maturity and weather conditions. This results in more efficient harvesting and superior olive oil quality
Intelligent water resource management	In a landscape where water conservation is pivotal, AI empowers smarter water utilization through sophisticated irrigation systems. These systems meticulously monitor and regulate water usage, curbing wastage and bolstering sustainability in grape cultivation	In a context where water is a precious resource, AI enables more efficient water use through smart irrigation systems. These systems monitor and control water usage precisely, reducing waste and increasing sustainability in olive tree cultivation

(continued)

Table 4.3 (continued)

	Wine case study	Olive oil case study
Improvement in quality	AI assumes a critical role in assuring wine quality by discerning patterns and traits that directly influence its excellence. This meticulous scrutiny guarantees the production and marketing of only the highest-grade wines	AI also plays a crucial role in the quality control of olive oil, allowing the detection of patterns and characteristics that directly affect its quality. This ensures that only the highest-quality olive oil is produced and marketed
Decision support	By processing extensive datasets, AI offers invaluable insights to winemakers, aiding in informed decisions regarding vineyard management, production forecasts, and strategic market approaches	With the ability to analyze large volumes of data, AI provides valuable insights to farmers, helping them make informed decisions about plantation management, production forecasts, and market strategies

Table 4.4 Fact sheet

Fieldwork	January through February 2024
Sample size	80 surveyed
Sample type	Convenience sampling
Survey type	Structured online questionnaire
Sampling error	5.8% assuming p = q = 0.5 and a confidence level of 95%

review of the literature. A convenience sample was used (non-probabilistic sampling procedure). When it is difficult to obtain a complete sampling, convenience sampling is suitable (Mercadé et al. 2017, 2018). The fieldwork was carried out between January and February of 2024 with a participation of 80 individuals, for a confidence level of 95% (and $p = q = 0.5$) and an increase in data error for the estimate of the proportion of 5.8%. The next table shows a summary of the information regarding the data collection and the technical matters of the sample (Table 4.4).

Data Analyses and Discussion

To analyze the differences in perceptions between experts, academics, and agriculture professionals, a covariance analysis (ANCOVA) has been carried out. The variance due to the individual differences is estimated from the regression between the dependent variable and the covariable. The scores in the dependent variable are statistically adjusted to the covariable. Finally, an ANOVA is performed on these adjusted scores (Tabachnick and Fidell 2007). Thus, the analysis controls the effect of the covariable, so that it eliminates the variation due to the mismatch of the ANOVA error.

Table 4.5 AI technologies applied in the agricultural sector

Variables	Experts and Academics	Agriculture professionals	F	p
Precision agriculture	4112	3153	102,323	0.000***
Monitoring and disease detection	3780	3526	24,762	0.000***
Predictive analytics	4211	3631	15,626	0.000***
Robotic farming	4354	3614	67,235	0.000***
Supply chain optimization	4366	3500	88,432	0.000***
Crop breeding and genomics	4321	3571	82,533	0.000***
Farm management systems	4234	3568	55,425	0.000***

* = $p < 0.1$; ** = $p < 0.05$; *** = $p < 0.01$

Table 4.5 analyzes the adjusted means F statistics and p-value for each group.

Table 4.5 displays the adjusted means, F statistics, and p-values. The analysis shows that there are statistically significant differences in all AI technologies applied to agriculture (p-value < 0.01 in all cases), always showing a higher score in experts and academics responses.

As for AI impacts in the agriculture sector, a four-point Likert scale, that rate 1 as "None", 2 as "Low", 3 as "Moderate", and 4 as "High" impact, was used. The t-student test has been carried out to verify whether there are significant differences between experts, academics, and agriculture professionals (Table 4.6).

In Table 4.6, it can be seen that experts and academics sample have higher scores in all the ítems analyzed, namely the higher impacts are in harvest optimization, intelligent water resource management, in precision in agricultural management, in decision support, and improvement in quality.

Finally, to analyze whether there is a correlation between AI and its influence on efficiency and decision support, a Cramer's test was applied (Table 4.7).

The results indicate that AI has an influence on both the variable's efficiency and decision support according to the perceptions of the participants in the study.

4.7 Recommendations for Public Policies to a Potential Promotion of AI in Agriculture

Public policies play a crucial role in fostering the integration of artificial intelligence in agriculture across Portugal. In this context, emerging from this study, next are presented some recommendations that can promote AI adoption in agriculture:

Table 4.6 Perceptions about AI impacts in the agriculture sector

Variables	Experts and academics mean	Agriculture professionals mean	t-student	*p*
Increased production efficiency	2.14	1.43	1.930	0.052*
Precision in agricultural management	2.48	1.95	3.743	0.000***
Harvest optimization	2.68	1.71	1.512	0.080*
Intelligent water resource management	2.58	1.40	7.644	0.000***
Improvement in Quality	2.05	1.77	1.732	0.03**
Decision Support	2.25	1.81	4.945	0.001***

* $= p < 0.1$; ** $= p < 0.05$; *** $= p < 0.01$

Table 4.7 AI impacts on efficiency and decision support

	Cramer's test	*p*
AI influences efficiency in agriculture	0.373	0.000***
AI contributes to effective decision support	0.368	0.00***

* $= p < 0.1$; ** $= p < 0.05$; *** $= p < 0.01$

1. Research and Development Funding: Government funding initiatives aimed at research and development in AI technologies specific to agriculture can incentivize innovation. Grants, subsidies, and research partnerships can encourage the development of AI tools tailored to the needs of farmers, processors, and marketers.
2. Education and Training Programs: Establishing educational programs and training initiatives focused on AI applications in agriculture can enhance the skill set of farmers and agricultural professionals. Workshops, courses, and skill development programs can facilitate the understanding and effective use of AI tools in farming practices.
3. Data Infrastructure Support: Building a robust data infrastructure that enables the collection, sharing, and analysis of agricultural data is essential. Policies supporting the creation of data-sharing platforms and frameworks for standardized data collection can aid in the development of AI-driven solutions for agriculture.

4. Regulatory Frameworks and Standards: Clear regulatory guidelines and standards specific to AI applications in agriculture are necessary. These regulations should ensure ethical AI use, data privacy, and interoperability of AI systems, fostering trust and confidence in adopting AI technologies.
5. Financial Incentives for Adoption: Providing financial incentives such as tax breaks, subsidies, or low-interest loans for farmers and producers adopting AI technologies can accelerate their uptake. These incentives can offset initial investment costs and encourage broader adoption of AI-driven solutions.
6. Collaboration and Knowledge Sharing: Encouraging collaboration between government bodies, research institutions, industry stakeholders, and tech innovators can facilitate knowledge sharing. Platforms for collaboration, exchange of best practices, and promoting partnerships can drive AI innovation in the agricultural sector.
7. Promotion of Sustainable Practices: Policies promoting the use of AI to improve sustainability in agriculture could be incentivized. AI-driven precision agriculture techniques can optimize resource utilization, reduce environmental impacts, and align with sustainability goals, thereby qualifying for support and recognition.
8. Support for Market Access: Facilitating market access for AI-enabled products and services related to agriculture is crucial. Policies that encourage the use of AI for quality assurance, traceability, and branding can enhance market competitiveness and consumer trust.

The process of integrating new practices and technologies within regulatory frameworks involves several key phases: policymaking, implementation, and evaluation; and the next Fig. 4.2 represents how these policy recommendations can be integrated into the public policy dimensions.

The policymaking phase involves the development of new policies or the revision of existing ones to accommodate advancements in agricultural practices and technologies. Policymakers, including government agencies, agricultural experts, and stakeholders, collaborate to identify emerging needs, assess available technologies, and formulate regulations/legislations that promote innovation in this case in the agriculture sector.

After the policies are formulated, it is needed to implement them by the agricultural sector, providing training and technical assistance, and establishing mechanisms for compliance monitoring and enforcement. Agricultural associations and research institutions facilitate the adoption of new practices and technologies. Moreover, incentivizing compliance through subsidies, grants, or tax incentives can encourage widespread adoption of AI and implementation.

The dissemination based on the evaluation process involves monitoring key performance indicators, such as changes in productivity and new practices and technologies implemented. This phase helps policymakers understand the effectiveness of the policies to make necessary adjustments to enhance their impact.

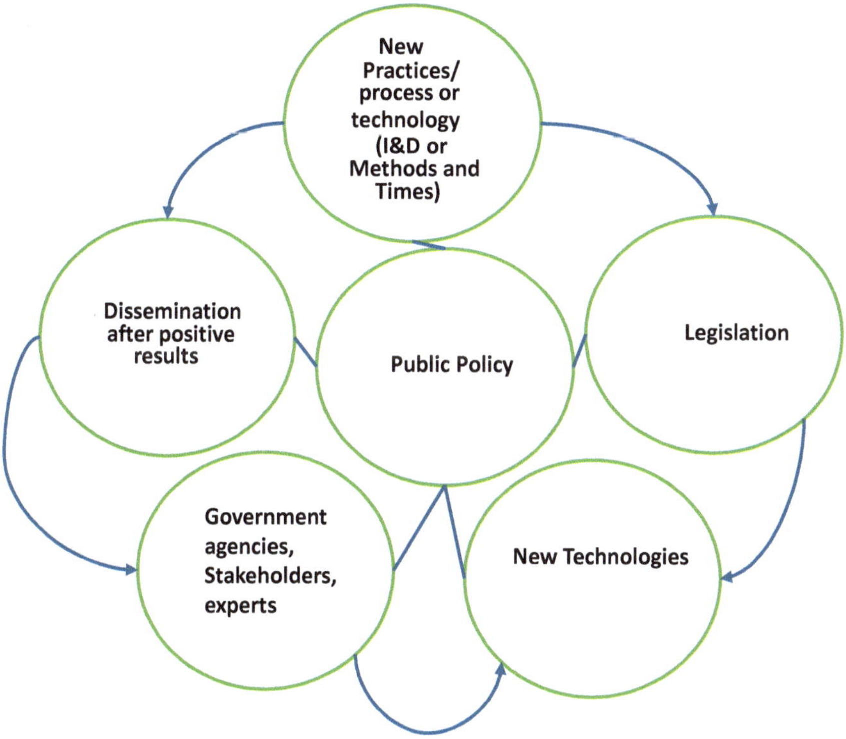

Fig. 4.2 Public policy dimensions

4.8 Conclusion, Limitations, and Future Research

AI's integration into agriculture represents a significant opportunity to address the sector's challenges while enhancing productivity and sustainability. By leveraging AI technologies such as precision agriculture, crop monitoring, predictive analytics, robotic farming, and supply chain optimization, farmers can make data-driven decisions, optimize resource use, and streamline operations. Furthermore, AI-driven solutions have the potential to reshape agricultural business models, fostering innovation and sustainability across the entire value chain.

Despite the promising potential of AI in agriculture as shown by this research, several challenges persist. The existing literature often lacks comprehensive real-world case studies demonstrating the tangible benefits of AI technologies on farms. Additionally, there is a shortage of AI-driven business models tailored to the agricultural sector. Moreover, the integration of AI into agriculture requires addressing technical, infrastructural, and regulatory challenges. Furthermore, the adoption of AI technologies may face resistance due to concerns about data privacy, cybersecurity, and the displacement of traditional farming practices. Finally, the scalability and accessibility of AI solutions in agriculture remain a concern, particularly for smallholder farmers.

Regarding the main limitations of this research, there is a sampling bias, because the demographic groups are underrepresented, and this limits the generalizability of the findings. As this topic is not very well-studied, there are no adequate scales already validated. The case studies can offer a higher in-depth exploration of the application of AI to agriculture, but there is little information available and a low number of studies.

Future research in the field of AI in agriculture should aim to advance the understanding of AI's implications for the sector. This includes conducting more comprehensive case studies that evaluate the real-world impact of AI technologies on farm productivity, sustainability, and profitability. Moreover, there is a need for research focused on developing AI-driven business models tailored to the unique needs and challenges of the agricultural sector. Additionally, interdisciplinary research collaboration is essential to address technical, infrastructural, and regulatory challenges hindering the adoption of AI in agriculture. Furthermore, future studies should explore the scalability and accessibility of AI solutions, particularly in the context of smallholder farmers. Overall, continued research in this area is crucial to unlocking the full potential of AI in transforming the agriculture sector.

References

Bagnoli C, Dal Mas F, Massaro M (2019) The 4th industrial Revolution: business models and evidence from the field. Int J E-Serv Mobile Applications 11(3):34–47

Bagnoli C, Massaro M, Dal Mas F, Demartini M (2018) Defining the concept of business model. searching for a business model framework. Int J Knowl Syst Sci 9:48–64

Biancone P, Brescia V, Lanzalonga F, Alam GM (2022) Using bibliometric analysis to map innovative business models for vertical farm entrepreneurs. Brit Food J 124(7):2239–2261

Biloslavo R, Bagnoli C, Edgar D (2018) An eco-critical perspective on business models: the value triangle as an approach to closing the sustainability gap. J Clean Prod 174:746–762

Blackmore S, Abson JM, Fischer J, Hanspach J (2015) Identifying and assessing barriers to the adoption of sustainable intensification practices in European agriculture: Insights from farmers, advisors and experts. Land Use Policy 47:98–111. https://doi.org/10.1016/j.landusepol.2015.03.005

Bogomolov A, Nevezhin V, Larionova M, Piskun E (2021) Review of digital technologies in agriculture as a factor that removes the growth limits to human civilization. In E3S Web Conferebce, 247. https://doi.org/10.1051/e3sconf/202124701074.

Castellini S, Pignatti E, Rizzoli AE, Torricelli M (2016) A review of the remote sensing of natural resources in agriculture. Precision Agric 17(6):654–681. https://doi.org/10.1007/s11119-016-9478-4

Cavazza A, Dal Mas F, Paoloni P, Manzo M (2023) Artificial intelligence and new business models in agriculture: a structured literature review and future research agenda. Br Food J 125(13):436–461. https://doi.org/10.1108/BFJ-02-2023-0132

Chiles RM, Broad G, Gagnon M et al (2021) Democratizing ownership and participation in the 4th industrial revolution: challenges and opportunities in cellular agriculture. Agric Hum Values 38:943–961. https://doi.org/10.1007/s10460-021-10237-7

Davies WJ, Ward SE, Wilson AW (2020) Can crop science really help us produce more and betterquality food while reducing the global environmental footprint of agriculture? Front Agric Sci Eng 7(1):28–44. https://doi.org/10.15302/J-FASE-2019299

Doerge TA (1999) Soil fertility basics. J Crop Prod 2(1):169–187. https://doi.org/10.1300/J144v02n01_09

Dwivedi YK, Hughes L, Ismagilova E, Aarts G, Coombs C, Crick T, Duan Y, Dwivedi, R., Edwards J, Eirug A, Galanos V, Ilavarasan PV, Janssen M, Jones P, Kumar Kar A, Kizgin H, Kronemann B, Lal B, Lucini B, Medaglia R, Le Meunier-FitzHugh K, Le Meunier-FitzHugh LC, Misra S, Mogaji E, Kumar Sharma S, Bahadur Singh J, Raghavan V, Raman R, Rana NP, Samothrakis S, Spencer J, Tamilmani K, Tubadji A, Walton P, Williams MD (2021) Artificial Intelligence (AI): multidisciplinary perspectives on emerging challenges, opportunities, and agenda for research, practice and policy. Int J Inform Manage 57:101994

FAO (2022) The state of food and agriculture. Food and Agriculture Organization of the United Nations

Ferguson AR, Burras CL (2009) The corn suitability rating revisited. Agron J 101(4):972–979. https://doi.org/10.2134/agronj2008.0217

Gebbers R, Böhner HJ (2010) Remote sensing and GIScience in precision agriculture: an overview. ISPRS J Photogramm Remote Sens 65(6):658–665. https://doi.org/10.1016/j.isprsjprs.2010.09.003

Hansen JW (2002) Realizing the Potential benefits of climate prediction to agriculture: issues, approaches, challenges. Agric Syst 74(3):309–330. https://doi.org/10.1016/S0308-521X(02)00025-1

Khan MA, Qazi S (2012) Role of water management in sustainable agriculture. Water Manage. IntechOpen, pp 1–13. https://doi.org/10.5772/37069

Kok JN, Boers EJW, Kosters WA, van der Putten P, Poel M (2002) Artificial intelligence: definition, trends, techniques, and cases. UNESCO. Available at: https://unesdoc.unesco.org/ark:/48223/pf0000128588

Komarek AM, De Pinto A, Smith VH (2020) A review of types of risks in agriculture: what we know and what we need to know, Agric Syst 178:102738

Li Y, Wang Q (2017) Robotic Agriculture: a promising field for precision farming. J Integr Agric 16(12):2773–2780. https://doi.org/10.1016/S2095-3119(17)61739-7

Mercadé-Melé P, Molinillo S, Fernández-Morales A (2017) The influence of the types of media on the formation of perceived CSR. Span J Mark-ESIC, 21(S1):54–64. https://doi.org/10.1016/j.sjme.2017.04.003

Mercadé-Melé P, Molinillo S, Fernández-Morales A, Porcu L (2018) CSR activities and consumer loyalty: the effect of the type of publicizing medium. J Bus Econ Manag 19(3):431–455

Mohanty SP, Hughes DP, Salathé M (2016) Using deep learning for image-based plant disease detection. Front Plant Sci 7:1419. https://doi.org/10.3389/fpls.2016.01419

Panpatte S, Ganeshkumar C (2021) Artificial intelligence in agriculture sector: case study of blue river technology. In: Lecture notes in networks and systems: proceedings of the second international conference on information management and machine intelligence. Springer Singapore, pp 147–153

Qin Z, Zhang M, Liu S, Li D, Song Z, Li Z (2019) A review of sensing technologies for precision agriculture crop monitoring. Adv Crop Sci Technol 7(4):1–7. https://doi.org/10.15406/acst.2019.07.00303

Rao M, Chhabria R, Gunasekaran A, Mandal P (2018) Improving competitiveness through performance evaluation using the APC model: a case in micro-irrigation. Int J Prod Econ 195:1–11

Saad MH, Hamdan NM, Sarker MR (2021) State of the art of urban smart vertical farming automation system: advanced topologies, issues and recommendations. Electronics 10(1422):1–40. https://doi.org/10.3390/electronics10121422

Rejesus RM, Zamora JP, Reyes JD (2003) Factors influencing the adoption of modern rice technologies among farmers in the Philippines. Food Policy 28(3):285–301. https://doi.org/10.1016/S0306-9192(03)00049-2

Singh G, Mann RS, Kaur H, Sharma K (2014) Decision support system for agriculture: a review. Int J Comput Appl 98(15):20–26. https://doi.org/10.5120/17182-1397

Tabachnick BG, Fidell LS (2007) Using multivariate statistics, 5th edn. Pearson/Allyn & Bacon

Tao F, Qi Q, Zhao X (2018) Sustainable agriculture and food supply chain: complexity, technology and innovation. Comput Electron Agric 147:108–115. https://doi.org/10.1016/j.compag.2018.03.007

Toniolo K, Masiero E, Massaro M, Bagnoli C (2019) Sustainable business models and artificial intelligence. Opportunities and challenges. In: Matos F, Vairinhos V, Salavisa I, Edvinsson L, Massaro M (eds) Knowledge, people, and digital transformation: approaches for a sustainable future. Springer, Cham, pp 103–117

United Nations (2015) Transforming our world: the 2030 agenda for sustainable development. United Nations.

Yela Aranega A, Ferraris A, Baima G, Bresciani S (2022) Guest editorial: sustainable growth and development in the food and beverage sector. British Food J 124(8):2429–2433

Zhang J, Song Q, Cregan PB, Jiang G-L, Song Q (2016) Genome-wide association study, genomic prediction and marker-assisted selection for seed weight in soybean (glycine max). Theor Appl Genet 129(1):117–130. https://doi.org/10.1007/s00122-015-2625-2

5 Wine Tourism as a Promoter of Talha Wine and Sustainability Practices in the Baixo Alentejo Region

Sónia Vieira, Fernanda Pereira, João Emanuel Martins, and Maria José Sousa

Abstract

Corporate sustainability is fundamental for the survival and competitiveness of companies in the long term. The European Union has proposed that the Corporate Sustainability Reporting Directive (CSRD) achieves the targets of the Sustainable Development Goals and promotes corporate sustainability reporting, which will be mandatory for listed companies from 2025, based on activity in 2024. The transformation in the business world promotes sustainability in the production of traditional products, such as talha wine, ensuring quality and authenticity. Talha wine, a tradition in Alentejo and Ribatejo brought by the Romans more than two thousand years ago, uses the technique of vinification in clay amphorae, giving the wine unique characteristics of aroma and flavor. Linked to the history and tradition of the Alentejo region, especially in the municipality of Vidigueira, talha wine contributes to the development of the region's wine sector. The study aims to analyze the importance of sustainability practices for traditional certified producers of talha wine in the municipality

S. Vieira (✉)
Beja Polytechnic Institute, Beja, Portugal
e-mail: sonia.vieira@ipbeja.pt

CiTUR, Algarve, Portugal

F. Pereira
Scientific Council of the Portuguese Association of Business Ethics, Lisbon, Portugal
e-mail: fernanda.pereira@ipbeja.pt

J. E. Martins
EDIA, SA, Beja, Portugal
e-mail: jmartins@edia.pt

M. J. Sousa
University of Lisbon, Lisbon, Portugal
e-mail: maria.jose.sousa@iscte-iul.pt

M. J. Sousa et al. (eds.), *Economic Sustainability Potentiated by Artificial Intelligence*,
Contributions to Environmental Sciences & Innovative Business Technology,
https://doi.org/10.1007/978-981-96-8089-4_5

of Vidigueira, Vila de Frades, through a questionnaire based on the existing literature, with the aim of showing that these practices are valued and contribute to the sustainability of companies and the territory.

Keywords

Sustainability • Industry 5.0 • Social • Governance • Talha wine

5.1 Introduction and Conceptualization

To understand the most important concepts associated to talha wine, the research begins with a search for relevant articles based on the words: Portugal + talha + wine or ancient wine. The search was made on the title, abstract, and keywords. Ten articles were found and will be the base for the following analysis of the literature published in Scopus database.

The VOSviewer application was then used to create a network and clusters for organizing literature and identifying key themes and connections between different studies. The representation created is the following (Fig. 5.1):

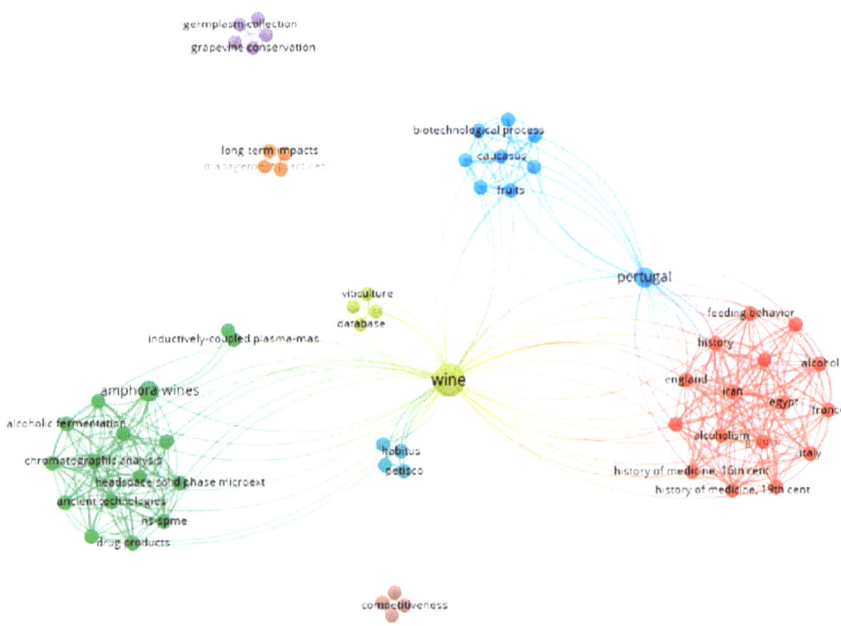

Fig. 5.1 Network of research attributes and concepts organized by clusters

By visualizing these clusters, a clearer understanding of the landscape of traditional and sustainable winemaking practices was achieved, ranging from historical and cultural significance to scientific analysis and regional studies, as follows.

Talha wine is a winemaking technique that dates back to the Roman era, when *talha* began to be used for fermentation and storage of wine (Martins et al. 2018; Beaven 1974). This method reflects the adaptation of the inhabitants of the region to the Mediterranean environment, where the scarcity of wood and the abundance of clay have shaped a "clay civilization," a term used by scholars such as Orlando Ribeiro (Ribeiro 1987). The technique has remained almost unchanged over two millennia, symbolizing a unique cultural continuity in the world of wine (Cabrita et al. 2018). This technique consists of fermenting and storing the wine in large clay containers (talha), giving the wine unique characteristics (Guasch-Jané et al. 2017; Reddy and Reddy 2012). This process is traditionally associated with family and community practices, having a great cultural and historical value (Ferreira et al. 2020).

Researchers highlight the importance of talha wine as cultural heritage. Its production recovers ancestral techniques and reinforces regional identity, also attracting wine tourism focused on authentic experiences (Cabrita et al. 2018; Radu 2019). Oenological studies indicate that talha give the wine unique organoleptic properties, such as freshness and a distinct aromatic profile (Martins et al. 2018; Reddy & Reddy 2012). During the Roman occupation, talha became a central element in the production of wine, taking advantage of the properties of clay, which helped in the controlled oxygenation of the liquid and gave distinct characteristics to the final product (Guasch-Jané et al. 2017; Beaven 1974). These containers were essential not only for the fermentation process but also for the conservation of the wine, allowing large-scale production, especially in regions such as Vila de Frades, considered the "capital" of talha wine (Cabrita et al. 2018).

Over the centuries, the practice of talha wine has evolved from a utilitarian necessity to a symbol of regional identity. Even with the introduction of modern technologies, this tradition remained alive in rural communities, where it was passed down from generation to generation (Gonçalves & Martins 2022). In recent years, talha wine has undergone a resurgence, attracting oenophiles and tourists in search of authenticity and historical oenological experiences (Ferreira et al. 2020). The recognition of its cultural importance has led to initiatives such as the application of talha wine to UNESCO (2022) Intangible Cultural Heritage, an effort that involves several institutions and aims to protect and promote this ancestral practice (Martins et al. 2018; Radu 2019). The traditional production of talha wine begins with the choice of grapes, often from old non-irrigated vines, which contributes to the diversity of grape varieties and unique characteristics of the wine (Gonçalves and Martins 2022). The process includes the spontaneous fermentation of the crushed grapes inside the talha, where the skins float and create a "blanket" that must be regularly dipped in the must (Cabrita et al. 2018). After fermentation, the wine is naturally filtered through the solid masses at the bottom of the talha, giving it a distinctive organoleptic profile (Ferreira et al. 2020; Reddy and Reddy 2012).

Currently, in addition to preserving traditions, talha wine is becoming an example of sustainability and innovation. On the one hand, production follows ecological methods, minimizing the use of chemicals and maximizing the use of local resources (Cabrita et al. 2018; Radu 2019). On the other hand, new producers have been exploring modern techniques to preserve the authenticity of wine while adapting it to global markets (Ferreira et al. 2020). This balance between tradition and modernity strengthens the position of talha wine as a culturally and environmentally relevant product in the contemporary market (Martins et al. 2018). The celebration of this heritage includes events such as Vitifrades, a festival dedicated to talha wine in Vila de Frades, which promotes local culture and attracts visitors from all over the world. These initiatives help to revitalize the region's economy and perpetuate a legacy that connects the present to the past in a vibrant and innovative way (Cabrita et al. 2018; Warren 2006).

The production process of talha wine is an ancestral practice, closely linked to the Alentejo tradition, and is characterized by artisanal and natural methods (Cabrita et al. 2018; Ferreira et al. 2020). The talha, large clay containers are carefully cleaned and internally coated with a mixture of beeswax and olive oil, which helps seal the pores of the clay and prevents liquid leaks. This coating also influences the sensory profile of the wine (Gonçalves and Martins 2022). The grapes, usually from traditional grape varieties of the region such as Antão Vaz, Aragonez, and Trincadeira, are harvested by hand to ensure the selection of the best (Cabrita et al. 2018; Reddy and Reddy 2012). The grapes are crushed, and the must (juice, pulp, skins, and seeds) is placed directly in the talha. The fermentation process is spontaneous, using natural yeasts present in the grapes and in the environment (Beaven 1974; Reddy and Reddy 2012). Fermentation lasts between 8 and 15 days, depending on the ambient temperature and the ripeness of the grapes (Cabrita et al. 2018). During fermentation, a solid layer is formed on the surface, composed of skins and seeds (the so-called "capitella"). This layer is removed regularly to allow the must to circulate and ensure homogeneous fermentation (Ferreira et al. 2020).

After fermentation, the wine remains in the talha, where natural clarification occurs. The solid particles descend to the bottom and act as a natural filter. The maturation period can vary, but it is usually 2–6 months (Martins et al. 2018). The wine is removed from the talha through small holes near the bottom of the container, without the need for artificial filtration. The liquid is transferred directly to bottles or other jars for further storage (Cabrita et al. 2018). Talha wine is sometimes sold in bottles that highlight its authenticity and artisanal production method (Gonçalves and Martins 2022). This process is characterized by minimal intervention, allowing the wine to reflect the natural conditions of the harvest and terroir. In addition, it has great cultural value, being preserved by small producers and traditional wineries in the Alentejo (Ferreira et al. 2020).

The production of talha wine can be understood as a sustainable model due to its low technological intervention and the use of local materials such as the clay used in talha (Cabrita et al. 2018; Radu 2019). This model has advantages in three dimensions of sustainability: low ecological impact, as the use of clay hoists eliminates the need for heavy machinery and reduces electricity consumption compared to industrial methods (Reddy and Reddy 2012); circularity of materials, as the clay used to manufacture the jars can be recycled or reused (Gonçalves and Martins 2022); and regenerative agriculture, as many producers resort to organic and biodynamic farming practices, minimizing the use of agrochemicals (Ferreira et al. 2020).

Valuing niche markets, talha wine attracts consumers looking for artisanal, natural, and sustainably sourced products, allowing for higher profit margins (Martins et al. 2018). Sustainable wine tourism also promotes local economic development, with wineries producing talha wine encouraging wine tourism (Cabrita et al. 2018; Radu 2019). Maintenance of traditions is another key aspect, as artisanal production contributes to the preservation of techniques passed down through generations, strengthening community ties (Gonçalves and Martins 2022). Additionally, cooperatives and local projects involving talha wine often train small producers, integrating them into the market (Ferreira et al. 2020).

Although it is a traditional practice, talha wine has been integrated into innovative strategies that balance authenticity and modernization. Advances in temperature monitoring and microbiological quality ensure greater consistency and safety in production batches (Cabrita et al. 2018). Some wineries are experimenting with the use of talha coated internally with natural wax or food polymers to diversify the wine's profile (Ferreira et al. 2020). Eco-labels, sustainable packaging, and sustainable tradition storytelling are explored to appeal to premium markets (Radu 2019). Additionally, talha wine has received organic and Protected Designation of Origin (PDO) certifications, increasing its global visibility (Martins et al. 2018).

5.2 Innovation, Development and Scientific Research

Modern oenological studies evaluate how talha materials, clay type, and fermentation processes impact the chemical and organoleptic profile of wine, promoting a dialog between science and tradition (Beaven 1974; Warren 2006). In Portugal, several innovative projects on talha wine have been carried out, both with European and private funding.

Located in the Cooperative Winery of Vidigueira, Cuba and Alvito, Casa das Talhas is a center for the valorization and promotion of talha wine. This project combines tradition and innovation, promoting wine tourism and the preservation of ancestral production methods. The project was recognized in national initiatives such as the Intermarché National Production Award, demonstrating its cultural and sustainable impact. Similarly, the Municipality of Vidigueira leads the application of the traditional method of production of talha wine to Intangible Cultural Heritage of Humanity by UNESCO (2022). This initiative also includes the creation of the "Talha Wine Route," promoting wine tourism and preserving the region's historical and cultural heritage.

Another notable project, Generations of Talha, located in Vila de Frades, promotes the certification of wine produced in talha. In addition, it includes guided tours of the vineyards and the history of the production method. It is an example of how innovation can be applied to enhance tradition and attract consumers. These projects stand out for combining tradition and innovation, promoting cultural and economic sustainability, and many have relied on European support and private partnerships for their implementation.

The combination of sustainability and innovation in talha wine suggests potential for expanding production sustainably while maintaining authenticity. Strengthening the appeal in international markets, especially in niches focused on natural wines, and integrating blockchain technologies for origin tracing and sustainability practices are also recommended.

Talha wine is more than a relic of the past; it is a model of how ancestral practices can be integrated into modern contexts to promote sustainability and innovation. At a time when consumers and markets are increasingly demanding environmental responsibility and authenticity, this wine represents a tangible response to the needs of the industry and communities. Its symbiotic relationship between tradition and innovation makes it a reference not only for the wine sector but also for other industries that seek to balance cultural preservation and technological progress. The future of talha wine looks promising, with a growing role on the global scene, while continuing to be an example of how the past can inspire sustainable solutions for the present and the future.

5.3 Methodology of the Empirical Study

5.3.1 Objectives of the Study

In the specific context of this investigation, we sought to develop a case study that focused on certified producers of talha wine. The case study is an empirical research approach that investigates phenomena directly, as they present themselves in reality, without the intervention of the researcher. This method involves an in-depth analysis of a particular case, assuming that it is possible to understand the phenomenon through detailed observation of a single case (Costa et al. 2013).

In view of the above, this study analyzed the importance of sustainability practices in wine tourism, especially for certified producers of talha wine in the municipality of Vidigueira, Vila de Frades, with the aim of showing that these practices are valued and contribute to the sustainability of companies and the territory. Different dimensions of sustainability were considered.

Therefore, the objectives proposed for the present study were the following:

Main objective: Show that sustainability practices are valued by the sector and contribute to the progress of companies and the territory. To respond to the main objective of this research, specific objectives were outlined for the three pillars of sustainability.

For specific objective 1, which seeks to respond to the environmental pillar, the following objective was outlined: Analyze the sustainability practices developed by producers at an environmental level.

For specific objective 2, which aims to respond to the social pillar, the following objective was outlined: Examine the integration of sustainability practices in the daily lives of employees, in the involvement of producers with the community and in the evaluation and sharing of the company's performance, aiming to increase the awareness and commitment to sustainable actions.

For specific objective 3, which seeks to respond to the economic pillar, the following objective was outlined: Investigate the potential of producers in relation to business goals and inspirations, their involvement in the implementation of processes, and the sustainability practices adopted in integration with the market and the value chain.

The choice of research methodology must be aligned with the research questions, the nature of the object of study and the type of answers intended to be obtained (Augusto 2014). In this context, the quantitative approach proved to be the most appropriate to achieve the objectives established at the beginning of this investigation. For data collection, a questionnaire survey was used as an instrument. The choice of the quantitative method was considered because the questionnaire is a simple and direct tool for obtaining data, without causing major inconvenience to the respondent and because of its similarity to the method used in the literature review.

The criterion for choosing the location was Baixo Alentejo, municipality of Vidigueira, specifically the parish of Vila de Frades as it is the capital of talha wine, due to its geographical proximity, its specific characteristics at different levels in the production of talha wine, for personal interest and appreciation of the region.

To choose the target population, certified companies producing talha wine in the parish were selected. After speaking with public entities in the region under study, it was realized that there were around five dozen talha wine producers, twenty of which are qualified to carry out wine tourism activities, and only nine are certified. The sample consists of only nine certified producers specializing in the product under study, emerging from a single case study, with the purpose of collecting, analyzing, and describing relative empirical evidence.

The questionnaire survey was designed based on a measurement scale adapted to the study context, more specifically an adaptation of Grace's (2011) social responsibility checklist. In the first phase, the questionnaire was pre-tested, which was validated by a sample of three respondents. The final version of the questionnaire was based on the results of this pre-test validation. The final questionnaire model was applied online, via Google forms, to companies producing certified talha wine in the parish of Vila de Frades, during the month of November 2024. A total of seven valid responses were obtained from the interviewed entrepreneurs, which can be analyzed in Table 5.1.

The questionnaire was produced using a literature review on the topic under study, adapting the question to the present investigation; in this sense, it was

Table 5.1 Technical data sheet of the study

Type of questions	Closed-ended with a single answer
Geographic scope	Portugal, Municipality of Vidigueira, Parish of Vila de Frades
Sample	9 producers
Contact method	Email, Google Forms, and in person
Response rate	77.7% (7 responses)
Fieldwork period	November 2024
Information processing	Computer processing using SPSS Version 29

divided into eight dimensions (Grace 2011), which encompass the model of the three pillars of sustainability:—How to start;—Company involvement;—Collaborators;—Environment;—Society;—Market;—Suppliers; and finally, evaluating and sharing the company's performance. Responses to items in each dimension were given on a 6-point qualitative scale, according to the participant's degree of agreement with each statement ("Completed"; "In progress"; "Could be improved"; "Worth considering"; "Out of question"; and "Not applicable").

Table 5.2 presents the connection between the specific objectives, the dimensions addressed in the literature review (Grace 2011), and the model of the three pillars of sustainability.

Table 5.2 Specific objectives, literature dimensions, and three pillar model

Specific objective	Dimension of literature (Grace 2011)	Model 3 pillars of sustainability
Specific objective 1: Analyze the sustainability practices developed by producers at an environmental level	Environment	Environmental
Specific objective 2: Examine the integration of sustainability practices in the daily lives of employees, in the involvement of producers with the community and in the evaluation and sharing of the company's performance, aiming to increase awareness and commitment to sustainable actions	Collaborators; colaboradores; society; evaluating and sharing company performance	
Specific objective 3: Investigate the potential of producers in relation to business goals and inspirations, their involvement in the implementation of processes, and the sustainability practices adopted in integration with the market and the value chain	How to start	Economic

Each dimension of the questionnaire aimed to analyze reports of producers' practices in specific areas. To this end, a set of statements relating to each dimension was prepared, covering the different pillars of sustainability under analysis, which made it possible to assess the level of practice and knowledge of producers. All questions formulated to answer the specific objectives contributed, as a whole, to achieving the main objective of this research.

5.3.2 Characterization of the Case Study Territory

Baixo Alentejo, a Portuguese statistical sub-region (NUTS III), is part of the Alentejo region and the district of Beja. It is located between Central Alentejo (to the north), Spain (to the east), the Algarve (to the south), and Alentejo Litoral (to the west). With an area of 8543 km^2, it is characterized by vast plains, dry climate, and fertile soils. Its economy is based on agriculture, livestock, and tourism (INE 2021), being a predominantly rural region.

In 2021, the resident population was around 115,237 inhabitants, resulting in a population density of 13.5 inhabitants/km^2 (INE 2021). The region encompasses 13 Alentejo municipalities: Aljustrel, Almodôvar, Alvito, Barrancos, Beja, Castro Verde, Cuba, Ferreira do Alentejo, Mértola, Moura, Odemira, Serpa, and Vidigueira. Economic activity diversifies, with emphasis on mining, forestry, hunting, agriculture, and livestock and the respective processed products (Cimbal 2024).

The municipality of Vidigueira, in the district of Beja, extends between Beja and Évora, occupying an area of 314.88 km^2 and divided into four parishes: Selmes, Pedrógão, Vidigueira, and Vila de Frades. With around 7000 inhabitants, the municipality is bordered to the north by Serra do Mendro, to the east by the Guadiana River and, to the south, by an extensive plain. Its history and identity are shaped by the interaction between the mountains, the river and the plain (CMV 2024). Vidigueira's economy is based on agriculture, in a rural and predominantly agricultural context. However, soil quality varies between parishes. In Vila de Frades and Vidigueira, the clayey and granite soils, of medium to low fertility, favor vineyard cultivation, making this area the main wine region in the municipality (Municipality of Vidigueira, 2023).

The tertiary sector has been growing, particularly in public services, education, social action, and commerce. Tourism also contributes significantly, exploring the scenic, environmental, cultural, and gastronomic heritage. The secondary sector, with small agri-food and beverage industries, has less impact on the local economy. Despite the decline in employment over the years, the primary sector still employs around 19% of the active population, remaining relevant to the municipal economy.

The parish of Vila de Frades, with 25.58 km^2 and 799 inhabitants (INE 2021), has a history that dates back to Roman times, with archeological remains that testify to its past. The ancient production of wine in carvings, a technique inherited from the Romans, gives it the title of capital of carving wine.

5.4 Data Analysis and Results

Certified producers of talha wine from the Vidigueira region, parish of Vila de Frades, were the target selected for the study, obtaining a response from 7 companies, all of them very heterogeneous in their constitution, but with the same objective relatively to the business area.

To respond to the objectives outlined, a set of questions were previously raised based on a questionnaire survey, which allowed the following conclusions to be drawn about the specific objectives outlined.

Regarding specific objective 1, "analyze the sustainability practices developed by producers at an environmental level," it is important to mention that the environmental sustainability policy has been gaining significant attention among certified producers of talha wine, reflecting a growing awareness and responsibility in the face of environmental challenges.

According to the analysis of the data collected, there is a significant trend toward the adoption of sustainable practices, with a particular focus on the areas of natural resource management, climate change mitigation, and biodiversity preservation.

Concern about climate change is one of the main priorities for producers, with 71.4% indicating their intention to develop or already implementing practices to mitigate the effects of these changes.

Regarding biodiversity, producers are taking significant steps to assess and reduce the environmental impact of their operations. Although a large proportion have already implemented practices related to the protection of biodiversity, a significant proportion (57.1–71.4%) are also planning to adopt or expand these measures, demonstrating a proactive approach to preserving ecological balance.

The efficient use of resources is a central concern for talha wine producers. With regard to energy, 85.7% of producers are already developing sustainable practices, which reflects a clear desire to reduce the ecological footprint associated with production.

Water management is another crucial point in producers' environmental policy, with the entire sample (100%) implementing or developing practices to optimize the consumption of this essential resource.

With regard to soil management, 85.6% of producers are already developing measures to prevent soil deterioration, such as conservation practices or the adoption of more sustainable cultivation techniques.

Waste management also stands out among producers' environmental practices, with the majority of respondents reporting the implementation of effective management practices. Although forest protection and fighting forest fires are not a central priority for the majority, 57.2% of producers demonstrate significant concern about preserving forests and mitigating fire risks, which have a direct impact on the environment and in its productions.

With regard to the creation of products and services, 57.2–71.4% of producers indicate that they are implementing environmental practices in the creation of their products, although a portion (28.6–42.9%) recognize that there is room for improvement in this aspect.

Finally, sustainable purchasing is a practice that 57.2% of producers have already implemented or are developing. However, a considerable part (28.6%) believes that there is still room to improve the incorporation of sustainable criteria in their acquisitions.

The talha wine market, with a strong component of tradition and innovation, seems to be increasingly aware of the need to align its processes with the principles of sustainability.

To respond to specific objective 2, "examine the integration of sustainability practices in the daily lives of employees, in the involvement of producers with the community and in the evaluation and sharing of the company's performance, aiming to increase awareness and commitment to sustainable actions," analysis of the data gathered reveals a series of practices being developed and implemented, with a particular focus on employee involvement, their ongoing training, and the social impact of their operations.

One of the main concerns of talha wine producers is the integration and involvement of employees in the company's dynamics. Although most companies are small in size, which limits some formal practices such as the creation of an "employee guide," 57.1% of producers show concern about integrating new employees.

Regarding diversity and inclusion, the majority of producers have already implemented or are developing practices in this area, which demonstrates a growing awareness of the importance of an inclusive and diverse work environment. This trend is also reflected in the professional development of employees, with a robust training and education practice that covers between 57.2 and 85.8% of producers.

Transparency in relationships with employees is a widely adopted practice, with 71.4–85.7% of producers indicating that they develop or are developing this practice.

Furthermore, recognition and reward programs for employees are becoming a reality in 57.1% of companies, although they are still in the development phase.

The health, safety, and well-being of employees are practices with disparate development, depending on the size of the companies. For many producers, especially smaller ones, there is a lack in these areas, but there is a visible interest in improving working conditions, as evidenced by the desire to implement improvements in specific aspects. In terms of work-life balance, 57.2% of producers are already developing or completing practices that promote this.

With regard to producers' involvement in society, there is a growing appreciation of social responsibility. More than 57.2–71.4% of producers are already implementing or plan to actively engage in socially responsible practices, demonstrating their commitment to the communities in which they operate. Donating products and services is one of the most common patronage and solidarity practices among producers, reflecting their direct contribution to social causes.

Volunteering also appears as a developing practice, with many producers considering or implementing programs that encourage their employees to participate in volunteer activities. Likewise, supporting education in the adjacent region, with a focus on local initiatives, is a growing practice, with a significant number of producers developing or planning programs that benefit community training and development.

Finally, evaluating and sharing company performance in terms of social responsibility is a practice that is under development or being considered by many producers. This openness to transparency and communication of the social impact of its activities reveals a growing commitment to the continuous improvement of social practices and sharing results with its stakeholders.

Finally, when we analyze specific objective 3, which "investigates the potential of producers in relation to business goals and inspirations, their involvement in the implementation of processes, and the sustainability practices adopted in integration with the market and the value chain," it is clear that the economic policy of sustainability in certified producers of talha wine demonstrates a growing commitment to incorporating ethical and responsible principles into their business practices.

Analysis of the data obtained from the sample reveals a concern with aligning operations with a sustainable vision and developing strategies that promote social responsibility.

The concern with defining a vision, mission, and values is a reality for 71.4% of talha wine producers. This focus, in the company's strategic guidelines, reflects the growing importance attributed to the clarification of long-term principles and objectives, which not only guide the economic growth of companies, but also ensure their integration with social and environmental sustainability practices. By developing these guidelines, producers show a clear intention to position themselves in the market with a solid identity, which values responsibility and commitment to the future.

The development of the company's code of ethics and conduct is being developed by 71.4% of certified producers, which denotes the growing concern in establishing clear standards of behavior and responsibility within the organization.

Communication also plays an important role for talha wine producers, especially with regard to social responsibility. Most producers demonstrate clear care in the way they communicate their practices and social commitments, seeking to reinforce the image of a responsible company, aware of its impact, and committed to sustainability.

Sustainability in the supply of products and services is another important aspect for talha wine producers. A total of 71.4% of producers already implement sustainable practices in the creation and marketing of their products, with 14.3% still in the development phase.

Regarding the choice of suppliers, most producers have already developed or are developing practices that ensure that their suppliers also respect sustainability criteria.

This aspect is crucial in creating a sustainable value chain, where social and environmental responsibility is integrated throughout the company's operations, from production to distribution. The selection of suppliers with responsible practices ensures that the company's operations are aligned with its ethical and environmental commitments, promoting a positive impact throughout the production chain.

After validating the specific objectives, it is possible to have a deeper view of the starting question of this investigation, which gave rise to the following main objective, "show that sustainability practices are valued by the sector and contribute to the progress of companies and the territory."

In terms of environmental sustainability policy, talha wine producers demonstrate a growing commitment to adopting practices that minimize environmental impact and guarantee the long-term sustainability of their operations.

Sustainability in the supply of products and services is another important aspect for talha wine producers. A total of 71.4% of producers already implement sustainable practices in the creation and marketing of their products, with 14.3% still in the development phase.

Regarding the choice of suppliers, most producers have already developed or are developing practices that ensure that their suppliers also respect sustainability criteria.

This aspect is crucial in creating a sustainable value chain, where social and environmental responsibility is integrated throughout the company's operations, from production to distribution. The selection of suppliers with responsible practices ensures that the company's operations are aligned with its ethical and environmental commitments, promoting a positive impact throughout the production chain.

After validating the specific objectives, it is possible to have a deeper view of the starting question of this investigation, which gave rise to the following main objective, "show that sustainability practices are valued by the sector and contribute to the progress of companies and the territory."

In terms of environmental sustainability policy, talha wine producers demonstrate a growing commitment to adopting practices that minimize environmental impact and guarantee the long-term sustainability of their operations. Although there are areas that are still developing or improving (such as forest protection and creating more sustainable products/services), most producers are already implementing significant strategies for resource management, mitigation of changes climate change, and the preservation of biodiversity.

In terms of social sustainability policy, talha wine producers show a responsible and increasingly structured approach, which encompasses the well-being of employees, the promotion of inclusive practices and professional development, as well as involvement with society. Although there are areas where practices are still developing or require improvement (such as support for employee health and safety and the reform program), the ongoing effort to integrate social sustainability into the operations of these companies is evident.

Finally, with regard to the economic policy of sustainability, talha wine producers show a clear commitment to economic responsibility, reflecting a growing adaptation to the demands of a more conscious and demanding market. The implementation of practices such as the development of vision, mission and values, the creation of a code of ethics and conduct, transparent communication, the offering of sustainable products, and the careful choice of suppliers are proof that producers are integrating sustainability strategically in its operations.

In conclusion, talha wine producers have demonstrated a growing commitment to environmental, social, and economic sustainability practices. Although there are areas for improvement, it is clear that these practices have a positive impact on the operations and progress of companies, as well as the territory. In this way, sustainable strategies value the competitiveness of producers and contribute to a more balanced and lasting development of the sector.

This case study concerns the talha wine production sector, with particular emphasis on the municipality of Vidigueira, in the parish of Vila de Frades. The small number of producers makes this study highly specific and strongly conditioned by the different size and business practice of each producer.

5.5 Conclusions

Sustainability has become unavoidable in all agricultural and agro-industrial practices, which implies a new stance and attitude on the part of businesspeople, as can be seen in the results obtained in the questionnaires administered.

In the environmental pillar, most producers adopt practices for resource management, energy efficiency, and soil and biodiversity preservation. In creating more sustainable products and services, there is still room for improvement.

In the social pillar, there is progress in valuing employees, especially in the area of training. Community involvement and social responsibility actions are growing. With regard to health and safety at work, producers face challenges resulting from current legislation.

In the economic pillar, the integration of sustainability principles into products and the value chain is evident. Economic and financial sustainability is being integrated into business strategies, with the definition of codes of ethics and communication reinforcing transparency and trust in the sector.

The majority of carving vineyard producers have ongoing sustainability practices and reveal an assumed concern aligned with the goals outlined to achieve the Sustainable Development Goals.

The fact that Vinho da Talha is a vehicle for promoting wine tourism in the region, contributing directly and effectively to the increase in wealth generated in the Territory, makes this artisanal and ancient product a diamond in the rough that the region must take care of and preserve, as an element of great importance for sustainable development and territorial cohesion, as per the objective of the Integrated Territorial Development Strategy for Baixo Alentejo.

A recommendation for future research would be to analyze the relevance of the issues from the perspective of technological wineries in the region, other players in the same sector, and other related regions.

Reports and Projects on Sustainability, Innovation Regarding Talha Wine

European Commission Report on the European Green Deal and Sustainable Food SystemsDiscussion on sustainability goals in agricultural production, including oenological practices. Link to the report: https://agriculture.ec.europa.eu/farm-fork-strategy-and-biodiversity-strategies_en.

Farm to Fork Strategy—European Commission This paper explores the integration of sustainability into the EU agricultural sector, with implications for wine production. Read the full document: https://ec.europa.eu/food/farm2fork_en.

LIFE VineAdapt Project Initiative that promotes sustainable practices in European vineyards. This project aims to adapt vineyards to climate change. Access the project: https://ec.europa.eu/life-vineadapt_en.

Sustainability strategies in the European wine sector Report on technological initiatives and practices with low environmental impact in the wine sector. See the study: https://ec.europa.eu/sustainable-wine-projects_en.

GreenWine—Study of sustainable agricultural practices A survey on how the transition to green practices is shaping the wine industry in Europe. See more details: https://greenwine-project.eu/.

Innovation in European Wine Production—Horizon 2020 Report on EU-funded projects in the wine sector, addressing innovation and sustainability. Read the report: https://ec.europa.eu/horizon2020-wine_innovation.

The impact of global warming on Alentejo wine Study carried out by Portuguese universities analyzing climate change and necessary adaptations in wine production. Access the study: https://www.alentejo-wine-climate-study.pt/.

Preservation of cultural heritage: Talha Wine as UNESCO Intangible Heritage A document that discusses the preservation of traditional practices, such as talha wine, and their cultural relevance. Read more about the topic: https://unesco.org/wine_cultural_heritage.

H2020 Winetech Project This project addresses emerging technologies in the wine sector, including sustainable practices and innovative production. Access the project: https://h2020-winetech.eu/.

Innovation and regional identity in Alentejo wine A regional study that explores the role of talha wine as a driver of innovation and cultural identity. See the study: https://www.alentejo-regional-wine-innovation.eu/.

References

Augusto A (2014) Metodologias quantitativas/metodologias qualitativas: Mais do que uma questão de preferência. Forum Sociológico

Beaven DW (1974) Wine through the ages. N Z Med J 79(507):611–615

Cabrita MJ, Martins N, Barrulas P, Garcia R, Dias CB, Pérez-Álvarez EP, Costa Freitas AM, Garde-Cerdán T (2018) Multi-element composition of red, white and palhete talha wines from Alentejo by ICPMS. Food Control 92:80–85. https://doi.org/10.1016/j.foodcont.2018.04.041

Cimbal (2024) Retrieved from https://cimbal.pt/pt/2837/vidigueira.aspx, 30 Sept 2024

CMV (2024). Retrieved from https://www.cm-vidigueira.pt, 30 Sept 2024.

Costa AS, Nascimento AV, Cruz EB, Terra LL, Siva MR (2013) O uso do método de estudo de caso na Ciência da Informação no Brasil. InCID: R. Ci. Inf. e Doc. 4(1):49–69

European Commission (2020) Green deal: sustainable wine production and EU funding opportunities. Recovered from https://ec.europa.eu

Ferreira CSS, Veiga A, Caetano A, Gonzalez-Pelayo O, Karine-Boulet A, Abrantes N, Keizer J, Ferreira AJD (2020). Assessment of the impact of distinct vineyard management practices on soil physico-chemical properties. Air Soil Water Res 13. https://doi.org/10.1177/1178622120944847

Gonçalves E, Martins A (2022) Efficient assessment and large-scale conservation of intra-varietal diversity of ancient grapevine varieties: case study Portugal. Plants 11(15). https://doi.org/10.3390/plants11151917

GRACE (2011) Primeiros Passos – Guia prático para a Responsabilidade Social das Empresas. Instituto Ethos, São Paulo

Guasch-Jané MR, Fonseca S, Ibrahim M (2017) The complete corpus of viticulture and winemaking scenes from the ancient Egyptian private tombs. In: XI International Congress of Egyptologists: Florence Egyptian Museum. Archaeopress, pp 254–259. https://doi.org/10.2307/j.ctv177tjnf.49

INE (2021 Retrieved from https://www.ine.pt/, 30 Sept 2024

International Organization of Vine and Wine (2021) Traditional winemaking techniques and their role in sustainable practices. OIV, Paris

Martins N, Garcia R, Mendes D, Costa Freitas AM, da Silva MG, Cabrita MJ (2018) An ancient winemaking technology: exploring the volatile composition of talha wines. LWT 96:288–295. https://doi.org/10.1016/j.lwt.2018.05.048

Municipality of Vidigueira (2023) Talha wine route and its socioeconomic impact. Vidigueira, Portugal

Radu L (2019) The wine sector of Romania, challenges and perspectives in the context of increasing competitiveness. In: Soliman KS (ed) Proceedings of the 33rd international business information management association conference, IBIMA 2019: education excellence and innovation management through vision 2020. International Business Information Management Association, IBIMA, pp 8192–8199. https://www.scopus.com/inward/record.uri?eid=2-s2.0-85074094787&partnerID=40&md5=3a29f6053cc54fdf26b8712447b7947d

Reddy LV, Reddy OVS (2012) Production and characterization of wine from mango (*Mangifera indica* L.) fruit juice. In: Handbook of plant-based fermented food and beverage technology, 2nd edn. CRC Press, Boca Raton, pp 249–272. https://doi.org/10.1201/b12055

Ribeiro O (1987) Civilization of clay in the Alentejo. Calouste Gulbenkian Foundation, Lisbon

UNESCO (2022) Application of talha wine to intangible cultural heritage of humanity. Retrieved from https://unesco.org

Warren V (2006) Yearning for the spiritual ideal: the influence of India on Western Dance 1626–2003. Dance Res J 38(1–2):97–114. https://doi.org/10.1017/S0149767700007403

The Impact of Artificial Intelligence (AI) and the Internet of Things (IoT) on the Sustainability of Globalized Businesses

6

André Luis Azevedo Guedes, Bruno Santos Cezario, Carlos Antônio da Silva Carvalho, Orlando Celso Longo, Vítor Gouvêa Andrezo Carneiro, and Lucio Fabio Cassiano Nascimento

Abstract

The growing adoption of artificial intelligence (AI) and the Internet of Things (IoT) drives digital transformation and the search for more sustainable business models. However, the literature still does not provide a comprehensive view of how these technologies are applied to promote sustainability in global businesses, especially in countries with substantial research and development investments, such as Japan, South Korea, USA, China, and parts of Europe. To fill this gap, this systematic literature review (SLR) seeks to identify the main

A. L. A. Guedes (✉) · B. S. Cezario · C. A. da Silva Carvalho · L. F. C. Nascimento
Postgraduate Program in Local Development (PPGDL), Centro Universitário Augusto Motta—UNISUAM, Rio de Janeiro, RJ, Brazil
e-mail: andre.guedes@gmail.com

B. S. Cezario
e-mail: brunoscezario@souunisuam.com.br

C. A. da Silva Carvalho
e-mail: carlos.carvalho@faa.edu.br

L. F. C. Nascimento
e-mail: lucionascimento@souunisuam.com.br

O. C. Longo
Postgraduate Program in Civil Engineering (PPGEC), Universidade Federal Fluminense—UFF, São Domingos, Niterói, RJ, Brazil
e-mail: orlandolongo@id.uff.br

V. G. A. Carneiro
Postgraduate Program in Defense Engineering, Military Institute of Engineering—IME, Rio de Janeiro, RJ, Brazil
e-mail: andrezo@ime.eb.br

M. J. Sousa et al. (eds.), *Economic Sustainability Potentiated by Artificial Intelligence*, Contributions to Environmental Sciences & Innovative Business Technology,
https://doi.org/10.1007/978-981-96-8089-4_6

trends, challenges, and opportunities at the intersection of AI, IoT, and sustainability with analyses through the VOSviewer software. The research mapped and analyzed the existing literature using Scopus, Web of Science, and CAPES Journals databases. The results include identifying best practices and advancing knowledge on the interface between technology and sustainability, contributing to the construction of a more sustainable future aligned with the Sustainable Development Goals (SDGs) of the United Nations (UN).

Keywords

Artificial intelligence • Internet of things • Sustainability • Globalization • Agenda 2030

6.1 Introduction

In recent years, technological advancements, such as the Internet of Things (IoT) and artificial intelligence (AI), have transformed how corporations operate and how society interacts with the world around them. These emerging technologies not only simplify and automate processes but also foster new opportunities for sustainable businesses, balancing economic growth with environmental preservation and social responsibility (Gajić et al. 2024). Furthermore, the urgency to promote sustainable practices and implement structural changes across several society sectors, while leveraging the interdependencies of the Sustainable Development Goals (SDGs) and engaging in strategic, technological, and political planning, underscores the importance of considering the connections between social actors and their nations (Lampropoulos et al. 2024).

IoT has the capacity to interconnect devices and software globally, enabling companies to monitor, analyze, and adjust their operations simultaneously, resulting in increased energy efficiency and waste reduction. Moreover, it encompasses a wide range of technological applications that facilitate interconnection and communication between diverse devices and systems through the Internet (Gajić et al. 2024). Simultaneously, AI provides unprecedented insights and automation, enhancing strategies for sustainable development by forecasting demands, optimizing resource usage, and aiding in more conscious, data-driven decision-making (Wang et al. 2021). Thus, the convergence of IoT and AI catalyzes the development of environmentally sustainable, intelligent, and resilient business models. By leveraging these technological advancements, many organizations can not only improve their outcomes but also amplify practices that benefit populations worldwide, contributing to a sustainable future (The Pact for the Future 2024).

In a globalized world, where climate changes and the depletion of natural resources have become critical concerns, technological innovation emerges as an indispensable ally in promoting a sustainable ecosystem (Rosário and Dias 2022). Many innovations, from IoT and AI to data analytics and renewable energy, are transforming how companies and governments address environmental issues, facilitating the creation of more sustainable and efficient solutions. These digital technologies are crucial in achieving the goals outlined in the 2030 Agenda for Sustainable Development (UN 2015), allowing for optimized resource usage, waste reduction, and real-time environmental impact monitoring, enabling more informed and data-driven decision-making (Vermesan et al. 2021).

Environmental and social issues have taken center stage in the global agenda. Sustainability has evolved from merely being a differentiator to becoming an essential component of long-term business success. Digital innovation has become commonplace for organizations, companies, and industries worldwide, driving the adoption of more sustainable practices and operational models, reducing environmental impact, adapting to changes, and remaining competitive in response to increasingly conscious consumer demands (Tran-Dang and Kim 2021). This movement extends beyond isolated initiatives. It involves structural changes, governance strategies, and a holistic approach that considers product life cycles, circular economy principles, and corporate social responsibility. Sustainable practices, while challenging, present significant opportunities for innovation and competitive advantage. Companies embracing this approach gain access to new markets, build stronger partnerships, and contribute to the development of a more balanced and responsible economy. Thus, corporate sustainability not only protects the environment but also strengthens the foundation for more resilient and inclusive economic growth, which is crucial for the future of businesses and the planet (The Pact for the Future 2024).

The use of technology in sustainable practices goes beyond simple solutions: it revolutionizes some sectors, such as agriculture, transportation, energy, and manufacturing, promoting clean production processes, circular economy principles, and intelligent waste management, while offering innovative solutions for sustainable resource management (Ramírez-Márquez et al. 2024). By integrating disruptive technologies into a vision of sustainable development, organizations worldwide are mitigating their environmental impact and opening doors to a more balanced and resilient future. In this context, the present study is aligned with the goals established in the UN's 2030 Agenda for Sustainable Development and directly linked to the global transformation outlined in the Pact for the Future created by the UN in 2024, aiming for a more equitable and prosperous world for all humanity.

6.2 Methodology

Initially, an extensive bibliographic search was conducted on the sites of Web of Science and Scopus, using the following keywords: sustainability, artificial intelligence, sustainable development goals, and Internet of Things. The recommendations of the Preferred Reporting Items for Systematic reviews and Meta-Analyses (PRISMA) (Moher et al. 2009) were adopted for the bibliographic search. The following filters were applied: scientific and review articles in English published in the last five years, peer-reviewed, and open access.

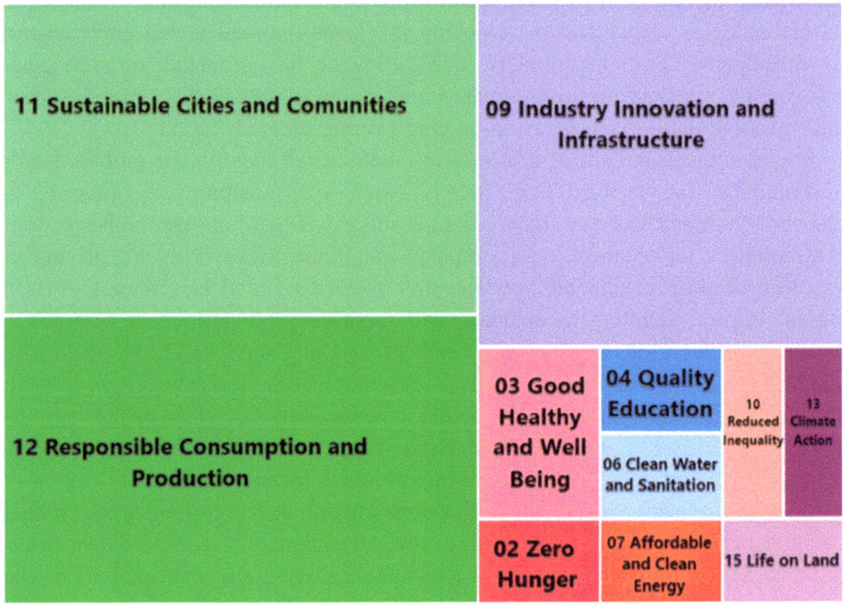

Fig. 6.1 Bibliographic research on the sustainable development goals

The obtained database is aligned with the following Sustainable Development Goals of the UN's 2030 Agenda (Fig. 6.1): Zero Hunger (SDG 2); Good Health and Well-Being (SDG 3); Quality Education (SDG 4); Clean Water and Sanitation (SDG 6); Affordable and Clean Energy (SDG 7); Reduced Inequalities (SDG 10); Sustainable Cities and Communities (SDG 11); Responsible Consumption and Production (SDG 12); Climate Action (SDG 13); Life on Land (SDG 15).

The bibliographic data from the found references were exported as a "ris" metadata file and processed using the VOSviewer 1.6.20 software (Perianes-Rodriguez et al. 2016). This system performs a quantitative bibliometric analysis to monitor published research, providing publication patterns related to a specific field of literature. The software enables the construction and visualization of bibliometric networks or knowledge clusters, which are groupings and interconnections among articles, allowing for the visualization of occurrences and facilitating the categorization of documents by keywords or authors. This approach results in a Bibliometric-Systematic Literature Review (B-SLR), characterized by the screening and selection of knowledge references to ensure the effectiveness and accuracy of the interpreted data (Rosário and Dias 2022).

Using this methodology, it was possible to map, analyze, and identify a network of co-occurrence relationships between the keywords of the selected articles, enabling the creation of a conceptual arrangement of a knowledge domain. This means that when keywords are repeated across the articles that compose the analyzed data, the concepts attributed to these keywords are connected. Figure 6.2

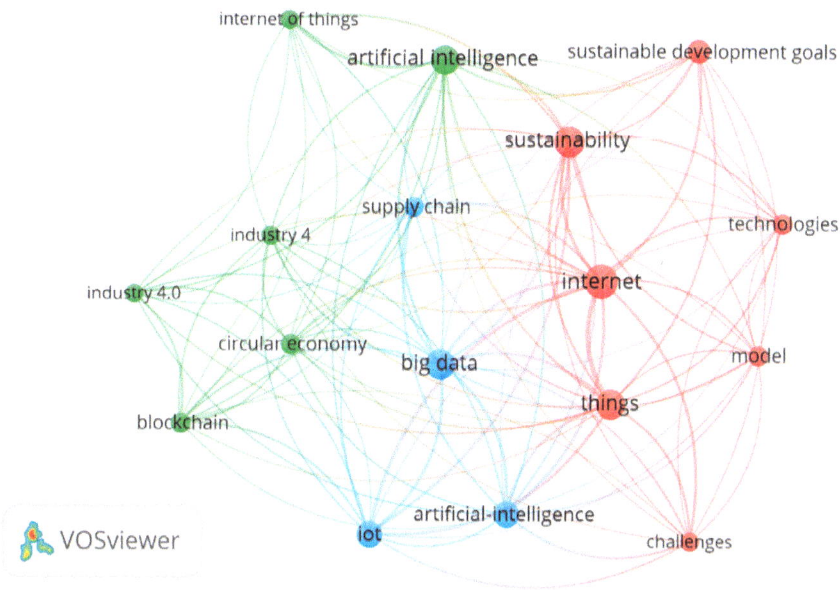

Fig. 6.2 Network of clusters

shows how the VOSviewer software creates keyword networks among the selected articles. The data is cross-referenced, and the research is divided into four main clusters, represented by colored circles. Each cluster groups keywords that share similarities and interactions among the articles. The connections between them occur through structural associations. The stronger the link, the greater the interaction between the topics. Figure 6.2 illustrates the co-occurrences of keywords found across all articles retrieved after the search vector query. It can be observed that they are grouped into a knowledge network, subdivided as follows: network 1 in green, network 2 in red, and network 3 in blue, respectively.

It is noted that the keywords used in the initial search are among the knowledge networks guided by the bibliometric study, aimed at identifying indicators to streamline and advance scientific knowledge (Rosário and Dias 2022). This indicates that the choice of keywords was effective in this research, as demonstrated in Fig. 6.3.

However, still employing quantitative methods to implement the eligibility criteria, and emphasizing the logic of selection, evaluation, and synthesis of the study, a search was conducted in the abstracts of the 43 articles using the keywords—sustainability, artificial intelligence, sustainable development goals, and Internet of Things—to identify articles relevant to the topic. At this stage, 42 items remained selected.

To ensure that all possible articles relevant to the study were included, a second round of mapping was performed on bibliographic materials containing at least two of the mentioned keywords, resulting in 33 articles. Considering the use of

Fig. 6.3 Keyword cluster network

the VOSviewer software and the systematic review, five references were added from other sources discussing the methodologies used and relevant to the topic, incorporating these references into the bibliographic database of this investigation. Finally, applying qualitative eligibility criteria, the abstracts and conclusions of these articles were read in full, resulting in the selection of 25 references. In addition, eight articles were discarded due to a lack of relevance to the topic. Figure 6.4 shows the PRISMA flowchart for this study.

6.2.1 The Role of IoT and AI in Achieving the SDGs in Business

Emerging technologies, such as artificial intelligence (AI) and the Internet of Things (IoT), when combined, have positioned themselves at the forefront, with robust solutions to address the global challenges outlined by the UN's Sustainable Development Goals (SDGs) (Lampropoulos et al. 2024). This synergy between digital technologies not only enhances operational efficiency but also drives data-driven decision-making to promote positive impacts in environmental, social, and economic areas (Vermesan et al. 2021). Thus, in a contemporary scenario, globalization has brought to light a series of challenges and opportunities for companies worldwide. In this context, digital technologies become an essential resource for organizations seeking to boost their performance while promoting the objectives of the 2030 Agenda (Wang et al. 2021). This integration enables innovative solutions to complex problems, balancing economic progress with socio-environmental responsibility.

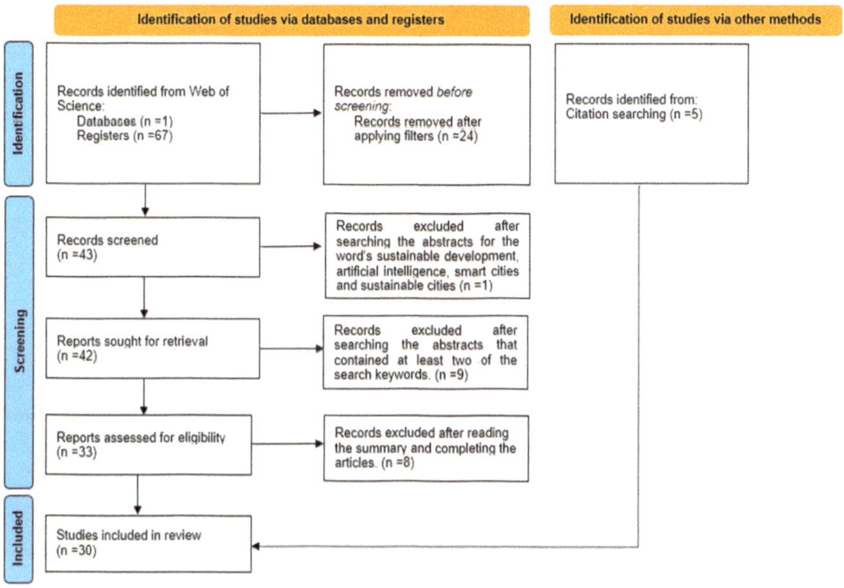

Fig. 6.4 PRISMA flowchart

6.2.2 Expanding Access to Clean Energy (SDG 7)

AI and IoT are key tools for global companies transitioning to clean and sustainable energy use (Ramírez-Márquez et al. 2024). IoT-equipped monitoring systems collect data on energy consumption, enabling AI algorithms to analyze and identify patterns to optimize energy use sustainably (Varriale et al. 2024). Additionally, combining AI and IoT facilitates the efficient integration of renewable energy sources, such as solar and wind, into corporate power grids.

Access to clean and affordable energy is one of the fundamental pillars of sustainable development, as defined by SDG 7 (UN 2015). Clean energy plays a critical role in reducing greenhouse gas emissions, combating climate change, and improving the quality of life in communities worldwide (Singh et al. 2022). One of the main tools for expanding access to clean energy is technological advancement. IoT-based systems monitor electrical grids in real-time, while AI algorithms analyze energy demand and optimize decision-making related to sustainable sources (Tran-Dang and Kim 2021). This combination ensures greater grid stability, reduces infrastructure costs, and democratizes access in remote areas.

Initiatives to decentralize energy generation and adopt microgrids and autonomous renewable energy systems enable communities without access to conventional grids to reliably and cleanly obtain energy (Ramírez-Márquez et al. 2024). These systems have a significant impact on rural areas and developing countries, where the lack of energy infrastructure poses a barrier to development. On the other hand, financing is a critical factor in expanding access to clean energy.

Fig. 6.5 SDG 7 of UN's 2030 agenda

Integration among governments, international organizations, and the private sector is essential to facilitate investments in research and the implementation of sustainable projects, supporting the development of a smart economy (Metelenko et al. 2024). Public policies, such as subsidies for adopting renewable sources and taxes on polluting energy, have the potential to accelerate this transition.

Efforts on education and awareness also play a vital role, as informed consumers are more likely to adopt clean technologies in their homes and support brands committed to sustainability. This movement encourages companies to invest in renewable sources and more efficient energy solutions. In this context, the integration of technological advancements, government initiatives, and community participation creates a conducive ecosystem for achieving SDG 7. Expanding access to clean energy is not just a challenge but an opportunity to build a more equitable, inclusive, and sustainable future for all (Fig. 6.5).

6.2.3 Promoting Innovation in Emerging Markets (SDGs 8 and 10)

The adoption of emerging technologies, such as AI and IoT, can foster inclusion and innovation in emerging economies (Lampropoulos et al. 2024). For instance, companies can use digital technology solutions to enhance intellectual capital in various regions, supporting economic growth in communities and addressing socio-environmental inequalities (Patalas-Maliszewska and Łosyk 2024). This not only contributes to sustainable economic growth but also reduces inequalities between countries and regions.

Promoting innovation in emerging markets is critical to driving sustainable economic growth, as defined by SDG 8, and reducing inequalities within and between countries, as proposed by SDG 10 (UN 2015). These markets present challenges and opportunities that demand creative approaches to stimulate socio-economic and environmental development, while also acting as catalysts in various sectors of emerging markets. The combination of artificial intelligence (AI) and the Internet of Things (IoT), for example, is transforming sectors such as agriculture, healthcare, and education (Ramírez-Márquez et al. 2024). In rural areas, AI can be used to detect pests and other threats to crops (Singh et al. 2022), while connected sensors enable real-time monitoring of weather conditions and soil quality, boosting the productivity of small-scale farmers and contributing to inclusive economic growth (Martos et al. 2021).

However, fostering innovation requires robust strategies for financial value aggregation and sustainability. Circular thinking, in partnership with governments, international organizations, and the private sector, plays a vital role in generating environmental and economic benefits by providing financial resources and enabling environmentally focused economic growth (Hoosain et al. 2020). Such initiatives help to create quality jobs and increase access to sustainable and innovative technologies.

Workforce training is also essential to sustaining green innovation in emerging markets (Gajić et al. 2024). Investments in technical and professional education, as well as upskilling programs in digital technologies, ensure that workers are prepared for the demands of an advanced global economy, transforming society and contributing to reduced inequalities (Cabanillas-Carbonell et al. 2023). However, digital inclusion remains a critical challenge in emerging markets, as they face inequalities in Internet access and digital technologies. To address this, governments and private partners must invest in affordable digital infrastructure and expand connectivity networks in underserved areas (Cabanillas-Carbonell et al. 2023).

Furthermore, innovation must be guided by principles of sustainability, in order to encourage business models focused on the circular economy, renewable energy, and efficient resource management. This approach promotes economic growth that respects environmental limits, aligning with global demands for responsible production (Pandey et al. 2022).

Finally, emerging markets offer a vast field for the creation of innovative solutions. Open innovation projects involving local communities, universities, and companies foster solutions tailored to the specific needs of each region. This not only promotes economic growth but also ensures that the benefits of innovation are broadly distributed, reducing social and economic inequalities. Thus, by integrating emerging technologies, strategic financing, and social inclusion, it is possible to create an innovation ecosystem that empowers emerging markets, contributing significantly to the objectives outlined by SDGs 8 and 10 (Figs. 6.6, and 6.7).

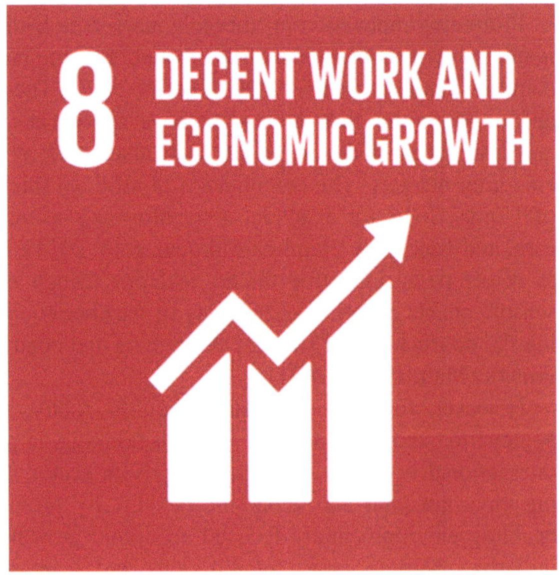

Fig. 6.6 SDG 8 of UN's 2030 agenda

Fig. 6.7 SDG 10 of UN's 2030 agenda

6.2.4 Encouraging Sustainable Production (SDGs 9 and 12)

In the industry, AI and IoT are drivers of sustainable development. Big data sensors establish new levels of intelligence to collect real-time data on efficiency and operational conditions. AI analyzes this information, enabling collaboration across sectors of society, identifying patterns, predicting maintenance needs, and reducing disruptions and waste (Al Maadeed and Ponnamma 2023). In addition, integrating these technologies with global supply chains optimizes productive resources, reduces waste, and minimizes environmental impacts, ensuring compliance with responsible consumption and production goals.

Sustainable production plays a fundamental role in the SDGs, particularly SDG 9, which addresses infrastructure, inclusive industrialization, and innovation; and SDG 12, which emphasizes responsible consumption and production patterns as part of the 2030 Agenda (UN 2015). Encouraging sustainable production requires innovative solutions that combine technological advancement with socio-environmental responsibility. In this context, attention must be given to the transition to a circular economy, a critical pillar of sustainable production (Akram et al. 2022). This approach aims to maximize resource utilization and minimize waste, promoting solutions, such as recycling, material reuse, and durable product design. Leading companies are investing in initiatives to close production and consumption cycles, contributing to the goal of responsible consumption and production (SDG 12).

However, for these initiatives to be effectively implemented, it is crucial to establish public–private partnerships and support small- and medium-sized enterprises, assisting in the creation of governmental policies that advocate for the adoption of clean technologies, while training programs prepare workers to operate more sustainably (Gajić et al. 2024). Another relevant aspect is the increase in transparency in global supply chains. Technologies such as blockchain, combined with AI and IoT, can trace the origin and environmental impact of each stage of production, ensuring that practices are ethical and aligned with the SDGs (Singh et al. 2022).

Moreover, promoting awareness among consumers and investors is of utmost importance. Growing demand for eco-friendly products and ways to enhance corporate responsibility are driving companies to adopt more sustainable strategies, demonstrating that the integration of profit and socio-environmental responsibility is not only achievable but also profitable in the long term (Figs. 6.8, and 6.9).

6.2.5 Mitigating the Carbon Footprint in Transportation and Logistics (SDG 13)

Reducing the carbon footprint in the transportation and logistics sectors has become a fundamental priority in the globalized world, aligned directly with Sustainable Development Goal 13 (UN 2015), which aims to combat climate change and its impacts. This goal seeks to reduce greenhouse gas (GHG) emissions and promote practices that contribute to the transition toward a more sustainable, resilient, and inclusive economy.

In a globalized world, transportation plays a crucial role but is also one of the main sources of greenhouse gas emissions. Artificial intelligence (AI) and the Internet of Things (IoT) are revolutionizing the logistics sector through innovative solutions, such as real-time fleet tracking, route optimization, and the use

Fig. 6.8 SDG 9 infographic of UN's 2030 agenda

Fig. 6.9 SDG 12 of UN's 2030 agenda

of smarter autonomous vehicles (Vermesan et al. 2021). These innovations minimize fossil fuel consumption and reduce carbon emissions, aligning with the 2030 Agenda objectives for addressing climate change.

Digital technologies, as the combination of AI and IoT, are playing a pivotal role in this context by building "smart connectivity" (Cabanillas-Carbonell et al. 2023). The usability of these technologies enables autonomous vehicles to operate with reliability and transparency. Real-time information allows for consumption control, helping optimize processes for sustainable green mobility (Vermesan et al.

Fig. 6.10 SDG 13 of UN's 2030 agenda

2021). In this way, they aid in streamlining production processes, reducing waste, predicting corrective maintenance needs, and mitigating environmental impacts.

The use of AI technologies and big data analytics is becoming a crucial tool for optimizing management and efficient technological decision-making. Through data-sharing algorithms, it is possible to calculate the most efficient vehicle routes, considering travel time, weather conditions, and traffic (Gohari et al. 2022). Moreover, these emerging technologies help to optimize cargo loads, maximize vehicle utilization, and reduce unnecessary trips. Such planning decreases fuel consumption and CO_2 emissions.

However, the carbon footprint in transportation and logistics is a complex issue, given the diversity of transportation modes and the unique characteristics of each local supply chain. Transitioning to low-carbon alternatives and implementing effective mitigation measures are not simple tasks. Nonetheless, it is essential to identify the key challenges that must be addressed to achieve global emissions reduction and sustainability goals (Fig. 6.10).

6.2.6 Enhancing Waste Management (SDGs 11 and 15)

Globalized businesses generate massive volumes of waste, and inadequate management of these materials can cause irreversible environmental damage. IoT offers tools to address this issue. Smart bin sensors can monitor the accumulation of waste, while AI determines the best methods for disposal or recycling (Abbas et al. 2023). These technologies help to minimize environmental impact and protect ecosystems (SDG 15).

Efficient waste management is essential for promoting more sustainable cities (SDG 11) and protecting terrestrial ecosystems (SDG 15) (UN 2015). Population growth, accelerated urbanization, and increasing consumption place unprecedented pressure on waste management systems, making the adoption of innovative and sustainable solutions imperative (Paiva et al. 2021). One of the main challenges is the high generation of urban solid waste, which often overwhelms existing infrastructure and negatively impacts the environment. Recycling and material reuse emerge as fundamental solutions. Investments in technologies for automated waste sorting, using IoT sensors and AI algorithms, can significantly improve the efficiency of recycling processes.

Implementing a circular economy is also vital. This model encourages material reuse, durable product design, and waste reduction in production chains (Pandey et al. 2022). Companies and governments can collaborate to create financial incentives and regulations that promote practices aligned with this paradigm.

In cities, sustainable resource management through digital technologies is becoming a reality for Society 5.0 (Ramírez-Márquez et al. 2024). IoT-based systems monitor waste volume in containers in real time and optimize collection routes, reducing costs and carbon emissions associated with transportation. Furthermore, digital platforms that properly encourage citizens to separate waste contribute to the overall improvement of the system.

Environmental education and adoption of green practices play a critical role in this process (Gajić et al. 2024). Awareness campaigns on the importance of reducing, reusing, and recycling help to change individual and collective behaviors. Schools, community organizations, and companies can collaborate to foster a culture of environmental responsibility from an early age.

Waste management also has a direct impact on preserving terrestrial ecosystems. Improper waste disposal directly affects fauna and flora, pollutes rivers and soil, and compromises biodiversity. Emerging digital technologies can be utilized to monitor degraded or polluted areas and provide effective solutions to prevent and remediate environmental damage (Varriale et al. 2024).

On the other hand, legislation plays a fundamental role in establishing standards for waste management. International agreements and policies are reshaping how society and companies approach sustainability in their operations (Pandey et al. 2022). Additionally, public–private partnerships can accelerate the development of innovative solutions, sharing costs and responsibilities among the involved sectors. Collaboration between governments, companies, and civil society organizations is essential to integrate technological solutions into the realities of each region.

Improving waste management goes beyond mitigating environmental impacts. It ensures healthy living conditions in cities and protects terrestrial ecosystems, ensuring that current and future generations enjoy a cleaner and more balanced planet (Pandey et al. 2022) (Figs. 6.11, and 6.12).

6.2.7 Transparency and Corporate Accountability (SDG 16)

The use of digital technologies helps companies to ensure transparency and accountability in corporate governance, thanks to advancements in monitoring capabilities in a globalized world (Young et al. 2020). IoT-equipped systems can

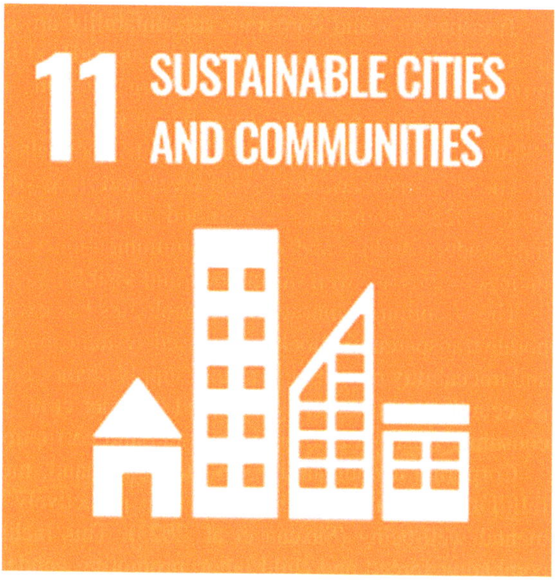

Fig. 6.11 SDG 11 of UN's 2030 agenda

Fig. 6.12 SDG 15 of UN's 2030 agenda

monitor industrial processes, ensuring regulatory compliance, while AI analyzes data to identify potential risks of corruption or embezzlement. Furthermore, technological solutions can enhance communication with stakeholders, reinforcing companies' commitments to peace, justice, and effective institutions.

Transparency and corporate accountability are central elements in promoting peaceful, just, and inclusive societies, as outlined by SDG 16 (UN 2015). These principles strengthen public trust in organizations and stimulate ethical and sustainable business practices. Transparency involves ensuring that key information about operations, environmental impact, labor relations, and sustainable financial practices is made available in a clear and accessible manner to society (Kumar et al. 2022). Companies committed to this principle use sustainability reports, independent audits, and direct communication with stakeholders to ensure that their activities are well-understood and visible.

The adoption of innovative technologies has expanded the possibilities for corporate transparency. Blockchain-based systems, for instance, enable the monitoring and traceability of transactions and supply chains, ensuring that critical information is securely and immutably recorded (Vernier et al. 2021). This not only increases consumer trust but also prevents fraud and corruption.

Corporate accountability, on the other hand, means that companies not only fulfill their legal obligations but also proactively promote social and environmental well-being (Saxena et al. 2023). This includes respecting human rights, combating forced and child labor, promoting gender equality, and contributing to the development of communities where they operate. Integrating corporate social responsibility into business strategies has become essential for modern companies. Programs for private social investment, environmental sustainability initiatives, and adherence to global frameworks (The Pact for the Future 2024) demonstrate organizations' commitments to creating positive impacts.

Strong corporate governance is another pillar of accountability. Diverse and ethically aligned boards of directors play a vital role in ensuring compliance with organizational guidelines (Saxena et al. 2023). Internal monitoring mechanisms and the implementation of codes of conduct ensure that business decisions are made according to ethical standards.

The impact of transparency and corporate accountability is reflected across various areas. Environmentally, companies that disclose greenhouse gas emissions data and adopt environmental management practices are more likely to be recognized as leaders in combating climate change (The Pact for the Future 2024). Socially, those who invest in dignified working conditions and diversity promotion contribute to more equitable societies.

The role of consumers and investors is equally significant. By demanding greater transparency and accountability from corporations, they drive the adoption of ethical practices on a global scale. Corporate evaluation platforms based on ESG (Environmental, Social, and Governance) increasingly influence investment decisions, pressuring companies to implement improvements (Saxena et al. 2023). Thus, promoting transparency and corporate accountability creates the conditions for more sustainable businesses, fosters social well-being, and protects the environment. These elements are decisive for achieving SDG 16, ensuring that companies and organizations act as positive agents in building a more ethical and just world (Fig. 6.13).

Fig. 6.13 SDG 16 of UN's 2030 agenda

6.3 Conclusion

The combination of artificial intelligence and the Internet of Things is redefining globalized businesses, helping companies to address sustainable development goals while enhancing their competitiveness and efficiency. However, these technologies must be implemented ethically and equitably, considering cybersecurity and digital inclusion for all markets. Thus, AI and IoT not only drive economic progress but also ensure a fairer and more sustainable future. These emerging digital technologies raise environmental, social, and governance initiatives, representing a fundamental and significant step toward corporate sustainability in a globalized world, as well as toward achieving the United Nations Sustainable Development Goals (SDGs).

This research provides insights into how IoT and AI, combined with other technologies, such as blockchain and big data, transform data collection, validation, and analysis, in general, enabling the generation of more accurate, reliable, and transparent information. It also highlights that IoT and AI help address socio-environmental challenges related to inconsistent data quality, promoting standardization and integration across various areas in the pursuit of sustainable development.

The study emphasizes that AI enables predictive analysis and pattern recognition in complex data, supporting informed decision-making and responsible investment. Additionally, combining this technology with blockchain ensures security, transparency, and immutability in data-related processes, providing stakeholders with greater confidence in corporate reports. IoT, when associated with big data, handles large volumes of information from different sources, consolidating them into meaningful insights for sustainable business practices.

The research also noted that the positive impact of AI technologies and IoT devices extends beyond corporate practices: they encourage social innovation, mitigate environmental risks, improve social equity, and promote ethical governance. However, implementation faces challenges, such as high initial costs and the need for highly skilled professionals. Overcoming these obstacles will require collaboration between the public and private sectors, support from robust regulatory policies, and continuous investment in technological innovation.

In summary, the incorporation of emerging digital technologies from Industry 4.0 and 5.0 represents an opportunity to enhance corporate efficiency, but also a crucial commitment to sustainable development and the creation of shared value for society as a whole. Ongoing exploration of these technologies will help shape a resilient and responsible future for both businesses and society.

References

Abbas Q, Ahmad G, Alyas T, Alghamdi T, Alsaawy Y, Alzahrani A (2023) Revolutionizing urban mobility: IoT-enhanced autonomous parking solutions with transfer learning for smart cities. Sensors 23:8753. https://doi.org/10.3390/s23218753

Akram SV, Malik PK, Singh R, Gehlot A, Juyal A, Ghafoor KZ, Shrestha S (2022) Implementation of digitalized technologies for fashion industry 4.0: opportunities and challenges. Sci Program 2022:7523246. https://doi.org/10.1155/2022/7523246

Al-Maadeed MAA, Ponnamma D (2023) Fiber chemistry and technology: their contributions to shaping society 5.0. Discover Nano 18:115. https://doi.org/10.1186/s11671-023-03888-4

Cabanillas-Carbonell M, Pérez-Martínez J, Zapata-Paulini J (2023) Contributions of the 5G network with respect to poverty (SDG1), systematic literature review. Sustainability 15:11301. https://doi.org/10.3390/su151411301

Gajić T, Petrović MD, Pešić AM, Conić M, Gligorijević N (2024) Abordagens inovadoras na gestão hoteleira: integrando inteligência artificial (IA) e a internet das coisas (IoT) para melhorar a eficiência operacional e a sustentabilidade. Sustainability 16:7279. https://doi.org/10.3390/su16177279

Gohari A, Ahmad AB, Rahim RBA, Supaat ASM, Abd-Razak S, Gismalla MSM (2022) Involvement of surveillance drones in smart cities: a systematic review. IEEE Access 10:56611–56628. https://doi.org/10.1109/ACCESS.2022.3177904

Hoosain MS, Paul BS, Ramakrishna S (2020) The impact of 4IR digital technologies and circular thinking on the United Nations sustainable development goals. Sustainability 12:10143. https://doi.org/10.3390/su122310143

Kumar S, Sharma D, Rao S et al (2022) Past, present, and future of sustainable finance: insights from big data analytics through machine learning of scholarly research. Ann Oper Res 21:4410. https://doi.org/10.1007/s10479-021-04410-8

Lampropoulos G, Garzón J, Misra S, Siakas K (2024) The role of artificial intelligence of things in achieving sustainable development goals: state of the art. Sensors 24:1091. https://doi.org/10.3390/s24041091

Martos V, Ahmad A, Cartujo P, Ordoñez J (2021) Ensuring agricultural sustainability through remote sensing in the era of agriculture 5.0. Appl Sci 11:5911. https://doi.org/10.3390/app11135911

Metelenko N, Nikitenko V, Meniailo V (2024) Development of the smart economy as the main source of competitiveness and sustainable development. Baltic J Econ Stud 10(2):187–195. https://doi.org/10.30525/2256-0742/2024-10-2-187-195
Moher D, Liberati A, Tetzlaff J, Altman DG, PRISMA Group (2009) Preferred reporting items for systematic reviews and meta-analyses: the PRISMA statement. PLoS Med 6(7):e1000097. https://doi.org/10.1371/journal.pmed.1000097
Page MJ, McKenzie JE, Bossuyt PM, Boutron I, Hoffmann TC, Mulrow CD, Shamseer L et al (2021) The PRISMA 2020 statement: an updated guideline for reporting systematic reviews. BMJ 372:71
Paiva S, Ahad MA, Tripathi G, Feroz N, Casalino G (2021) Enabling technologies for urban smart mobility: recent trends, opportunities and challenges. Sensors 21:2143. https://doi.org/10.3390/s21062143
Pandey S, Twala B, Singh R, Gehlot A, Singh A, Montero EC, Priyadarshi N (2022) Wastewater treatment with technical intervention inclination towards smart cities. Sustainability 14:11563. https://doi.org/10.3390/su141811563
Patalas-Maliszewska J, Łosyk H (2024) Changes in sustainable development in manufacturing in cases of unexpected occurrences: a systematic review. Sustainability 16:717. https://doi.org/10.3390/su16020717
Perianes-Rodriguez A, Waltman L, van Eck NJ (2016) Constructing bibliometric networks: a comparison between full and fractional counting. J Inform 10(4):1178–1195. https://doi.org/10.1016/j.joi.2016.10.006
Ramírez-Márquez C, Posadas-Paredes T, Raya-Tapia AY, Ponce-Ortega JM (2024) Natural resource optimization and sustainability in society 5.0: a comprehensive review. Resources 13:19. https://doi.org/10.3390/resources13020019
Rosário AT, Dias JC (2022) Sustainability and the digital transition: a literature review. Sustainability 14:4072. https://doi.org/10.3390/su14074072
Saxena A, Singh R, Gehlot A, Akram SV, Twala B, Singh A, Montero EC, Priyadarshi N (2023) Tecnologias empoderadas ambientais, sociais e de governança (ESG): um cenário da indústria 4.0. Sustentabilidade 15:309. https://doi.org/10.3390/su15010309
Singh R, Akram SV, Gehlot A, Buddhi D, Priyadarshi N, Twala B (2022) Energy system 4.0: digitalization of the energy sector with inclination towards sustainability. Sensors 22:6619. https://doi.org/10.3390/s22176619
Singh R, Singh R, Gehlot A, Akram SV, Priyadarshi N, Twala B (2022) Horticultura 4.0: adoção de tecnologias da indústria 4.0 na horticultura para atender à agricultura sustentável. Appl Sci 12:12557. https://doi.org/10.3390/app122412557
The Pact for the Future (2024) Seventy-ninth session agenda item 123 strengthening of the United Nations system. www.un.org/pga/wp-content/uploads/sites/109/2024/09/The-Pact-for-the-Future-final.pdf
Tran-Dang H, Kim DS (2021) The physical internet in the era of digital transformation: perspectives and open issues. IEEE Access 9:164613–164631. https://doi.org/10.1109/ACCESS.2021.3131562
UN, United Nations (2015) 2030 agenda: sustainable development goal. https://sdgs.un.org/goals
UN, United Nations (2023) The sustainable development goals report 2023: special edition. https://unstats.un.org/sdgs/report/2023/
Varriale V, Cammarano A, Michelino F, Caputo M (2024) The role of digital technologies in production systems for achieving sustainable development goals. Sustain Product Consump 47:87–104. https://doi.org/10.1016/j.spc.2024.03.035
Vermesan O, John R, Pype P, Daalderop G, Kriegel K, Mitic G, Lorentz V, Bahr R, Sand HE, Bockrath S, Waldhör S (2021) Inteligência automotiva incorporada na tecnologia de veículos elétricos autônomos e compartilhados conectados para mobilidade verde sustentável. Front Fut Transp 2:688482. https://doi.org/10.3389/ffutr.2021.688482
Vernier C, Loeillet D, Thomopoulos R, Macombe C (2021) Adoption of ICTs in agri-food logistics: potential and limitations for supply chain sustainability. Sustainability 13:6702. https://doi.org/10.3390/su13126702

Wang K, Zhao Y, Gangadhari RK, Li Z (2021) Analyzing the adoption challenges of the internet of things (IoT) and artificial intelligence (AI) for smart cities in China. Sustainability 13:10983. https://doi.org/10.3390/su131910983

Young OR, Yang J, Guttman D (2020) Meeting cyber age needs for governance in a changing global order. Sustainability 12:5557. https://doi.org/10.3390/su12145557

7 The Role of Artificial Intelligence in Green Reverse Logistics: A Sustainability Perspective

Otacilio José Moreira, Maria Carolina Martins Rodrigues, Martius Vicente Rodrigues Y. Rodrigues, and André Luiz Azevedo Guedes

Abstract

Due to the population's high growth, man needs to find more efficient techniques to solve socio-environmental problems, such as reducing the generation of solid waste, using artificial intelligence (AI). This article aims to analyze and answer how using artificial intelligence (AI) in solid waste management enhances the efficiency and effectiveness of reverse logistics operations while promoting environmental sustainability through reuse, recycling, and remanufacturing. The concept of *green returns* is one of the main benefits of reverse logistics, ensuring the return of products that do not meet marketing standards. This is exploratory research based on a survey of bibliographic and documentary sources on solid waste management through reverse logistics, based on AI. Data were obtained from the WoS, Scopus, and Research Rabbit databases in the period between 2020 and 2024. The study aims to maximize efforts and contribute to the environmentally appropriate disposal of the waste generated, emphasizing the potential of its AI-based management, increasing the environmental and economic benefits of reverse logistics, and presents a proposal for an integrated management system that can contribute to achieving these objectives.

O. J. Moreira (✉) · M. V. R. Y. Rodrigues
Universidade Federal Fluminense, Niterói, Brazil
e-mail: otaciliojm@gmail.com

M. C. M. Rodrigues
Cinturs—Universidade do Algarve, Faro, Portugal

A. L. A. Guedes
UNISUAM e Universidade Federal Fluminense, Niterói, Brazil

M. J. Sousa et al. (eds.), *Economic Sustainability Potentiated by Artificial Intelligence*, Contributions to Environmental Sciences & Innovative Business Technology,
https://doi.org/10.1007/978-981-96-8089-4_7

Keywords

Artificial intelligence • Agriculture • Efficiency • Case studies • Quantitative analysis

7.1 Introduction

In the era of globalization, marked by increased consumption, and heightened competition, new concepts such as sustainable development first defined in the Brundtland Report (World Commission on Environment and Development 1987), as "development that meets the needs of the present without compromising the ability of future generations to meet their own needs." Concurrently, products' shorter lifecycle leads to challenges in managing post-consumption goods and addressing environmental issues due to improper disposal. Consequently, the concept of logistics has expanded to include reverse logistics, which facilitates the management of material flows from the point of consumption back to the point of origin, particularly for products that do not meet commercial standards, thereby reducing waste.

Reverse logistics (RL) according to Chouinard et al. (2005) and Kazemi et al. (2018) is a vital component of sustainable production; unlike traditional forward supply chains, where the customer is the endpoint, RL extends the product's lifecycle by managing its end-of-life (EoL) or even initiating a new production cycle. This approach provides opportunities to prolong product use, conserve resources, reduce waste, and create secondary markets and jobs in remanufacturing and recycling. The authors argue that RL is inherently more complex; their return streams can include products, subassemblies, and materials, which can reintegrate the forward supply chain at multiple points. RL encompasses a set of actions, procedures, and means to facilitate the collection and return of solid waste to the business sector, for either reuse in production cycles or alternative disposal. However, according to Bhowmik et al. (2024), reverse logistics generate positive results, when processes are efficient from the point of origin to final disposal, requiring meticulous planning and implementation. Bhowmik et al. (2024) also cite that the application of artificial intelligence in reverse logistics can significantly enhance the efficiency and sustainability of supply chain processes. The question that guides this study is: How can the application of artificial intelligence optimize reverse logistics processes, contributing to environmental sustainability and business efficiency, while minimizing waste and promoting the development of a circular economy?

The study's main objective is to analyze and answer how using artificial intelligence (AI) in solid waste management enhances the efficiency and effectiveness of reverse logistics operations while promoting environmental sustainability through reuse, recycling, and remanufacturing. The specific objectives are: to investigate the impact of the application of artificial intelligence (AI) in the process of collecting, sorting, and sorting solid waste in reverse logistics, evaluate how the technology improves the efficiency and accuracy of these operations; to analyze how AI can optimize the reverse supply chain, facilitating the reuse, recycling, and remanufacturing of products, and its role in reducing waste and maximizing material recovery and evaluating the role of AI in promoting environmental sustainability in reverse logistics operations, examining its contribution to reducing the carbon footprint, and strengthening greener and more circular business practices.

Currently, there is an urgent need to properly manage municipal solid waste to reduce environmental impact and promote sustainable development while minimizing adverse environmental effects. Therefore, this requires a robust waste management policy and the commitment of social institutions to maintain it. According to Aubakirova (2024) and Bhowmik et al. (2024), AI plays a critical role in improving green reverse logistics, optimizing processes, improving companies' performance, and reducing waste. The integration of AI with environmental processes further strengthens the association between waste minimization and business performance, promoting a more sustainable and circular economy. Future research should focus on exploring new AI techniques and their application in various industries to further advance the field of reverse logistics.

The study is justified by the urgent need for technological solutions that can mitigate the environmental impacts resulting from inadequate disposal and excessive consumption. With the growing production of urban solid waste and the pressure on natural resources, it is essential to find ways to optimize the recovery and reuse of materials, promoting a more sustainable production cycle.

The importance of the topic lies in the fact that the application of AI can transform reverse logistics operations, making them more efficient and effective, by reducing waste, improving recycling, and increasing product remanufacturing. In addition, AI has the potential to contribute significantly to environmental sustainability, by enabling smarter and more accurate waste management, promoting the circular economy, and reducing the carbon footprint of industrial operations. Thus, the study contributes to advances both in the field of sustainability and technology applied to environmental processes.

7.2 Theoretical Framework

Nowadays, the supply chain management environment deals with a global volume of waste at an alarming rate; as a result, there is a growing awareness of the urgent need to improve recycling efforts and optimize resource use to promote environmental sustainability and minimize the negative impacts of waste generation. According to Bali et al. (2022), as consumer awareness of environmental issues grows and ecological regulations evolve, organizations are pressured to adopt green practices through end-of-life (EoL) and end-of-use (EoU) processes; reverse logistics (RL) has emerged as a key focus in supply chain management

(SCM). However, integrating reverse logistics into the supply chain is a complex task, requiring a multifaceted approach. Reverse logistics focuses on the efficient and sustainable handling of products and materials after their initial use. This section provides a comprehensive overview of the definition and importance of reverse logistics, its role in the circular economy and sustainability, and the challenges faced in traditional reverse logistics processes.

7.2.1 Definition and Importance of Reverse Logistics

The Retailers Manual (1869) addressed the first studies on the concept of returning goods, Ferber et al. (1960), and Guiltinan and Nwokoye (1975); although the term used was reverse logistics flows. In the following years, according to Davis et al. (1995) and Patankar and Mitra (1995), businesses introduced various return policies, including money-back guarantees. For Duong et al. (2022), the origins of product returns (PR) date back to the 1960s and 1970s, when PR was quietly integrated into business tactics and strategies. The term *reverse logistics* is found in Murphy et al. (1995) who are among the first authors to use it, defining it as the use of a marketing channel to promote the flow of products from customer to producer, in contrast to the normal flow. The term ***reverse distribution***, which is interchangeable with reverse logistics, was used in practice by Carter and Ellram (1998), Jayaraman et al. (2003) and Dyckhoff et al. (2004). Giuntini and Andel (1995) defined reverse logistics as the materials management obtained from customers by an organization, without specifying the direction of resources. Stock and Lambert (2001) define reverse logistics as planning and managing the return of products and materials to the place of origin, providing a closed cycle of production and consumption. The term *product recovery management* (PRM) created by Thierry et al. (1995) encompasses all tasks, under the responsibility of a manufacturing company, related to the management of all products, components, and materials used and discarded. The author's goal with the PRM was to reduce waste and regain maximum economic and ecological value. Rogers and Tibben-Lembke (1999) consider reverse logistics as the planning, implementing, and controlling of the reverse flow of materials efficiently and cost-effectively, and related information from consumers to manufacturers to recover value or dispose of it appropriately.

Reverse logistics is the process of managing the return, disposal, and reuse of products and materials, according to Prajapati et al. (2019a, b), including the return, repair, and refurbishment of products, or their disposal at the end of their useful life. Carter and Ellram (1998) and Senthil (2014) emphasize the importance of reverse logistics for value recovery and waste management, contributing to business sustainability. The main actors in the network structure in reverse logistics are consumers, collection centers, treatment plants, and markets (Chileshe et al. 2018; Waqas et al. 2018; Zhang et al. 2022).

Consumers, collection centers, treatment plants, and markets are the major players in a network structure in reverse logistics (Chileshe et al. 2018; Waqas et al.

2018; Zhang et al. 2022). According to Arroyo et al. (2023), there is an increasing focus on using reverse and green logistics to reduce the environmental impact of logistics operations, minimize waste, and maximize the value of products and materials through efficient planning and management.

7.2.2 Understanding Reverse Logistics and Sustainability

Since the first industrial revolution, rapid economic and technological advancements have significantly enhanced living standards. However, according to Bao and Lu (2020) and Ahmed et al. (2020), this growth has also been responsible for global waste generation and resource depletion. Inefficient waste handling and unsustainable use of resources lead to environmental degradation, resource scarcity, and economic losses. As urbanization and industrialization continue to accelerate, the volume of waste generated globally is increasing at an alarming rate. In response to these challenges, there is a growing recognition of the need for more sustainable waste and resource management practices. Governments, businesses, and communities are increasingly adopting strategies that promote the circular economy, where waste is minimized, and resources are kept in use for as long as possible. This shift toward sustainability, according to Julianelli et al. (2020) and Sun et al. (2022), not only addresses environmental concerns but also creates opportunities for economic growth and innovation. *Reverse logistics* (RL) has emerged as a key solution to these challenges, optimizing material use and addressing resource depletion by enabling the recovery and repurposing of goods. To Shah et al. (2019), this approach not only enhances environmental sustainability but also improves operational efficiency and financial performance. For the authors, companies converting waste materials into valuable resources and addressing environmental challenges are promoting economic growth. Reverse logistics (RL) enables the retrieval, recycling, remanufacturing, and repurposing of products while reducing waste and easing the burden on natural resources. This not only decreases the environmental impact of waste disposal but also extends the material's lifecycle, fostering sustainability in both industrial and consumer practices.

The main motivation for establishing reverse logistics stated by Keh et al. (2012) is to improve customer service and satisfaction and meet sustainable requirements. Competitive advantage now extends beyond manufacturing and sales to include associated services.

In addition to its environmental benefits, RL increases the efficiency of companies' value chains by reintroducing recovered goods into the production process or reselling them, thereby unlocking previously untapped value from products that would otherwise be discarded. This optimization of the value chain reduces production costs, improves resource utilization, and boosts profitability. Furthermore, RL encourages businesses to innovate in their product design and processes, aligning them more closely.

The next point will analyze the challenges involved in sustainable reverse logistics adoption.

7.2.3 Challenges in Traditional Reverse Logistics Processes

By enabling the efficient return and repair of products, reverse logistics help businesses reduce waste, preserve resources, and minimize their environmental footprint. Additionally, RL can significantly improve a company's reputation and customer satisfaction by offering seamless and effective return and repair services, which are increasingly valued by environmentally conscious consumers. However, the successful implementation of a reverse logistics program is complex. In accordance with Waqas et al. (2018) and Zhang et al. (2022), demands close coordination among various departments and partners, including suppliers, logistics service providers, and repair facilities to overcome challenges. Keh et al. (2012) revealed that the synergies between reverse logistics and the supply chain help mitigate the operational and economic challenges outlined in the literature. The authors conclude that the "*closed-loop*" supply chain model, by integrating forward and reverse logistics, allows companies to simultaneously overcome the main challenges: economic benefits through savings and revenue generation, environmental responsibility through compliance with legislation and preservation of resources, and social challenges of creating or preserving jobs.

The complexity of devolution control and retrieval services in supply chain management, a shortage of skilled professionals, low environmental awareness, and an insufficient understanding of legislation are significant challenges to sustainability in developing countries. In the words of Dabees et al. (2023), overcoming these challenges requires a concerted effort to educate, empower, and engage all stakeholders in promoting sustainable practices.

AI-driven innovations enable companies to enhance corporate social responsibility, improve customer satisfaction through efficient return services, and strengthen competitive advantage in a market increasingly focused on sustainability. The following point will explore these benefits in detail.

7.2.4 AI-Driven Sustainability Benefits in Reverse Logistics

While reverse logistics offers substantial benefits, its successful implementation requires a well-coordinated effort involving multiple stakeholders and the strategic use of technology to overcome the inherent complexities. Moreover, the integration of advanced technologies, such as artificial intelligence (AI) and Internet of Things (IoT), can further enhance the efficiency of reverse logistics operations. These technologies can optimize routes, predict demand for returns, and automate sorting and repair processes, thereby reducing costs and improving overall performance. In accordance with Aubakirova (2024) and Bhowmik et al. (2024), their studies highlight the interconnected nature of technology, human resource management, and firm performance in driving sustainable supply chain practices. The integration of AI and big data analytics, coupled with effective human resource management practices and strong firm performance, can result in significant improvements in waste management and overall corporate environmental performance.

As mentioned by Kumar et al. (2023) and Adesoga et al. (2024), artificial intelligence (AI) is emerging, offering significant improvements in efficiency and decision-making, and stands out for responding better in less time. Its applications vary by industry and show the potential to transform repetitive and time-consuming tasks. The author emphasizes that AI's main objectives in reverse logistics include optimizing collection networks and improving the identification, inspection, and sorting of returned products. It is advantageous in companies that manage a limited range of products and multiple recovery paths, increasing efficiency in sorting and determining the appropriate destination for each item.

7.3 Methodology

Despite the growing academic focus on integrating AI into various phases of reverse logistics (RL), according to Bhowmik et al. (2024), the technological advancement in this area remains unclear, highlighting a significant research gap. To address this, the study uses, initially, a narrative review of the literature that as highlighted by Ferrari (2015) has an important role in continuing education providing readers with up-to-date knowledge about a specific topic or topic. Additionally, Hart (2018), Chaney (2021), and Sukhera (2022) mention that a narrative review allows for a critical analysis of the existing literature and identifies gaps in the production of studies, providing current information on the subject, and stimulating new research. Greenhalgh et al. (2018) discuss the importance of narrative reviews to identify trends and guide subsequent empirical studies. The authors further argue that this approach is especially useful when dealing with emerging topics, such as the use of AI in LR.

A systematic search was conducted across the Web of Science, ScienceDirect, and Research Rabbit databases to identify documents published between 2019 and 2024. Levy and Ellis (2006) emphasize that utilizing multiple databases in systematic searches enhances the robustness of reviews, ensuring comprehensive and thorough coverage of the subject matter. The keywords used in the search were: reverse logistics, artificial intelligence, solid waste management, and environmental sustainability.

Following the narrative review, the research utilizes case studies that complement the research to provide a comprehensive understanding of the current state of AI-enabled LR. Yin (2018) defines the case studies methodology as a recommended strategy to explore contemporary phenomena in depth, especially in contexts where the boundaries between the phenomenon and the context are blurred. For Eisenhardt and Graebner (2007), case studies can provide valuable insights by capturing the dynamics of the implementation process of emerging technologies, such as AI in LR. And, Miles et al. (2014) state that the use of case studies as a qualitative analysis tool allows a detailed understanding of the impacts and challenges faced by businesses when adopting new technologies. The research enriches the existing literature by integrating AI within LR frameworks and demonstrating the potential of AI for waste reduction and sustainability. The study contributes to understanding how AI can optimize LR processes while offering insights into the environmental impacts of operational decisions in manufacturing.

7.3.1 AI in Reverse Logistics Operations:

The application of AI in reverse logistics operations offers several benefits, from optimizing collection networks and improving identification and sorting processes to improving decision-making and automating repetitive tasks. Davenport and Ronanki (2018) argue that AI is essential for modernizing supply chain management, offering automated and analytical solutions to complex problems such as reverse logistics. Van der Laan and Salomon (1997) suggest that the use of advanced technologies, including AI, for companies can significantly improve efficiency, and reduce waste, which are essential elements for a sustainable approach to reverse logistics.

As the availability of data continues to increase, the potential for AI to transform reverse logistics will only grow, becoming an essential component of any modern supply chain strategy (Table 7.1).

7.4 Case Studies

7.4.1 Real-World Examples of Companies Successfully Implementing AI in Green Reverse Logistics

Here are some cases where artificial intelligence (AI) has been applied in solid waste management and reverse logistics (Table 7.2).

These use cases illustrate how AI can be effectively applied to improve the efficiency and effectiveness of solid waste management and reverse logistics, promoting environmental sustainability through the reuse, recycling, and remanufacturing of products.

Table 7.1 Samples of AI in reverse logistics and their objectives

Operations	Objective	Authors
Waste classification and sorting	AI-driven waste segregation, automated sorting, and recycling processes	Qu (2022), Lilhore et al. (2023), Nasir and Al-Talib (2023)
Returns management	AI-powered return predictability, routing optimization, and inventory management	Chouinard et al. (2005), Kazemi et al. (2018), Bhowmik et al. (2024)
Repair, refurbishment, and remanufacturing	AI applications in assessing product condition, optimizing repair/refurbishment processes, and predicting remanufacturing potential	Aziz et al. (2021), Rahito and Azman (2019), Xiao and Huang (2024)
End-of-life product management	AI in disassembly, material recovery, and disposal processes	Wang et al. (2021), Daniyan et al. (2021)
Resource efficiency	AI's role in reducing waste, conserving resources, and improving process efficiency	Arun et al. (2024), Lanzalonga et al. (2024), Islam et al. (2024)
Emission reduction	AI applications for minimizing the carbon footprint of reverse logistics operations	Mukherjee et al. (2024), Kannan et al. (2012)
Social impacts	AI's contribution to job creation, skill development, and improved working conditions in reverse logistics	Morandini et al. (2023), Wilson et al. (2022)
Demand forecasting	AI-driven demand prediction for returned, refurbished, or recycled products	Krapp et al. (2013), Callialp et al. (2017)
Network optimization	AI algorithms for optimizing the location and capacity of reverse logistics facilities	Kangi et al. (2024), He et al. (2024)
Route optimization	AI-powered dynamic routing for collection, transportation, and delivery of returned or recycled items	Dabo and Hosseinian-Far (2023), Mtetwa (2024)

Fonte: Author creation

7.5 Future Trends and Research Directions

7.5.1 Emerging AI Technologies and Their Potential Applications in Green Reverse Logistics

By integrating AI and machine learning into their operations, according to Riahi et al. (2021), companies can achieve significant improvements in efficiency, accuracy, and speed, ultimately leading to better business outcomes in reverse logistics

Table 7.2 Samples of real-world case studies

Project	Organization	Summary	Technology	Results
Waste classification and separation	ZenRobotics	AI-equipped robots to classify and sort construction and demolition waste	Computer vision and machine learning	Separation of wood, metal, and plastic increases efficiency and recycling accuracy
Waste classification and separation	Eco Urbis			
Optimization of waste collection routes	Rubicon global	Real-time data analysis of the volume of waste and traffic to determine the most efficient routes for collection	Machine learning and real-time data analysis	Reduced costs and carbon emissions
Optimization of waste collection routes	Eco logic			
Monitoring and forecasting of waste	Waste management, Inc	Monitoring and forecasting the generation of waste in different areas	IoT sensors and machine learning	Better resource allocation and planning
Monitoring and forecasting of waste	Sustentare			
Automation of waste sorting centers	Projeto reciclo	Identification and separation of different types of recyclable materials with high precision	Computer vision and machine learning	Increased efficiency and quality of recycling
Automation of waste sorting centers	Tomra systems			
Reverse logistics management	Recicla+	Managing reverse logistics program, from collection to repair, and reuse of electronic equipment	Machine learning and predictive analytics	Better use of transport modal and identification of components that can be reused or recycled
Crop optimization	Stara	Agricultural machinery and implements	Artificial intelligence system	Operational costs were reduced by 30% and customer satisfaction improved by 25%

Fonte: Author creation

and other supply chain processes, ending in a competitive advantage in the market. Authors believe that (AI) methodologies have the potential to help decision-making in reverse logistics processes by offering sophisticated capabilities in data analysis, trend identification, and forecasting. These methodologies empower businesses in accordance with Yu (2022) and Kuznar and Lorenc (2023), to derive meaningful insights from extensive datasets, uncovering patterns that may not be readily apparent through conventional approaches. The author found that AI can predict trends by examining historical and real-time data, facilitating more informed and proactive decision-making. This capability, according to Didast et al. (2024), is essential in reverse logistics, where anticipating demand, streamlining operations, and reducing risks can greatly enhance performance and efficiency. Furthermore, AI-powered tools can improve decision support and aid in more precise scenario planning, enabling organizations to maintain competitiveness and adapt to rapidly evolving environments.

7.5.2 The Role of AI in Advancing the Circular Economy and Closed-Loop Supply Chains

According to Govindan et al. (2015), Kazemi et al. (2018) and Wilson et al. (2022), reverse logistics is a key driver of a closed-loop supply chain; the authors recognize these practices as environmentally friendly, helping to make conventional supply chains more sustainable. Reverse logistics is considered the core component of the closed-loop supply chain (CLSC) due to its crucial role in facilitating the return and recovery of products after their initial use. The authors found that the traditional supply chain process, which often emphasizes the direct movement of goods from raw materials to the end consumer, transforms into a closed-loop supply chain when AI-driven reverse logistics is incorporated. According to Bhattacharya et al. (2024), reverse logistics adds to the reverse flow processes, activities such as product return, recycling, remanufacturing, and proper disposal. This integration improves resource efficiency and minimizes waste, contributing to more sustainable supply chain operations.

According to Singh (2023), the integration of artificial intelligence (AI) in circular economies accelerates the transition to sustainable systems, delivering environmental, economic, and social benefits aligned with global sustainability objectives. These technologies facilitate the creation of new approaches to designing products that are more sustainable and that consider, from the outset, easy recycling and reuse. Geissdoerfer et al. (2017) complement this view, stating that **the circular economy** is based on a strategy that ignites the continuous reuse of materials and products. From this perspective, AI is essential to coordinate circular economy strategies, as it enables more efficient waste sorting and the automatic identification of materials that can be recycled or remanufactured, optimizing reverse logistics processes and significantly reducing waste.

Bhattacharya et al. (2024) argues that by closing the loop with the widespread implementation of AI techniques in reverse logistics which can close the loop in supply chains, organizations can promote a significant reduction in the use of virgin resources, energy consumption, and the amount of waste sent to landfills. This allows companies to recover value from end-of-life products, reduce environmental impact, and align their operations with sustainability goals, while potentially generating new revenue streams and reducing costs associated with sourcing raw materials.

7.6 Conclusion

The present study highlights the relevance and positive impacts of the integration of artificial intelligence (AI) in reverse logistics for solid waste management, with a focus on promoting environmental sustainability. The use of AI proves crucial in modernizing and optimizing reverse logistics operations, improving efficiency and accuracy in waste collection, sorting, and sorting processes. The study employed a narrative literature review combined with case studies to examine the practical applications of AI in depth, highlighting both the benefits and challenges faced by companies. The findings suggest that AI technology can enhance operational performance and support the transition to a circular economy by aligning with global sustainability goals and facilitating closed-loop supply chains.

The study revealed that the application of AI in reverse logistics enables more efficient and sustainable waste management, reducing waste and optimizing the reuse and recycling of materials. AI's ability to perform advanced data analysis and generate reliable predictions facilitates decision-making, allowing businesses to quickly adapt to changing market demands and environmental policies. In addition, AI has proven effective in scenario planning, which is crucial for managing complex operations and anticipating any risks. The implementation of AI not only promotes a significant reduction in the use of virgin resources but also helps to minimize environmental impact by reducing the amount of waste destined for landfills, generating new revenue opportunities through the recovery of value from end-of-life products.

Despite the significant contributions of this study, some limitations must be considered. The technological complexity and cost of implementing AI systems can pose challenges for small and medium-sized businesses, which may not have the resources to invest in advanced solutions. In addition, integrating AI into existing processes may face cultural resistance within organizations, which may require time to adapt practices and develop specific technical skills. The study is also limited by the dependence on secondary data and the lack of access to some proprietary information of the companies analyzed, which may restrict the scope of the conclusions.

For future research, it is recommended to explore the application of AI in specific reverse logistics sectors, such as electronics and plastics, where recycling challenges are more complex. Analyzing different business models that adopt AI and their impacts on financial and environmental sustainability can provide valuable insights. It is also suggested to investigate the ethical and regulatory aspects associated with the use of AI, focusing on the implications for data privacy and technological governance. In addition, the development of methodologies that evaluate the return on investment of AI technologies applied to reverse logistics can provide additional support for the decision-making of companies that wish to adopt these solutions.

References

Adesoga TO, Olayaa OP, Onuma EP, Ajavi OO, Olagunju OD (2024) Review of reverse logistics practices and their impact on supply chain sustainability. Int J Sci Res Arch 12(02):162–168

Ahmed Z, Asghar MM, Malik MN, Nawaz K (2020) Moving towards a sustainable environment: the dynamic linkage between natural resources, human capital, urbanization, economic growth, and ecological footprint in China. Resour Policy 67:101677. https://doi.org/10.1016/j.resourpol.2020.101677

Arroyo LAB, Barreto CADLS, Vasquez OBS, Nicola RJV (2023) The importance of reverse logistics and green logistics for sustainability in supply chains. J Bus Entrep Stud 7(4):46–72

Arun V, Krishna-Rao-Patro E, Anusuya Devi VS (2024) AI based prediction algorithms for enhancing the waste management system: a comparative analysis. ES Web Confer 552:01052

Aubakirova D (2024) Directions for using big data analytics in logistics management. Develop Manag 23(1):27–36

Aziz NA, Adnan NAA, Wahab DA (2021) Component design optimization based on artificial intelligence in support of additive manufacturing repair and restoration: current status and future outlook for remanufacturing. J Clean Product 296:126401. https://doi.org/10.1016/j.jclepro.2021.126401

Bali S, Gunasekaran A, Aggarwal S (2022) A strategic decision-making framework for sustainable reverse operations. J Clean Product 381:135058. https://doi.org/10.1016/j.jclepro.2022.135058

Bao Z, Lu W (2020) Developing efficient circularity for construction and demolition waste management in fast emerging economies: lessons learned from Shenzhen. China. Sci Total Environ 724:138264. https://doi.org/10.1016/j.scitotenv.2020.138264

Bhattacharya A, Srivastava S, Majumdar A (2024) Circular supply chains in manufacturing—Quo vadis? Accomplishments, challenges and future opportunities. Bus Strategy Environ 33(5):4397–4423. https://doi.org/10.1002/bse.3702

Bhowmik O, Chowdhury S, Ashik JH, Mahmud GMI, Khan MM, Hossain NUI (2024) Application of artificial intelligence in reverse logistics: a bibliometric and network analysis. Supply Chain Anal 7:100076. https://doi.org/10.1016/j.sca.2024.100076

Callialp F, Vayvay O, Temucin T (2017) Importance of artificial intelligence in reverse logistics: the state of the art. In: Sixth international conference on advances in applied science and environmental technology—ASET 2017. Institute of Research Engineers and Doctors, pp 1–5

Carter CR, Ellram LM (1998) Reverse logistics: a review of the literature and framework for future investigation. J Bus Log 19(1):85–102

Chaney MA (2021) So you want to write a narrative review article? J Cardioth Vascul Anesth 35:3045–3049. https://doi.org/10.1053/j.jvca.2021.06.017

Chileshe N, Rameesdeen R, Hosseini MR, Martek I, Li HX, Panjehbashi-Aghdam P (2018) Factors driving the implementation of reverse logistics: a quantified model for the construction industry. Waste Manag 79:48–57. https://doi.org/10.1016/j.wasman.2018.07.013

Chouinard M, D'Amours S, Aït-Kadi D (2005) Integration of reverse logistics activities within a supply chain information system. Comput Ind 56(1):105–124. https://doi.org/10.1016/j.compind.2004.07.005

Dabees A, Barakat M, Elbarky SS, Liasec A (2023) A framework for adopting a sustainable reverse logistics service quality for reverse logistics service providers: a systematic literature review. Sustainability 15:1755. https://doi.org/10.3390/su15031755

Dabo A-AA, Hosseinian-Far A (2023) An integrated methodology for enhancing reverse logistics flows and networks in industry 5.0. Logistics 7(4):97. https://doi.org/10.3390/logistics7040097
Daniyan I, Muvunzi R, Mpofu K (2021) Artificial intelligence system for enhancing product's performance during its life cycle in a railcar industry. Proced CIRP 98:482–487. https://doi.org/10.1016/j.procir.2021.01.138
Davenport and Ronanki (2018) argue that AI is essential for modernizing supply chain management, offering automated and analytical solutions to complex problems such as reverse logistics
Davis S, Gerstner E, Hagerty M (1995) Money back guarantees in retailing: matching products to consumer tastes. J Retail 71(1):7–22
Didast FZ, Nassih R, Elbachir IA (2024) Artificial intelligence and logistics: recent trends and development. Preprints. https://doi.org/10.20944/preprints202410.2141.v1
Duong QH, Zhou L, Meng M, Nguyen TV, Ieromonachou P, Nguyen DT (2022) Understanding product returns: a systematic literature review using machine learning and bibliometric analysis. Int J Product Econ 243:108340. https://doi.org/10.1016/j.ijpe.2021.108340
Dyckhoff H, Lackes R, Reese J (eds) (2004) Supply chain management and reverse logistics, pp 91–112. https://doi.org/10.1007/978-3-540-24815-6
Eisenhardt KM, Graebner ME (2007) Theory building from cases: opportunities and challenges. Acad Manag J 50:25–32. https://doi.org/10.5465/amj.2007.24160888
Ferber R, Hoagland JH, Lazer W (1960) The retailer's manual of 1869. J Market 24(3):59–60. https://doi.org/10.1177/002224296002400311
Geissdoerfer M, Savaget P, Bocken NMP, Hultink EJ (2017) The circular economy: a new sustainability paradigm? J Clean Product 143:757–768. https://doi.org/10.1016/j.jclepro.2016.12.048
Giuntini R, Andel T (1995) Advance with reverse logistics: part 1. Transp Distr 36(2):73–75
Govindan K, Soleimani H, Kannan D (2015) Reverse logistics and closed-loop supply chain: a comprehensive review to explore the future. Eur J Oper Res 240(3):603–626. https://doi.org/10.1016/j.ejor.2014.07.012
Greenhalgh T, Thorne S, Malterud K (2018) Time to challenge the spurious hierarchy of systematic over narrative reviews? Eur J Clin Invest 48(6):1–6. https://doi.org/10.1111/eci.12931
Guiltinan JP, Nwokoye NG (1975) Developing distribution channels and systems in the emerging recycling industries. Int J Phys Distr 6(1):28–38
Hart C (2018) Doing a literature review: releasing the research imagination (2nd ed). SAGE Publications
He M, Li Q, Wu X (2024) A novel multi-level reverse logistics network design optimization model for waste batteries considering facility technology types. J Clean Product 467:142966
Islam Z, Ahmed A, Alfify MH et al (2024) The impact of artificial intelligence on environment and sustainable development in India. Educ Admin Theory Pract 30(5):1850–1856
Jayaraman V, Patterson R, Rolland E (2003) The design of reverse distribution networks: models and solution procedures. Eur J Oper Res 150(1):128–149. https://doi.org/10.1016/S0377-2217(02)00497-6
Julianelli V, Caiado RGG, Scavarda LF, Cruz SPMF (2020) Interplay between reverse logistics and circular economy: critical success factors-based taxonomy and framework. Resour Conser Recycl 158:104784. https://doi.org/10.1016/j.resconrec.2020.104784
Kangi F, Pasandideh SHR, Mehdizadeh E, Soleimani H (2024) The optimization of an integrated forward and reverse logistics network based on routing and cross-docking strategy. J Appl Res Ind Eng 11(2):251–282. https://doi.org/10.22105/jarie.2023.368851.1510
Kannan D, Diabat A, Alrefaei M et al (2012) A carbon footprint based reverse logistics network design model. Resour Conser Recycl 67:75–79
Kazemi N, Modak NM, Govindan K (2018) A review of reverse logistics and closed loop supply chain management studies: a bibliometric and content analysis. Int J Product Res 24:1–24. https://doi.org/10.1080/00207543.2018.1471244
Keh P, Rodhain F, Meissonier R, Llorca V (2012) Financial performance, environmental compliance, and social outcomes: the three challenges of reverse logistics. Supply Chain for Int J 3(3):26–39. https://doi.org/10.1080/16258312.2012.11517296

Krapp M, Nebel J, Sahamie R (2013) Using forecasts and managerial accounting information to enhance closed-loop supply chain management. Or Spect 35(4):975–1007
Kumar KR, Melchior R, Madhuri A, Shireesha B (2023) The role of artificial intelligence in decision-making processes. Appl Res Artif Intell Cloud Comput 6(9):31–46
Kuznar M, Lorenc A (2023) A hybrid method for technical condition prediction based on AI as an element for reducing supply chain disruptions. Appl Sci 13(22):12439. https://doi.org/10.3390/app132212439
Lanzalonga F, Marseglia R, Irace A, Biancone PP (2024) The application of artificial intelligence in waste management: understanding the potential of data-driven approaches for the circular economy paradigm. Manag Dec 10:1733
Levy Y, Ellis TJ (2006) A systems approach to conduct an effective literature review in support of information systems research. Inform Sci Int J Emerg Transdiscip 9:181–212
Lilhore UK, Simaiya S, Dalal S, Damaševičius R (2023) A smart waste classification model using hybrid CNN-LSTM with transfer learning for sustainable environment. Multimedia Tools Appl 32:1–25. https://doi.org/10.1007/s11042-023-16677-z
Miles MB, Huberman AM, Saldaña J (2014) Qualitative data analysis: a methods sourcebook, 3rd edn. SAGE Publications, New York
Morandini S, Fraboni F, De Angelis M et al (2023) The impact of artificial intelligence on workers' skills: upskilling and reskilling in organizations. Inform Sci Int J Emerg Transdiscip 26:039–068
Mtetwa C (2024) Reverse logistics and waste reduction alternatives. In: Munuhwa S (ed) Advances in civil and industrial engineering. IGI Global, pp 345–373. https://doi.org/10.4018/979-8-3693-3755-4.ch013
Mukherjee S, Nagariya R, Mathiyazhagan K, Baral MM, Pavithra MR, Appolloni A (2024) Artificial intelligence-based reverse logistics for improving circular economy performance: a developing country perspective. Int J Logist Manag 35(6):1779–1806. https://doi.org/10.1108/IJLM-03-2023-0102
Murphy PR, Poist RF, Braunschweig CD (1995) Role and relevance of logistics to corporate environmentalism. Int J Phys Distrib Log Manag 25(2):5–19
Nasir I, Al-Talib GAA (2023) Waste classification using artificial intelligence techniques: literature review. Techn Roman J Appl Sci Technol 5:49–59
Patankar JG, Mitra A (1995) Effects of warranty execution on warranty reserve costs. Manag Sci 41(3):395–400. https://doi.org/10.1287/mnsc.41.3.395
Prajapati H, Kant R, Shankar R (2019a) Bequeath life to death: state-of-art review on reverse logistics. J Clean Product 211:503–520. https://doi.org/10.1016/j.jclepro.2018.11.187
Prajapati H, Kant R, Shankar R (2019b) Prioritizing the solutions of reverse logistics implementation to mitigate its barriers: a hybrid modified SWARA and WASPAS approach. J Clean Product 240:118219. https://doi.org/10.1016/j.jclepro.2019.118219
Qu D (2022) Application of artificial intelligence in waste classification management at university. In: Sharma H, Vyas VK, Pandey RK, Prasad M (eds) Proceedings of the international conference on intelligent vision and computing (ICIVC 2021). ICIVC 2021. Proceedings in adaptation, learning and optimization, vol 15. Springer, Cham. https://doi.org/10.1007/978-3-030-97196-0_27
Rahito WDA, Azman AH (2019) Additive manufacturing for repair and restoration in remanufacturing: an overview from object design and systems perspectives. Processes 7(11):802. https://doi.org/10.3390/pr7110802
Riahi Y, Saikouk T, Gunasekaran A, Badraoui I (2021) Artificial intelligence applications in supply chain: a descriptive bibliometric analysis and future research directions. Exp Syst Appl 173:114702. https://doi.org/10.1016/j.eswa.2021.114702
Rogers DS, Tibben-Lembke RS (1999) Going backwards: reverse logistics trends and practices. Reverse Logistics Executive Council
Senthil S, Venkatachalam S, Baskaran R (2014) Reverse logistics: a key to sustainable business practices. Int J Eng Technol 6(2):123–130. https://doi.org/10.15623/ijret.2014.0323031

Shah S, Dikgang G, Menon S (2019) The global perception of industry 4.0 for reverse logistics. Int J Econ Manag Syst 4:103–107
Singh JP (2023) Artificial intelligence in circular economies: a pathway to sustainable resource management. Int J Sci Res 12(12):1128–1131
Stock JR, Lambert DM (2001) Strategic logistics management (4th ed). McGraw-Hill/Irwin
Sukhera J (2022) Narrative reviews: flexible, rigorous, and practical. J Grad Med Educ 14(4):414–417. https://doi.org/10.4300/JGME-D-22-00480.1
Sun X, Yu H, Solvang WD (2022) Towards the smart and sustainable transformation of reverse logistics 4.0: a conceptualization and research agenda. Environ Sci Pollut Res 29(46):69275–69293
Thierry M, Salomon M, Nunnen J, Wassenhove L (1995) Strategic issues in product recovery management. California Manag Rev 37(2):114–135
Van der Laan E, Salomon M (1997) Production planning and inventory control with remanufacturing and disposal. Eur J Oper Res 102(2):264–278. https://doi.org/10.1016/S0377-2217(97)00108-2
Wang L, Liu Z, Liu A, Tal F (2021) Artificial intelligence in product lifecycle management. Int J Adv Manufact Technol 114(3–4):771–796. https://doi.org/10.1007/s00170-021-06882-1
Waqas M, Dong Q, Ahmad N, Zhu Y, Nadeem M (2018) Critical barriers to implementation of reverse logistics in the manufacturing industry: a case study of a developing country. Sustainability 10(11):4202. https://doi.org/10.3390/su10114202
Wilson M, Paschen J, Pitt L (2022) The circular economy meets artificial intelligence (AI): understanding the opportunities of AI for reverse logistics. Manag Environ Qual Int J 33(1):9–25
World Commission on Environment and Development (Org.) (1987) Our common future. Oxford University Press, Oxford
Xiao J, Huang K (2024) A comprehensive review on human–robot collaboration remanufacturing towards uncertain and dynamic disassembly. Manufact Rev 11:17
Yin RK (2018) Case study research and applications: design and methods (6th ed). SAGE Publications
Yu H (2022) Modeling a remanufacturing reverse logistics planning problem: some insights into disruptive technology adoption. Int J Adv Manufact Technol 123:4231–4249. https://doi.org/10.1007/s00170-022-10387-w
Zhang X, Zou B, Feng Z, Wang Y, Yan W (2022) A review on remanufacturing reverse logistics network design and model optimization. Processes 10(1):84. https://doi.org/10.3390/pr10010084

Renewable Energy Innovations and Artificial Intelligence: A Scientific Mapping Study

8

Donaji Jiménez-Islas, Miriam Edith Pérez-Romero, José Álvarez-García, and María de la Cruz del Río-Rama

Abstract

The development and application of renewable energy technologies have increased in recent years due to the need for energy and the government's drive to combat climate change. The use of renewable energy presents advantages and challenges that in most cases are solved through product and service innovation. A topical tool used in renewable energy applications is artificial intelligence (AI), which helps address the challenges of the clean energy sector. The objective of the work is to identify the patterns of innovation in renewable energy and artificial intelligence, and a bibliometric analysis was performed to determine the publications, citations, authors, keywords, subjects, journals, countries, and dynamics of collaboration in the field of study. The data were obtained from publications indexed in the Scopus database. Information is displayed in graphs and maps that allow to visually identify trends in the field of knowledge, and tools such as excel solver, VOSviewer, Sankey, and chord diagrams will be

D. Jiménez-Islas · M. E. Pérez-Romero
TecNM/ITS de Huichapan, Huichapan, Mexico
e-mail: djimenez@iteshu.edu.mx

M. E. Pérez-Romero
e-mail: mperezgp@alumnos.unex.es; meperez@iteshu.edu.mx

M. E. Pérez-Romero
University of Extremadura, Badajoz, Spain

J. Álvarez-García (✉)
Universidad de Extremadura, Cáceres, Spain
e-mail: pepealvarez@unex.es

M. de la Cruz del Río-Rama
University of Vigo, Ourense, Spain
e-mail: delrio@uvigo.es

M. J. Sousa et al. (eds.), *Economic Sustainability Potentiated by Artificial Intelligence*, Contributions to Environmental Sciences & Innovative Business Technology,
https://doi.org/10.1007/978-981-96-8089-4_8

used. The rate of publications has grown exponentially from 2018 to 2023, having a rate of 0.039 $years^{-1}$, also the type of review document has the highest TC/TP ratio, and the most cited author is Tabor, Daniel P. The subject engineering, computer sciences, and energy are the three main fields of knowledge, and the most productive countries in publications are India, the United States, and China. The top 25 documents with the highest number of citations (91.7%) correspond to quartile 1 journals. The keywords related to artificial intelligence are machine learning, internet of things, learning systems, decision support systems, and big data applied to renewable energy. This work may be of interest to students, researchers, academics, and decision-makers in companies related to the renewable energy sector and is the first study to implement bibliometric analysis in combination with innovation and artificial intelligence.

Keywords

AI • Renewable energy • Innovation • Bibliometric • Scopus • Green energy

8.1 Introduction

Society's need for comfort has led to a continuous increase in energy demand and dependence on fossil fuels. Major countries are seeking to provide energy that is compatible with the planet and meet the Sustainable Development Goals (SDGs), specifically number 7 (Minas et al. 2024). This gradual transition is key not only for energy security, but also for addressing interconnected challenges such as climate change, poverty, and food security. Renewable sources such as solar, wind, and geothermal improve energy security and reduce dependence on finite fossil fuels (Bashiru et al. 2024).

Renewable energy sources, including solar, wind, and hydropower, are vital for energy security and economic development, especially in G20 countries, where collaborative efforts are crucial for deployment (Dixit 2024). Despite the momentum of renewables globally, challenges such as market failures, limited access to information in some countries, and infrastructure constraints hinder the widespread adoption of renewable technologies (Bashiru et al. 2024; Hassan et al. 2024).

AI can address complex problems by simulating human reasoning and decision-making processes, making it valuable in clinical settings and beyond (Pollock et al. 2024). Artificial intelligence (AI) technology has evolved rapidly in recent decades and has been widely applied in energy management systems, renewable energy systems, and smart grids, providing powerful tools for design, control, fault diagnosis, and simulation in modern industrial power systems (Chander and Gopalakrishnan 2024). AI algorithms are powerful tools that analyze historical data to predict energy consumption patterns, enabling better resource allocation (Barkavi et al. 2024). AI improves operational efficiency by identifying and mitigating energy losses during distribution; in addition, various investment instruments can support the implementation of AI in the energy sector (Kuzmenko et al. 2024). Despite the promise of technological tools, merging AI systems with existing energy infrastructure can be complex and requires specialized skills (Olatunde et al. 2024).

The integration of AI in renewable energy will improve resource management, optimize raw materials, improve the sustainability of the source, reduce emissions to the environment, and increase the efficiency of the system by establishing predictive maintenance that extends the life of the infrastructure, in addition to monitoring data in real time to have greater reliability of energy systems (Nguyen et al. 2024; Onwusinkwue et al. 2024).

The deployment of AI in energy management systems addresses the challenges of balancing intermittent renewable energy generation with variable demand. It improves resource forecasting, predicts electricity market dynamics, and facilitates intelligent decision-making, thus ensuring a more stable and reliable power supply (Adewuyi et al. 2023). AI frameworks use multi-agent systems that employ machine learning for predictive analytics, enabling real-time decision-making to optimize load distribution and manage distributed energy resources (DER) effectively, and these systems assimilate data from sensors and smart meters, enabling dynamic adjustments to energy supply in response to fluctuations in demand and generation levels (Dasari et al. 2024). AI algorithms can forecast renewable energy generation patterns and align them with energy demand, optimizing the use of clean energy sources. This capability improves grid stability by effectively managing the interplay between intermittent generation and variable demand for a constant power supply (Muhammad et al. 2023).

It is necessary to investigate the extent of IA implementations in renewable energies, taking into consideration how researchers in the field of knowledge conduct their lines of work. The evolution of scientific knowledge in recent years has increased at an accelerated pace, where authors seek to disseminate the results of their research (Álvarez-García et al. 2020; Jiménez-Islas et al. 2024).

Given the technological changes and innovations, it is proposed to quantify different aspects related to the area of knowledge through bibliometric indicators.

Complete descriptive information on many facets of research publications, authors, universities, and nations is frequently included in a thorough bibliometric analysis. Depending on the research goals, it might also contain details about publication kinds, collaborative networks, citation patterns, and more (Hassan and Duarte 2024). A more thorough overview of the impact of research can be obtained by combining traditional and alternative measurements, which makes it easier to evaluate for financing and academic advancements (Berger 2022). Also, to promote research in particular subjects, they can identify understudied regions and offer chances to enhance data analytic skills, create educational initiatives, and encourage multidisciplinary collaborations using methods like bibliometrics.

Although there is a diversity of literature on the benefits of IA in renewable energy, it is also possible to identify some research gaps, which leads to the following research questions:

RQ-1: At what speed do authors publish research results on the topic of innovations in renewable energy and AI? RQ2: In what type of document do authors publish their work and how are papers cited in the fields of renewable energy and IA? RQ3: What are the countries' areas of research work? RQ4: Which countries are prominent in the number of publications, and what is their citation influence by other authors? RQ5: What is the impact of the publications on the scientific community through the quartile in which the journals are placed? RQ6: Which are the most cited publications? RQ7: Who are the leading authors in the fields of renewable energy and IA? RQ8: What are the keywords used by the authors in their publications, and how has the topic of renewable energy and IA evolved?

In view of the above, the purpose of this work is to carry out a bibliometric analysis through information obtained from the Scopus database for the topic of renewable energy, innovation, and artificial intelligence.

The format of this document is as follows: The theoretical justifications, the necessity for additional study in this area, earlier research, and the research topics are all presented in Sect. 8.1. The methods used to collect and analyze the data in order to address each of the study topics are then described in Sect. 8.2. Section 8.3, which is divided up into subsections, presents the findings. A discussion is presented in Sect. 8.4, and the paper's findings are presented in Sect. 8.5.

8.1.1 Review of the Scientific Literature

Energy conversion performance, storage, and distribution efficiency have greatly improved with the incorporation of artificial intelligence (AI) into renewable energy systems. Artificial intelligence (AI) technologies, such as machine learning and predictive analytics, improve sustainability and operational efficiency across a range of renewable energy sources, including biomass, solar, wind, and hydropower.

Some authors identify the key role of technological innovation.

- Current AI technologies can optimize energy production and improve system performance in some renewable energy sources such as solar and wind. They improve energy storage and distribution through predictive analytics, ensuring efficient power generation and balancing supply and demand, thus supporting smart grid developments (Manuel et al. 2024).
- Dibie (2024) concludes that AI and cloud technology offer immense potential to revolutionize renewable energy systems, but their deployment must balance technological progress with ethical considerations.
- Katkar et al. (2024) describe that predicting the production of renewable energy is significantly influenced by artificial intelligence (AI). The importance of AI-powered solutions in large-scale integration of renewable energy sources, such as solar panels, has not been well studied in multi-energy systems, despite this.

- With an emphasis on increasing the efficiency of renewable energy sources, the authors examine AI-driven solar energy generation and smart grid integration. While AI inclusion in smart grids helps with demand-side management and energy storage optimization, methods such as support vector regression and artificial neural networks enhance solar irradiance forecast (Wen et al. 2024).
- Abdur et al. (2024) report that AI enhances renewable energy systems through nine identified techniques, improving power generation, forecasting, and optimization. It addresses challenges like load balancing, cyberattack prevention, and efficiency, while offering superior accuracy and adaptability compared to conventional models, paving the way for future advancements.
- Dawodu et al. (2024) explore how AI enables predictive maintenance of renewable energy infrastructures through the use of machine learning algorithms to analyze large amounts of sensor data and historical performance. By anticipating equipment failures, reducing downtime, extending the lifespan of infrastructure, and improving dependability, this method saves a lot of money.

8.2 Materials and Methods

The generation of knowledge at a global level has increased in recent years as a response to the opening of information and the need for researchers to express their contributions and contribute to the scientific community. The analysis of information requires tools that allow summarizing and forecasting the main trends in the field of knowledge. Bibliometric analysis evaluates the scientific literature by analyzing publication trends, citation counts, and journal contributions (Tyasi et al. 2024). It was proposed to analyze the information through bibliometric indicators. The search, selection, and analysis of the information were carried out between October 2024 and November 2024. Documents for the bibliometric analysis were retrieved from the Scopus database, and no language restrictions were used. The keyword criteria defined were as follows: ((TITLE-ABS-KEY ("renewable energ*") OR TITLE-ABS-KEY ("green energ*") AND TITLE-ABS-KEY ("innovat*") AND TITLE-ABS-KEY ("artificial intelligence") AND PUBYEAR > 2009 AND PUBYEAR < 2024).

Different growth models (linear, exponential, quadratic) have been applied to analyze publication data, and the findings suggest that linear growth may better describe certain periods (Behrens and Lankenau 2006). The evolution of scientific publications in a specific area of growth is a multifactorial indicator, and it is important to establish a way to estimate the annual growth rate. In this work, we implemented the adjustment of data through the Gompertz equation and its normalization, Eqs. 8.1 and 8.2, respectively. Microsoft Excel was used to estimate the values of the variables of the Gompertz equation and normalization, and the "Solver" function was used for optimization. Regression data analysis was used with significance of $P = 0.05$, and the coefficient of determination R^2 was estimated (Jiménez-Islas et al. 2023).

$$P_{(y)} = ae^{-be^{-\mu t}} \tag{8.1}$$

$$\left(P_{(y)\mathrm{Scopus}} - P_{(y)\mathrm{Model}}\right)^2 \tag{8.2}$$

Here,

$P_{(y)}$ is the cumulative documents by year,
μ is the specific growth rate of publications
a is the asymptotic publications,
b is an integration constant related to initial publications,
t is the time in years,
$P_{(y)\mathrm{Scopus}}$ is the publications reported on Scopus database, and
$P_{(y)\mathrm{Model}}$ is the estimated publications by model.

Data were downloaded from the Scopus platform in.csv format and processed in Microsoft Excel. To construct the visual maps, Microsoft 365 Power Bi was used for chord map, Sankeymatic.com platform for Sankey map, and VOSviewer for keywords and countries map.

8.3 Results and Discussions

8.3.1 Publications

Technological innovation has evolved through various models, reflecting changes in the sources, dynamics, and foundations of innovation over decades (Xu and Chen 2023). The evolution of technological innovation is influenced by a number of factors, such as policy interventions, social entrepreneurship, and the integration of knowledge economies, which together enhance economic growth, competitiveness, and the development of innovative practices in different sectors and regions (Kumar et al. 2023). Business innovation is also enhanced by the integration of scientific knowledge, as firms that take advantage of published research tend to produce more patents (Chen et al. 2024) and gain competitive advantages that allow them to stay at the technological forefront. The relationship between scientific publications and technological innovation is evident in sectors such as renewable energy, where knowledge from scientific articles has driven advances in technology, an example of this is in wind turbines used in wind energy (Lacerda 2019). The evolution of scientific publications has two stages: the first until 2016 where the number of published documents is no more than 10 and the second stage where the IA is socialized and integrated with an exponential behavior in the number of publications in the area of study, Fig. 8.1. The publication data were fitted with the Gompertz model to describe their behavior, and the publication rate estimated by the model is 0.039 years^{-1} rate, with a fit of 0.977 coefficient of determination (R^2).

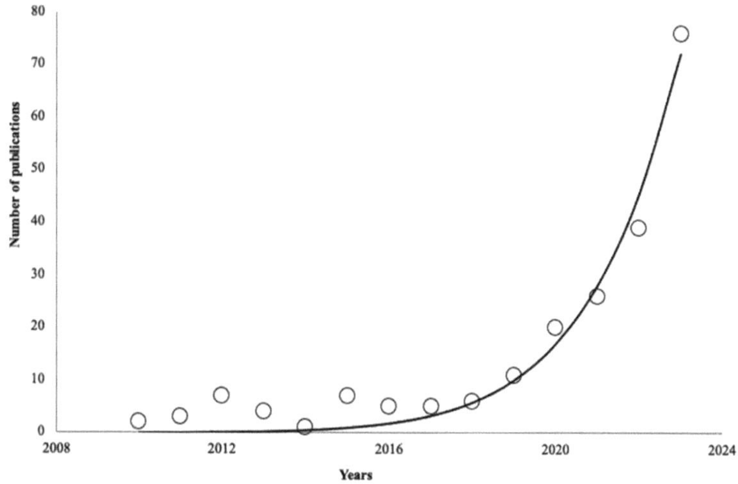

Fig. 8.1 Evolution of scientific publications on the subject of innovation in renewable energy and IA. ○ indicates Scopus data, and the solid line indicates data fitting by the Gompertz model

8.3.2 Document Type and Citation

The data obtained from the Scopus database are shown in Table 8.1. Knowing the type of publication of the authors serves as a basis for establishing who are the main authors and papers for the field of knowledge; in addition, articles focus on original investigations in specific scientific subjects, while review articles synthesize existing research, offering comprehensive analyses of current knowledge, identifying gaps, and proposing future research agendas, thereby enhancing understanding within a particular field of study (Chigbu et al. 2023). The data obtained from the Scopus database are shown in Table 8.1. 84.4% of the documents are articles, conference papers, and reviews. In this work, articles and review are the most cited documents.

Table 8.1 Document type and citations

Rank	Document type	TP	Ratio (%)	TC (2010–2023)	Ratio	TC/TP
1	Article	88	41.5	2412	57	27.4
2	Conference paper	69	32.5	381	9	5.5
3	Review	23	10.8	1304	30.8	56.7
4	Book	12	5.7	44	1	3.7
5	Book chapter	12	5.7	36	0.9	3
6	Conference review	6	2.8	0	0	0
7	Editorial	2	0.9	51	1.2	25.5

TC—total citations, TP—total publications

8.3.3 Most Frequently Cited Documents

Table 8.2 shows the top ten most cited documents. The paper with the highest number of citations (511) is "Accelerating the discovery of materials for clean energy in the era of smart automation" de Tabor et al. (2018); the document discusses the integration of AI, robotics, and automation to revolutionize the discovery of clean energy materials, significantly reducing development time and costs, thereby facilitating the transition to a low-carbon economy within the next 5–10 years. The second most cited paper is "Energy storage emerging: A perspective from the Joint Center for Energy Storage Research," which discusses the evolution of energy storage, highlighting advances in lithium-ion batteries, the emergence of diverse battery technologies for various applications, and strategic research efforts to meet the changing demands of energy storage (Trahey et al. 2020). Citations for some authors are a multifactorial result of the quality of the paper, and the analysis in Table 8.2 refers to the positive aspect of authors having more citations of their publications. Researchers' self-assessments of the quality of their work show a statistically significant relationship with citation counts, particularly with respect to scientific importance (Aksnes et al. 2023).

Table 8.2 Most frequently cited documents

Document title	Year	Journal	Cited by	References
Accelerating the discovery of materials for clean energy in the era of smart automation	2018	Nature Reviews Materials	511	Tabor et al. (2018)
Energy storage emerging: a perspective from the joint center for energy storage research	2020	Proceedings of the National Academy of Sciences of the United States of America	222	Trahey et al. (2020)
Application of artificial intelligence methods for hybrid energy system optimization	2016	Renewable and Sustainable Energy Reviews	221	Zahraee et al. (2016)
The impact of renewable energy, internet use and foreign direct investment on carbon dioxide emissions: a method of moments quantile analysis	2022	Renewable Energy	149	Awan et al. (2022)
A comparative investigation of maximum power point tracking methods for solar PV system	2016	Solar Energy	139	Gupta et al. (2016)

(continued)

Table 8.2 (continued)

Document title	Year	Journal	Cited by	References
Optimized control of DFIG-based wind generation using sensitivity analysis and particle swarm optimization	2013	IEEE Transactions on Smart Grid	136	Tang et al. (2013)
Artificial intelligence for smart renewable energy sector in Europe: smart energy infrastructures for next generation smart cities	2020	IEEE Access	125	Venayagamoorthy (2011)
Dynamic, stochastic, computational, and scalable technologies for smart grids	2011	IEEE Computational Intelligence Magazine	99	Venayagamoorthy (2011)
From lagging to leading? Technological innovation systems in emerging economies and the case of Chinese wind power	2013	Energy Policy	94	Gosens and Lu (2013)
An overview on the smart grid concept	2010	IECON Proceedings (Industrial Electronics Conference)	93	Cecati et al. (2010)

TC—total citations, TP—total publications

8.3.4 Top 15 Countries with the Highest Number of Publications and Citations

The top five most productive countries in terms of publications are India (36), United States (26), China (25), Italy (14), and Romania (11). As the aim of the researchers is to have their work consulted and cited by others, the nationality of the researchers with the highest citation impact was identified as United States with 1270 citations, China with 853 citations, Germany with 670 citations, Mexico with 579 citations, and India with 546 citations. Germany is the country with the highest CT/PT rate with 67, indicating that the documents are frequently cited. (Table 8.3).

According to the "*Precedence research*" platform (www.precedenceresearch.com), the global artificial intelligence (AI) in renewable energy market size will increase at a projected CAGR of 27.70% from 2023 to 2032. One of the key elements driving the advancement of global artificial intelligence (AI) in the renewable energy sector is the increasing need for electricity in developing and established areas. The increase in demand for smarter energy grids is also spurring the advancement and evolution of artificial intelligence (AI) in renewables in the global market. Another element that is driving the growing demand for artificial intelligence (AI) in renewables is the digital transformation of the energy sector. The Asia Pacific region takes leadership in AI in the expansion of the renewable

Table 8.3 Top 15 countries most prominent in publications and citations

Rank	Country	Documents	Citations	TC/TP
1	India	36	546	15.2
2	United States	26	1270	48.8
3	China	25	853	34.1
4	Italy	14	262	18.7
5	Romania	11	209	19.0
6	Germany	10	670	67.0
7	Turkey	10	257	25.7
8	Canada	9	351	39.0
9	United Kingdom	9	233	25.9
10	Saudi Arabia	9	445	49.4
11	Malaysia	8	295	36.9
12	Russian Federation	8	28	3.5
13	Poland	7	62	8.9
14	Portugal	7	57	8.1
15	Australia	7	73	10.4

TC—total citations, TP—total publications

energy sector. The energy distribution sector represents the fastest growing segment of artificial intelligence in the renewable energy sector in 2022. Energy distribution is one of the most important global tasks. The government is taking action to develop this sector. This is directly encouraging the development of global artificial intelligence (AI) in the renewable energy sector (Precedence research Report 2024). Early on, the majority of developed nations were seen to be the birthplaces of innovations in renewable energy. However, a number of developing nations, particularly China and India, have emerged as well-known pioneers in the use of renewable energy in recent years (Bayer et al. 2013; Elavarasan et al. 2019).

8.3.5 Top Ten Countries–Subjects

They were ranked in the top 10 countries and subjects, the data were processed on the Sankeymatic.com platform, and a Sankey map was generated. Figure 8.2 shows the relationship between the top 10 of different countries and the subject areas. The top three subjects where authors publish are engineering, followed by computer science and Energy. The countries with the highest contribution to the subject areas are India, the United States, and China.

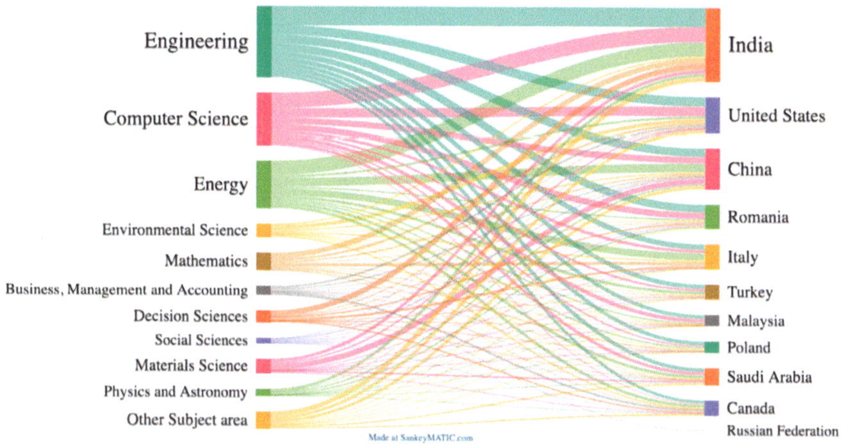

Fig. 8.2 Sankey map of scientific publications on the subject–countries of innovation in renewable energy and IA

8.3.6 Journals and Quartile of Publications

Academic journals are ranked according to several criteria, including the Q (quartile) indicator, which is an analytical approach used to evaluate scientific journals according to their impact factor and their affiliation to a certain scientific group. As the number of publications in the IA and renewable energy fields increases, authors seek to share their knowledge through journals with the best quality ratings and the widest global reach (Jiménez-Islas et al. 2024). Consequently, they have tended to publish in journals that are in quartiles 1 and 2 (Q1 and Q2). 62% of authors prefer to publish in journals with Q1 quartile, 12% in Q2 quartile, and 12% in Q3 quartile. The rest of the publications are in journals that do not have a quartile. Authors seek to publish in journals with global reach so that their results are read and cited, and for this reason, they seek to publish in Q1 quartile journals, Fig. 8.3. More than 50% of the papers published in Q1 journals belong to "Energies," "IEEE access," and Sustainability, and two of the journals belong to the MDPI publishing house.

The journal with one publication and the highest number of citations (511) is "Nature reviews materials," followed by "Renewable and sustainable energy reviews" with four publications and 343 citations.

8.4 Keywords

The information obtained from keyword mapping (Fig. 8.4) can guide future lines of research, helping to identify gaps and emerging trends (Sarudin et al. 2024). To carry out the keyword analysis, a map was created in VOSviewer (Fig. 8.4), in

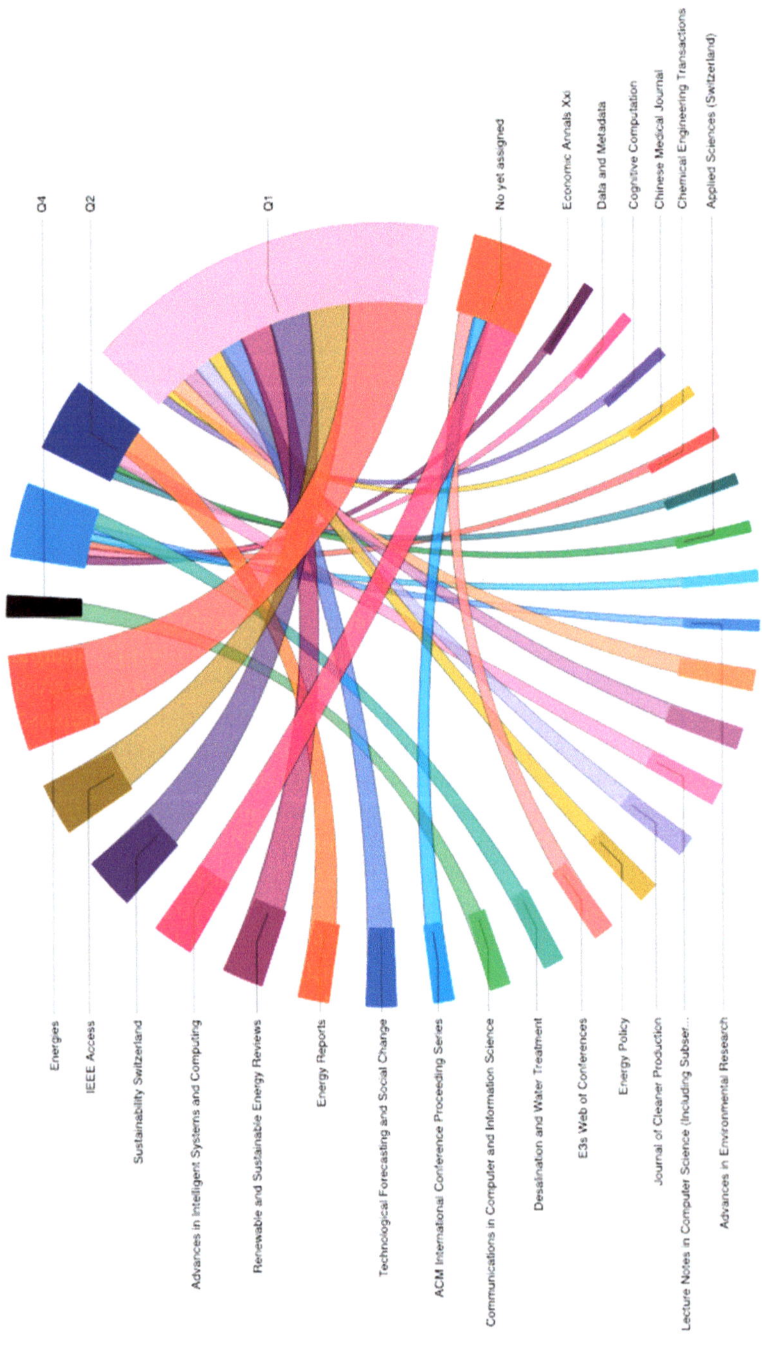

Fig. 8.3 Chord map of journals and quartiles

which four distinct clusters were obtained. At the center of the word map is the term "artificial intelligence" with occurrence of 114, and it is connected to most of the keywords of the different clusters. In the center of the word map is located the term "artificial intelligence" with occurrence of 114, and it is connected to most of the keywords of the different clusters, in the second place the words "renewable energy resources" and in third place "renewable energies." In the analysis, five clusters were identified, and the green cluster related to "renewable energy" is associated with the keywords "gas emissions," "energy systems," "decision support systems," "energy efficiency," "digital storage," "AI." The blue cluster related to "alternative energy" is associated with "innovation," "developing countries," "technological innovation system," "technological development," and others. The yellow cluster is highlighted by the connected words "smart grid," "smart power grids," "deep learning," "energy management," "electric power transmission," and "iot". The purple cluster is connected to all clusters; the word "artificial intelligence" is related to "machine learning," "fuzzy logic," "learning algorithms," "forecasting," and others. The red cluster is the one with the highest presence of artificial intelligence; the words related to "sustainable development" are "machine learning," "internet of things," "machine-learning," "energy," "big data," "blockchain," "data analytics," "engineering education," and "digital transformation." AI's role in optimizing solar and wind energy is expected to develop and innovate significantly in future. AI-driven solutions will play a key role in the sustainable growth of renewable energy in future, as evidenced by the trend toward hybrid learning approaches and the growing application of AI in technical and geographic assessments. It is anticipated that the incorporation of IoT and other cutting-edge technology will improve renewable energy systems' accuracy and efficiency even more. AI technology will be essential in resolving the intermittent nature of renewable energy sources and in streamlining grid operations as they develop further. Significant expansion and innovation are possible in the field of artificial intelligence in solar and wind energy optimization in future (Ohalete et al. 2023).

8.5 Discussion

The transformative role of AI in renewable energy includes improving wind speed forecasts, optimizing biofuel production, and improving energy efficiency. Collaborations between academia and industry accelerate sustainable energy transitions, emphasizing the need for innovation and investment in technology-driven solutions (Raman et al. 2024).

The main AI findings in renewable energy innovation have been identified by various authors. Predictive analytics enable better forecasting of energy demand, allowing for more effective resource allocation (Rane et al. 2024). AI technologies like machine learning and predictive analytics optimize energy efficiency, enhance grid management, and support renewable energy integration (Mohammadi Lanbaran et al. 2024). The combination of IoT devices and machine learning allows

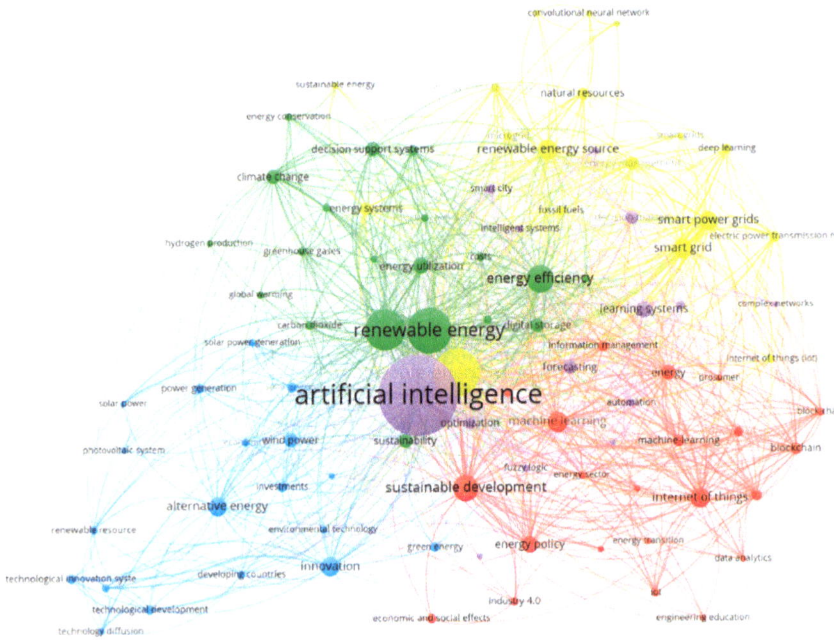

Fig. 8.4 Network

for real-time data collection and analysis, improving the precision of energy production forecasts in wind energy systems (Emexidis and Gkonis 2024). AI-driven predictive maintenance enhances operational efficiency and reliability in power systems, particularly by diagnosing faults in induction motors. This capability is essential for minimizing energy cuts and ensuring stable integration of renewable energy sources (Hasani et al. 2024).

While the integration of AI in renewable energy systems presents numerous benefits, challenges such as data privacy and integration complexity remain significant hurdles that need to be addressed for widespread adoption (Komala et al. 2024). AI is significantly transforming the research and development of new energy materials, particularly in the realm of battery technology. By utilizing advanced algorithms for the simulation and evaluation of chemical combinations, researchers can discover novel compounds that enhance battery efficiency, such as solid-state batteries.

In this work, the evolution in the number of publications from 2018 to 2023 had an exponential growth, Fig. 8.1. As the need for energy grows, renewable sources are a technically and economically viable option; however, they require optimization and improvement tools for their management, generation, and storage. IA is a tool that shows promise in renewable energy management and publications in databases are growing, possibly due to the interest of researchers, companies, and stakeholders. In the same vein, Eslava-Zapata et al. (2024) reported the interest of researchers from 2021 to 2023 in the use of artificial intelligence with renewable energy with the aim of promoting sustainable development aligned to the 2030 Agenda. In addition, the application of artificial intelligence has grown exponentially from 2016 to 2022 as demonstrated by Liu et al. (2022) report, which identified a 35-fold growth in AI publications in 2021 compared to the 2011 period. By fitting the data using the Gompertz model, it is possible to identify in this paper that the exponential growth in publications will continue, which is consistent with data reported by other authors.

Numerous bibliometric studies have shown that publications of various document kinds not only collect citations at varying rates, but also have varying average citation counts (Haunschild and Bornmann 2022; Wang 2013). Table 8.1 shows that articles and reviews are the type of document with the highest number of citations, which may be due to the fact that they are more widely available than other types of documents in the databases. Authors have a higher number of citations in articles and reviews because they are more easily accessible, as well as a higher number of TC/TP conversions; this information is also verified in Table 8.2.

The document with the highest number of citations is "Accelerating the discovery of materials for clean energy in the era of smart automation" (Tabor et al. 2018), and its relevance in the field of knowledge is due to the fact that it addresses the need to integrate advances in automation, robotics, and informatics with traditional chemistry and materials science. This integration is expected to transform research and development processes in both industry and academia, making them more efficient and effective.

The country with the highest productivity is India, but the United States has the highest number of citations. Government initiatives and regulations play a vital role in creating an enabling environment for innovation and investment in renewable energy technologies; the integration of smart grids, artificial intelligence (AI), Internet of Things (IoT), and big data is crucial to transform India's renewable energy landscape (Bhattacharya 2023). The authors also point to issues such as infrastructure readiness, data privacy concerns, and the need for qualified staff as significant barriers to successful integration.

The fusion of artificial intelligence (AI) with renewable energy technologies is driving significant innovations in the energy field. Artificial intelligence boosts the efficiency and sustainability of renewable energy systems by addressing challenges such as variability and intermittent, ultimately promoting a smoother transition to greener energy sources. Artificial intelligence algorithms anticipate energy production from renewable resources, optimizing the organization and allocation of resources (Salma et al. 2024). Artificial intelligence systems facilitate constant analysis of energy production and use, optimizing efficiency in operations (Lakshmi et al. 2024). Artificial Intelligence enables dynamic modifications to energy supply based on real-time consumption data, improving resource utilization (Komala et al. 2024). Artificial intelligence facilitates predictive maintenance of renewable infrastructure, decreasing downtime and operational costs (Singh and Kaur 2024). Challenges such as information privacy, complexity of integration, and demand for supporting regulatory frameworks remain crucial aspects for future progress.

8.6 Conclusions

The paper identifies the current state of innovation at the intersection of renewable energy and artificial intelligence (AI) through bibliometric analysis. This analysis highlights the evolution of publications, citations, and leading countries in this field, indicating a growing interest and relevance of the topic in the scientific community.

The study reveals an exponential growth in publications from 2016 to the present, suggesting that the topic is gaining significant traction. The Gompertz model was used to project future growth, indicating that this trend is likely to continue due to the growing importance of renewable energy and AI innovations. It was found that articles and reviews tend to receive more citations compared to other types of papers. This suggests that comprehensive reviews and original research in this area are highly valued by the scientific community, reflecting their importance in advancing knowledge.

The analysis indicates that India, the United States, and China are the most influential countries in terms of scientific publications in this field. These countries benefit from scientific globalization, which improves their access to information and opportunities for collaboration. The findings suggest that there are ample opportunities for researchers in the field of renewable energy and AI. The growth in publications and the emergence of relevant keywords indicate a vibrant area for future research and innovation.

One of the main limitations noted in the study is the exclusion of papers not indexed in the Scopus database. Future research could benefit from a comparative analysis across different databases to provide a more complete picture of the literature. The study employed bibliometric indicators to analyze publication trends and citation counts, using tools such as Microsoft Excel and VOSviewer for data visualization. This methodological approach underlines the importance of quantitative analysis in understanding research dynamics.

References

Abdur R, Pradipta B, Angona B, Abdullah, AN, Kishor DG, Roy G (2024) 10 present and future of AI in renewable energy domain: a comprehensive survey. https://doi.org/10.48550/arxiv.2406.16965

Adewuyi OB, Folly KA, Oyedokun DTO, Sun Y (2023) Artificial intelligence application to flexibility provision in energy management system: a survey. In: Manshahia MS, Kharchenko V, Weber GW, Vasant P (eds) Advances in artificial intelligence for renewable energy systems and energy autonomy. EAI/Springer innovations in communication and computing. Springer, Cham. https://doi.org/10.1007/978-3-031-26496-2_4

Aksnes DW, Piro FN, Fossum LW (2023) Citation metrics covary with researchers' assessments of the quality of their works. Quant Sci Stud 4(1):105–126. https://doi.org/10.1162/qss_a_00241

Álvarez-García J, Durán-Sánchez A, del Río-Rama MC, Simonetti B (2020) Big data and tourism research: measuring research impact. Qual Quant 57(S3):271–292. https://doi.org/10.1007/s11135-020-01044-z

Awan A, Abbasi KR, Rej S, Bandyopadhyay A, Lv K (2022) The impact of renewable energy, internet use and foreign direct investment on carbon dioxide emissions: a method of moments quantile analysis. Renew Energy 189:454–466. https://doi.org/10.1016/j.renene.2022.03.017

Barkavi P, Muthu-Meena S, Shapna S (2024) Artificial intelligent, energy and environment. Fut Trends Artif Intell 3:91–101

Bashiru O, Ochem C, Enyejo LA, Manuel HNN, Adeoye TO (2024) The crucial role of renewable energy in achieving the sustainable development goals for cleaner energy. Global J Eng Technol Adv 19(3):011–036

Bayer P, Dolan L, Urpelainen J (2013) Global patterns of renewable energy innovation, 1990–2009. Energy Sustain Develop 17(3):288–295. https://doi.org/10.1016/j.esd.2013.02.003

Behrens H, Lankenau I (2006) Wissenschaftswachstum in wichtigen naturwissenschaftlichen Disziplinen vom 17. bis zum 21. Jahrhundert. Ber Wiss 29(2):89–108

Bhattacharya D (2023) Smart grid, AI, IOT and big data: powering India's renewable energy revolution. Manag Account J 58(10):20

Cecati C, Mokryani G, Piccolo A, Siano P (2010) An overview on the smart grid concept. In: IECON 2010—36th annual conference on IEEE industrial electronics society. https://doi.org/10.1109/iecon.2010.5675310

Chander B, Gopalakrishnan K (2024) Integration of artificial intelligence techniques for energy management. Sustain Manag Electr Waste 24:1–46. https://doi.org/10.1002/9781394166923.ch1

Chen X, Mao J, Ma Y, Li G (2024) The knowledge linkage between science and technology influences corporate technological innovation: evidence from scientific publications and patents. Technol Forecast Soc Change 198:122985. https://doi.org/10.1016/j.techfore.2023.122985

Chigbu UE, Atiku SO, Du Plessis CC (2023) The science of literature reviews: searching, identifying, selecting, and synthesising. Publications 11(1):2. https://doi.org/10.3390/publications11010002

Dasari K, Raj VH, Nijhawan G, Kalra R, Pahwa S, Abdul-Zahra DS (2024) Integration of AI in distributed energy resource management for enhanced load balancing and grid stability. E3S Web Confer 505:03008. https://doi.org/10.1051/e3sconf/202450503008

Dibie E (2024) The future of renewable energy: ethical implications of AI and cloud technology in data security and environmental impact. J Adv Math Comput Sci 39(10):62–73. https://doi.org/10.9734/jamcs/2024/v39i101935

Dixit A (2024) The role of renewable energy in achieving sustainable development goals: a global perspective on G20 nations. Fut Trends Manag 3:91–96

Elavarasan R, Shafiullah G, Manoj Kumar N, Padmanaban S (2019) A state-of-the-art review on the drive of renewables in Gujarat, State of India: present situation, barriers and future initiatives. Energies 13(1):40. https://doi.org/10.3390/en13010040

Emexidis C, Gkonis P (2024) On the integration of internet of things and machine learning for energy prediction in the wind turbines. https://doi.org/10.20944/preprints202409.2351.v1

Eslava-Zapata R, Sánchez-Castillo V, Juaneda-Ayensa E (2024) Key players in renewable energy and artificial intelligence research. EAI Endorsed Trans Energy Web 82:11. https://doi.org/10.4108/ew.5182

Gosens J, Lu Y (2013) From lagging to leading? Technological innovation systems in emerging economies and the case of Chinese wind power. Energy Policy 60:234–250. https://doi.org/10.1016/j.enpol.2013.05.027

Gupta A, Chauhan YK, Pachauri RK (2016) A comparative investigation of maximum power point tracking methods for solar PV system. Sol Energy 136:236–253. https://doi.org/10.1016/j.solener.2016.07.001

Hasani A, Heydari H, Golsorkhi MS (2024) Enhancing microgrid performance with AI-based predictive control: establishing an intelligent distributed control system. IET Gener Transm Distrib 18(15):2499–2508. https://doi.org/10.1049/gtd2.13191
Hassan W, Duarte AE (2024) Bibliometric analysis: a few suggestions. Curr Probl Cardiol 49(8):102640. https://doi.org/10.1016/j.cpcardiol.2024.102640
Hassan Q, Viktor PJ, Al-Musawi T, Mahmood-Ali B, Algburi S, Alzoubi HM, Khudhair-Al-Jiboory A, Zuhair-Sameen A, Salman HM, Jaszczur M (2024) The renewable energy role in the global energy transformations. Renew Energy Focus 48:100545. https://doi.org/10.1016/j.ref.2024.100545
Haunschild R, Bornmann L (2022) Relevance of document types in the scores' calculation of a specific field-normalized indicator: are the scores strongly dependent on or nearly independent of the document type handling? Scientometrics 127(8):4419–4438. https://doi.org/10.1007/s11192-022-04446-y
Jiménez-Islas D, Pérez-Romero ME, García JÁ, Ventura-Cruz I (2023) Trend in publications related to biomethane using a bibliometric approach. Int J des Nat Ecodyn 18(3):565–572
Jiménez-Islas D, Pérez-Romero ME, Álvarez-García J, del Río-Rama MC (2024) Mapping scientific knowledge of renewable energy and tourism. Sustainability 16(15):6356. https://doi.org/10.3390/su16156356
Katkar AS, Shelar PV, Mene RU, Salve T (2024) Large-scale renewable integration powered by AI. Expl Artif Intell Solar Energy Integr 93:285–306. https://doi.org/10.4018/979-8-3693-7822-9.ch010
Komala CR, Basha MM, Farook S, Niranchana R, Rajendiran M, Subhi B (2024) Smart energy systems-integrated machine learning, IoT, and AI tools. Reshap Environ Sci through Mach Learn IoT 25:201–229. https://doi.org/10.4018/979-8-3693-2351-9.ch011
Kumar RK, Hoor A, Jain SK, Singh R, Kumar K, Kumar P, Chamaria A (2023) Role of technological innovation and its governance in entrepreneurial evolution. Int J Electr Govern Res 20(1):1–25. https://doi.org/10.4018/ijegr.335069
Kuzmenko O, Chorna V, Kozhura L (2024) Implementation of artificial intelligence in energy consumption calculations to reduce excess generation in the context of Ukraine's recovery. Baltic J Econ Stud 10(1):153–162
Lacerda SJ (2019) Linking scientific knowledge and technological change: lessons from wind turbine evolution and innovation. Energy Res Soc Sci 50:92–105. https://doi.org/10.1016/j.erss.2018.11.012
Lakshmi M, Rajkumar M, Dhanasakkaravarthi B, Saravanan S, Ramya J (2024) Advances in novel power generation technology and AI software tools. Optim Techn Hybrid Power Syst 58:426–454. https://doi.org/10.4018/979-8-3693-0492-1.ch019
Liu Y-X, Zhu C, Wu Z-X, Lu L-J, Yu Y-T (2022) A bibliometric analysis of the application of artificial intelligence to advance individualized diagnosis and treatment of critical illness. Ann Transl Med 10(16):854–854
Manuel HNN, Kehinde HM, Agupugo CP, Manuel CAN (2024) The impact of AI on boosting renewable energy utilization and visual power plant efficiency in contemporary construction. World J Adv Res Rev 23(2):1333–1348
Minas AM, García-Freites S, Walsh C, Mukoro V, Aberilla JM, April A, Kuriakose J, Gaete-Morales C, Gallego-Schmid A, Mander S (2024) Advancing sustainable development goals through energy access: lessons from the global south. Renew Sustain Energy Rev 199:114457. https://doi.org/10.1016/j.rser.2024.114457
Mohammadi Lanbaran N, Naujokaitis D, Kairaitis G, Jenciūtė G, Radziukynienė N (2024) Overview of startups developing artificial intelligence for the energy sector. Appl Sci 14(18):8294. https://doi.org/10.3390/app14188294
Muhammad A, Ishaq AA, Idris MB (2023) Artificial intelligence and machine learning for real-time energy demand response and load management. J Technol Innov Energy 2(2):20–29
Nguyen TH, Paramasivam P, Dong VH, Le HC, Nguyen DC (2024) Harnessing a better future: exploring AI and ML applications in renewable energy. JOIV Int J Inform vis 8(1):55

Ohalete NC, Aderibigbe AO, Ani EC, Ohenhen PE, Daraojimba DO, Odulaja BA (2023) AI-driven solutions in renewable energy: a review of data science applications in solar and wind energy optimization. World J Adv Res Rev 20(3):401–417

Olatunde TM, Okwandu AC, Akande DO, Sikhakhane ZQ (2024) Reviewing the role of artificial intelligence in energy efficiency optimization. Eng Sci Technol J 5(4):1243–1256

Onwusinkwue S, Osasona F, Ahmad IAI, Anyanwu AC, Dawodu SO, Obi OC, Hamdan A (2024) Artificial intelligence (AI) in renewable energy: a review of predictive maintenance and energy optimization. World J Adv Res Rev 21(1):2487–2799

Pollock JR, Moore ML, Smith JF, Elahi MA (2024) Artificial intelligence. Transl Orthop 85:305–308. https://doi.org/10.1016/b978-0-323-85663-8.00085-4

Precedence Research Report (2024). https://www.precedenceresearch.com/artificial-intelligence-in-renewable-energy-market

Purushottam KM (2024) Artificial intelligence to enhance energy management and distribution in smart grid communication networks. Tuijin Jishu/j Propul Technol 45(02):4366–4378

Raman R, Gunasekar S, Kaliyaperumal D, Nedungadi P (2024) Navigating the nexus of artificial intelligence and renewable energy for the advancement of sustainable development goals. Sustainability 16(21):9144. https://doi.org/10.3390/su16219144

Rane NL, Kaya Ö, Rane J (2024) Artificial intelligence, machine learning, and deep learning applications in smart and sustainable industry transformation. Artif Intell Mach Learn Deep Learn Sustain Ind 5:271

Salma B, Tarik J, Roa L (2024) Enhancing renewable energy systems with advanced artificial intelligence solutions. SCT Proceed Interdiscip Insights Innov 2:319

Sarudin ES, Ariffin WNM, Jamaian SS (2024) Mapping the landscape: a bibliometric analysis of staff scheduling optimization research trends and keywords evolution. Int J Res Innov Soc Sci 8(8):358–372

Singh M, Kaur G (2024) Case studies on AI-driven innovations in renewable energy, waste management, and resource conservation. Maint Sustain World Nexus Environ Sci AI 15:455–484. https://doi.org/10.4018/979-8-3693-6336-2.ch018

Tabor DP, Roch LM, Saikin SK, Kreisbeck C, Sheberla D, Montoya JH, Dwaraknath S, Aykol M, Ortiz C, Tribukait H, Amador-Bedolla C, Brabec CJ, Maruyama B, Persson KA, Aspuru-Guzik A (2018) Accelerating the discovery of materials for clean energy in the era of smart automation. Nat Rev Mater 3(5):5–20. https://doi.org/10.1038/s41578-018-0005-z

Tang Y, Ju P, He H, Qin C, Wu F (2013) Optimized control of DFIG-based wind generation using sensitivity analysis and particle swarm optimization. IEEE Trans Smart Grid 4(1):509–520. https://doi.org/10.1109/tsg.2013.2237795

Trahey L, Brushett FR, Balsara NP, Ceder G, Cheng L, Chiang Y-M, Hahn NT, Ingram BJ, Minteer SD, Moore JS, Mueller KT, Nazar LF, Persson KA, Siegel DJ, Xu K, Zavadil KR, Srinivasan V, Crabtree GW (2020) Energy storage emerging: a perspective from the joint center for energy storage research. Proceed Natl Acad Sci 117(23):12550–12557. https://doi.org/10.1073/pnas.1821672117

Tyasi TL, Ergin M, Mathapo MC (2024) A bibliometric analysis of the literature on goat breeding. F1000Research 13:451

Venayagamoorthy G (2011) Dynamic, stochastic, computational, and scalable technologies for smart grids. IEEE Comput Intell Mag 6(3):22–35. https://doi.org/10.1109/mci.2011.941588

Wang J (2013) Citation time window choice for research impact evaluation. Scientometrics 94(3):851–872. https://doi.org/10.1007/s11192-012-0775-9

Wen X, Shen Q, Zheng W, Zhang H (2024) AI-driven solar energy generation and smart grid integration a holistic approach to enhancing renewable energy efficiency. Int J Innov Res Eng Manag 11(4):55–66

Xu Q, Chen J (2023) Evolution of enterprise technological innovation model. Indig Innov Pathways Chin Charact 15:85–94. https://doi.org/10.1007/978-981-99-5199-4_5

Zahraee SM, Khalaji Assadi M, Saidur R (2016) Application of artificial intelligence methods for hybrid energy system optimization. Renew Sustain Energy Rev 66:617–630. https://doi.org/10.1016/j.rser.2016.08.028

9 Generative AI Applications in Finance Analysis: New Technological Platforms Serving Companies and Investment Funds

Younes Karrouk, Mario Arias Oliva,
and Luis-Maria Fornies Sanchez

Abstract

AI is transforming the activities of financial institutions while driving the growth of start-ups. By harnessing real-time market data, AI provides deep insights into investment trends and enables trades to be executed with remarkable speed and precision. Furthermore, AI solutions analyze complex transaction patterns to strengthen risk management in critical areas such as security, fraud detection, anti-money laundering, know your customer, and regulatory compliance. This integration of AI is propelling finance into a new era, where data-driven decision-making enhances efficiency, security, and user experiences. Against this backdrop, our research will explore how technology platforms leverage generative AI, machine learning, and advanced analytics to evaluate the sustainability and social impact of companies and investment funds.

Keywords

Artificial intelligence • Generative AI • Technology platforms • Financial system • Financial institutions

Y. Karrouk (✉)
Abdelmalek Essaadi University, Tetouan, Morocco
e-mail: y.karrouk@uae.ac.ma

M. A. Oliva
Complutense University of Madrid, Madrid, Spain
e-mail: mario.arias@ucm.es

L.-M. F. Sanchez
URJC-IUNIT, Madrid, Spain
e-mail: luis.fornies@iunit.edu.es

M. J. Sousa et al. (eds.), *Economic Sustainability Potentiated by Artificial Intelligence*, Contributions to Environmental Sciences & Innovative Business Technology,
https://doi.org/10.1007/978-981-96-8089-4_9

9.1 Introduction

Artificial intelligence (AI) is revolutionizing business across all sectors by amplifying predictive capabilities, streamlining decision-making, and automating tasks traditionally requiring human intelligence. The financial sector is no exception, with AI systems (AIS) gaining traction in critical sectors like credit, insurance, and asset management. By enhancing advanced analytics, AI provides financial institutions with significant competitive advantages, making it a cornerstone of modern finance. As a result, organizations are increasingly adapting their operations and restructuring their frameworks to fully harness the transformative potential artificial intelligence solutions.

The adoption of AI in financial services offers promising development possibilities for costumers and financial institutions alike but also introduces significant vulnerabilities and risks. These challenges impact financial operations, market balance, and the protection of customers legitimate interests, highlighting the urgent need for regulatory oversight. Regulators must ensure that AI systems are implemented in ways that uphold market integrity, promote fair competition, and safeguard consumer rights (Maple et al. 2023).

While discussions around regulating AI are not new, tangible progress has been limited. The multifaceted nature of AI complicates regulatory efforts, which often need to address specific applications. Defining the core risks and challenges posed by these technologies—collectively labeled as "artificial intelligence"—remains an ongoing struggle.

With earlier generations of IT systems, ethical and legal frameworks primarily centered on data responsibility and privacy protection. However, modern AI systems, including non-learning expert systems, demand a reevaluation of these considerations to address their broader and more varied implications.

Among the industries leveraging AI, the finance sector stands out due to its reliance on mathematical modeling. Whether applied to credit, insurance, or asset management, AI systems (AIS) can systematize tasks or augment human decision-making capabilities, making finance one of the most promising arenas for AI innovation (Antwi et al. 2024).

The adoption of AI to enhance natural intelligence, commonly known as augmented intelligence, has transformative potential in finance. However, as AI increasingly automates processes with minimal human involvement and influences decisions through algorithmic recommendations, it also introduces significant legal and ethical menaces for financial institutions and presents distinct management challenges.

AI systems (AIS) do not demand the same degree of oversight, as their risks to markets and consumers vary widely. For instance, the risks posed by an AIS used to determine insurance premiums are fundamentally different from those of an AIS designed to improve customer experience, such as automated appointment scheduling. Recognizing these distinctions is critical to developing appropriate regulatory and governance strategies. Globally, numerous initiatives by intergovernmental organizations, governments, industries, and academia aim to implement ethical guidelines for the responsible use of AI (Tidjon and Khomh 2022).

While discussions about framing AI usage are well-trodden ground, formal AI regulation remains in its infancy. Several challenges contribute to this, including the proliferation of responsible AI frameworks, competition among stakeholders to establish their ethical standards, the limited efficacy of the institutions promoting these frameworks, and the abstract nature of ethical principles. Translating these principles into actionable rules and integrating them into practical governance remains a complex and ongoing endeavor.

9.2 What's the Purpose of AI in Finance?

To address the risks posed by AI and ensure its responsible deployment, it is essential to first understand its applications in finance. AI is progressively utilized to enhance the efficiency and precision of operational workflows, support various aspects of the investment process, and enhance customer experiences. These technologies harness computational power and large datasets to identify patterns, structure non-traditional data sources, and automate support and decision-making processes efficiently (Apruzzese et al. 2022).

AI applications in finance can be categorized into four main functions:

1. Evaluation: Examples include consumer credit scoring and assessing the ESG footprint of investments.
2. Incentives: AI contribute to minimize behavioral risks and optimize the costing of insurance products.
3. Optimization: This includes applications such as portfolio development and workflow enhancement.
4. Information and Advice: AI improves customer service and provides personalized recommendations.

This categorization simplifies the classification of AI applications in finance, enabling a clearer understanding of their ethical implications and facilitating a targeted approach to addressing the challenges they pose.

9.3 The Challenges of Using AI Responsibly in Finance

The application of AI in finance generates significant concerns within consumers, who frequently feel insecure in a digital, algorithm-based setting, which they neither fully understand nor control. The risks associated with improper AI usage in finance can be examined along two dimensions: the harm inflicted on individuals and the broader societal impact on specific populations. Additionally, significant harm to an individual can cascade to larger populations through the widespread replication of algorithmic decisions.

Ethical concerns surrounding AI in finance are informed by documented cases of harm—such as discriminatory lending practices—and the creation of general AI ethical charters, which often fail to address finance-specific issues. Identifying harm, however, depends on recognizing both the experience of harm and the underlying AI processes that cause it. Unfortunately, consumers and citizens often lack the knowledge or tools to pinpoint the origins of harm or understand how AI systems contribute to it (Saeedi et al. 2021).

Financial institutions typically strive to understand public expectations to mitigate reputational risks, yet they often lack clarity on what these expectations involve. Public consultations on AI in finance are crucial for bridging this gap and understanding the informed expectations of consumers and citizens (Kuiper et al. 2022).

While financial institutions prioritize data protection and fairness, citizens are more concerned with maintaining their autonomy and capacity to give free and informed consent. In order to restore consumer autonomy, consumers demand explanations and transparency in algorithmic decision-making by enabling individuals to accept or challenge decisions. This aligns with financial institutions' interests in favoring explainable algorithmic models and maintaining control over their decision-making processes.

Beyond traditional concerns like human accountability and system reliability, consumers are particularly sensitive to issues such as surveillance, privacy invasion, and fairness. Here, fairness extends beyond avoiding bias in competitive markets to encompass eliminating discrimination based on irrelevant criteria and promoting social justice (Chohlas-Wood et al. 2023).

Although socioeconomic inequalities are not directly caused by AI systems (AISs) in finance, consumers caution against their potential to exacerbate these inequalities, as seen in areas like credit and health insurance. While achieving social justice is not the primary mission of financial institutions, they are increasingly called upon to ensure that AIS applications do not disproportionately amplify existing disparities (Alexander and Karametaxas 2021).

Over the past two decades, advancements in computing technologies, increased processing power, and the rapid expansion of data sources have spurred innovation across the finance sector. This transformation builds on the industry's longstanding reliance on mathematical modeling and algorithmic risk management techniques, extending their impact to every facet of finance, including insurance, asset management, investment, trading, and consumer interactions.

Machine learning and artificial intelligence have emerged as leading technologies in this evolution, drawing substantial interest from investors. These tools represent a natural progression of computational methods, leveraging vast datasets to identify patterns, create value, and organize non-traditional data.

ML and AI are now widely applied in the financial sector to enhance customer experiences, strengthen operational efficiency and accuracy, and optimize performance across various investment processes (Li et al. 2021).

This section provides an overview of ML and AI applications in financial services and investment, beginning with a clarification of the terms. Although ML and AI are often employed interchangeably, they reflect distinct but overlapping concepts.

Artificial intelligence refers to a multidisciplinary scientific field focused on replicating mechanisms underlying natural intelligence. This contains contributions from disciplines such as engineering, anthropology, science, neuroscience, psychology, mathematics, computer science, biology, and statistics. AI also encompasses applied fields that aim to replicate human cognitive abilities by performing tasks and making decisions at levels traditionally associated with human expertise.

The development of AI represents a collaborative integration of statistical methods and computer science, powered by extensive datasets and ever-increasing computational capabilities (Alasmri and Basahel 2022). These innovations have paved the way for a new era of technological advancement in finance, reshaping how institutions manage operations, engage with customers, and pursue strategic objectives.

Machine learning is a branch of computing science focused on designing algorithms and methods that enable computers to encode knowledge and perform complex tasks through a process akin to "learning" rather than relying solely on basic programming. These methods depend on large volumes of data, from which the system collects relevant information for the task. Various types of learning methods have been developed, often inspired by analogies to human cognitive processes. For example, some methods simulate reasoning and learning from experience (Prasad 2021).

From this perspective, ML is best understood as a framework that enables computers to process data and execute cognitive tasks. It encompasses a variety of tools and methods allowing systems to "learn" how to perform tasks ranging from simple to highly complex, often mimicking aspects of human intelligence. Rather than being merely a component of AI, ML is a computational approach and a branch of computer science that explores encoding knowledge and experience into systems through learning-like processes. As such, it serves as an enabler of AI rather than a subordinate field (Pramod et al. 2021).

A key aspect of ML and AI applications is their deployment into production environments. To take advantage of these technologies, models should be integrated to the systems or workflows that connect data to actionable decisions in commercial or operational contexts. This integration defines the concept concerning an AI system.

The AI system is a structured system designed to generate value—whether economic, social, or operational—by improving processes, automating tasks, or enhancing decision-making. These systems interact with their environments, such as data or human knowledge, process this data using methods and models, and produce decisions or predictions. When these outputs are contextualized, they enhance the performance and capacity of the system's operators. Figure 9.1 provides a visual representation of the components and relationships within an AI system.

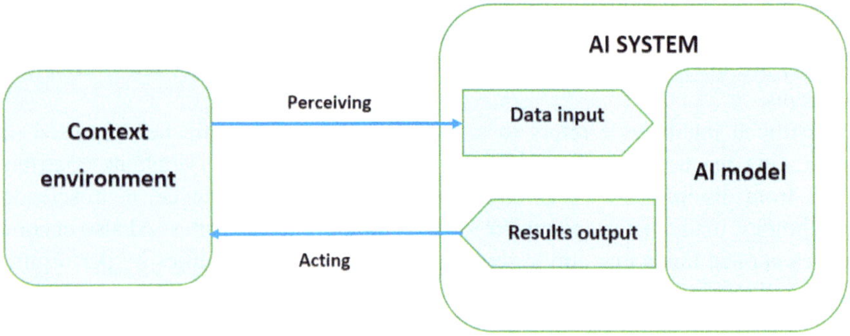

Fig. 9.1 Components and relationships within an AI system

The financial sector is increasingly adopting technologies broadly classified as artificial intelligence (AI) to drive digital transformation. These applications can be divided into two main categories: statistical AI and symbolic AI.

Statistical AI focuses on deriving value or enhancing process efficiency through data. It employs predictive algorithms and advanced modeling techniques to empower quantitative teams, automating or optimizing investment decision-making processes.

Symbolic AI, in contrast, emphasizes leveraging human knowledge and expertise by converting unstructured information into structured, actionable insights. This approach enables decision-making automation and optimization in areas not yet fully addressed by statistical AI, further broadening the scope for value creation.

Building on these foundational concepts, this discussion introduces a preliminary framework for categorizing AI applications in finance. It specifically examines the functions and roles of AI within financial services and investment, providing a structured approach to understanding how these technologies create value.

9.3.1 Applying AI in Finance and Investment: A Classification by Function

The implementation of AI and ML technologies in asset management and financial services has expanded rapidly in recent years. These technologies are now employed in multiple applications, from proofs of concept to fully integrated products, demonstrating varying levels of maturity. AI systems are applied across the entire value chain in these sectors, serving diverse objectives and employing different methodologies (Hoang and Wiegratz 2022).

To facilitate analysis and discussion, it is crucial to establish a taxonomy that organizes these applications. Existing taxonomies often classify AI systems based

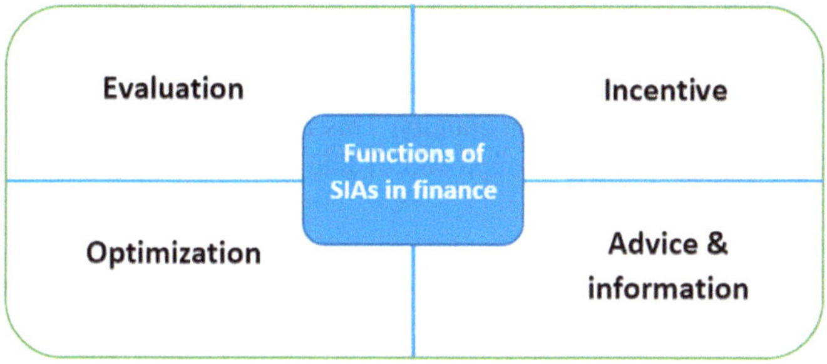

Fig. 9.2 Taxonomy of the main functions of AI systems in the financial sector

on their fundamental models and techniques, the data they utilize, or the specific business contexts in which they operate.

Here, we propose a taxonomy that is based on the function performed by an ML technology or AI within a broader AI system. In this framework, a model is defined as the application of an AI system within a corporate context, comprising multiple functions, each supported by specific technological components. Figure 9.2 illustrates this taxonomy, identifying four key functions of AI systems in the financial sector.

We illustrate each function with a dedicated use case, presented in a sidebar. These examples highlight how innovative AI systems address business challenges, creating value for companies and users through the integration of AI/ML models, data, and digital strategies. Notably, an AI system often fulfills multiple functions concurrently, showcasing its versatility and impact.

9.3.1.1 Evaluation

To generate quantitative measurements from data, an AI system carries out an evaluation function by using AI/ML models or technologies. These are then used at various levels within the system, contributing to its final output. The evaluation function offers complex realities to be represented as a finite set of enriched data points, which are then applied in subsequent stages of the system. This function can be seen in various use cases (Burnell et al. 2023).

Credit Scoring

In AI systems developed for credit adjudication, models analyze an individual's historical characteristics—such as transaction history, purchasing behavior, risk tolerance, and debt levels—and generate a score. This score is then used to evaluate the individual's risk profile, informing decisions on credit approval and determining loan terms. Similarly, scoring systems are widely utilized in the insurance sector to facilitate processes of underwriting (Abdoli et al. 2022).

Fraud Detection
In systems designed to detect fraudulent transactions, models analyze transaction histories, metadata, and personal data from user involvement to produce a qualitative assessment of transaction authenticity. This assessment is then used to decide whether a transaction should be approved or flagged for further investigation.

ESG Footprint
Models are used to quantitatively represent information from companies' ESG reports, derived from the original data sources. AI/ML models transform this unstructured data into metrics that can be incorporated into fundamentally quantitative investment processes through an AI system designed to integrate environmental, social, and governance (ESG) factors into an asset manager's investment process (Ahn et al. 2019).

Pricing Car Insurance Products
In an AI system designed to calculate car insurance premiums, models evaluate an individual's personal data, driving behavior, and driving history in real time to create a quantitative assessment of risk. This assessment is then used to determine the final premium rating within the system.

Case 1. Lenddo

Sector: Credit.

Application: Evaluation.

Secondary Application: Advice and information.

Context

Traditionally, establishing and maintaining a good credit score takes time and consistent financial activity. For creditors, the process of determining creditworthiness is also time-consuming, as it involves reviewing personal data. To offer an alternative to conventional credit scoring, Lenddo has created an application that leverages transaction data and non-financial information to provide credit access to customers with no prior credit history.

System Description and Objectives

Lenddo tackles the slow process of traditional credit systems by offering online and mobile services that generate credit scores for individuals without a credit history. By installing the Lenddo app on their smartphones, users enable Lenddo to calculate their credit scores through an analysis of their digital footprint. Lenddo also examines the number of selfies in a user's photo library, which may signal youthfulness, enabling the company to better segment its customer base.

System Benefits

Banks and credit unions can leverage LenddoScores to improve their assessment of loan default risks. By tapping into a wider array of customer data, banks utilizing this technology can extend credit to individuals who are often excluded from conventional credit systems.

Technology Used

The software employs natural language processing to examine users' social media activity and browsing behavior, identifying signs of responsibility or risk-taking. These data are then processed through predictive algorithms to generate a credit score, which helps evaluate the likelihood of a customer repaying a loan.

9.3.1.2 Virtuous Incentives

An AI system (AIS) performs an incentive function when it uses AI/ML models or technologies to generate interactions with end-users aimed at influencing their behavior in a particular direction. This influence encourages the adoption of the system, boosts its usage, or generates additional data that the system can leverage to enhance its performance and/or create value within the business context where the AI system is applied (Chen et al. 2022).

Here are a few examples:

Pricing Automobile Insurance Products

In an AI system designed to calculate car insurance premiums, AI/ML processes aim to influence users by offering personalized rewards or "nudges" that encourage safer driving behavior. This approach not only personalizes the pricing but also helps reduce overall losses.

Solution to Manage Current Accounts

In an AI system designed to help manage bank accounts, AI/ML processes recommend savings options, alert users when they exceed their budget, or suggest strategies for achieving financial goals. This function personalizes customers' financial expectations, encourages the adoption of new practices, proposes relevant products, and fosters sound financial literacy. By offering a comprehensive view of an individual's finances, the system promotes savings and investments through additional product recommendations (Jain and Vanzara 2023).

Case 2. Vitality

Sector: Insurance (Health).

Application: Incentive.

Secondary Application: Advice and information.

Connected Objects (COs) and Insurance Premiums

Connected objects (COs) offer significant potential for insurers seeking to accurately measure the parameters used in determining their customers' insurance premiums. In 2015, Life Insurance Company launched its first interactive life insurance policy based on COs: the Vitality program. By 2018, the company declared that they would exclusively offer dynamic life insurance plans.

Through the COs integrated into their plans, the company can set premiums based on data collected from devices used by their customers, in addition to their smartphones. Policyholders can earn discounts on premiums by meeting exercise targets tracked by these devices.

System Description and Objectives

In essence, customers are motivated to prioritize their physical health. They benefit from the natural rewards of being in good condition, along with the financial incentives provided by the company. In return, the company profits more from the specific customer while maintaining highly accurate actuarial calculations for each customer.

System Benefits

The value generated by this business model allows the company to pass those benefits back to their consumers in various forms. Customers are rewarded for their participation with perks such as receiving a free connected object (CO), discounts (up to 20% off their life insurance), and substantial savings from selected retailers.

The company also employs a gamification strategy to motivate customers to adopt healthy behaviors by rewarding them for their achievements.

Technology Used

Connected objects (COs) utilize machine learning (ML) to analyze data collected from sensors, and GPS through an application programming interface. Common ML techniques employed in this process include decision tree algorithms, instance-based learning methods (such as IB3), Bayesian classification, and multi-layer perception algorithms.

9.3.1.3 Optimization

An AI system performs optimization when it uses an AI/ML model or technology to find the best solution for a specific problem or process, taking predefined constraints into account. The aim is to achieve practical outcomes, improve efficiency, and incorporate additional data sources to boost the system's performance. This function also includes AI systems that streamline workflows by automating tasks, from data collection to decision-making. Below are examples that demonstrate optimization use cases.

Here are a few examples:

Portfolio Construction

In an AI system designed for portfolio construction, AI/ML processes optimize the allocation of securities within a portfolio to align with return objectives and risk tolerance constraints. These systems can uncover new opportunities in bond and equity markets, enhance diversification by identifying previously unnoticed correlations, and minimize errors in model parameterization and calibration. For instance, "model-free" algorithms, such as reinforcement learning, can identify an optimal portfolio that meets specific financial objectives with greater precision (Gupta 2023).

Moreover, AI enhances analysis and prediction by integrating diverse factors like macroeconomic conditions and economic cycles. It also incorporates unstructured data, including textual information from social media, blogs, news articles, and financial reports, through sentiment analysis techniques and advanced language processing. This allows for a more comprehensive and dynamic approach to portfolio optimization.

ESG Footprint

In an AI system designed to reduce the ESG footprint, AI/ML processes optimize portfolio allocations focusing on minimizing the environmental, social, and governance (ESG) impact. This function identifies new opportunities in bond and equity markets while ensuring the accuracy of ESG-related information disclosed by investments. By analyzing unstructured data, such as textual content in social media, financial reports, news articles, and blogs, AI can validate ESG claims through language processing and sentiment analysis techniques. This enables more informed decision-making while aligning investments with sustainability goals.

Workflow Optimization

In an AI system designed to optimize workflows, AI/ML processes focus on streamlining the execution of specific tasks. This function can be applied to reduce costs, automate procedures, or enhance production and employee productivity. It can be used in areas such as cash flow optimization, supply chain management, routing of interbank transfers and payments, logistics, or even selecting the most efficient brokers (Hambly et al. 2023).

Case 3. VITAL

Sector: Investment.

Application: Optimization.

Secondary Application: Consulting and information.

Context

VITAL (Validating Investment Tool for Advancing Life Sciences) was an algorithm created for the board of Deep Knowledge Ventures (DKV), a Hong Kong-based venture capital firm, in May 2014. Specializing in regenerative medicine, DKV used VITAL to assess potential investments in biotech start-ups. The algorithm was granted voting rights on investment decisions, similar to a human board member. VITAL was removed from the company in 2017.

System Description and Objectives

VITAL was designed to assist DKV's board in predicting which emerging technologies and companies were most likely to deliver a favorable return on investment. Given that nine out of ten biotech start-ups fail, VITAL proved valuable in identifying early warning signs of companies at risk, helping to flag potentially overly risky investments.

Model Benefits

VITAL's analyses gave Deep Knowledge Ventures (DKV) a strategic edge, positioning the firm as a leader in biotechnology and regenerative medicine. By leveraging AI, the company was able to make rational, data-driven business decisions, minimizing the influence of emotional biases that can affect human judgment. But, DKV's CEO also acknowledged the value of intuition, Therefore VITAL's insights were used to complement and support human intuition, rather than replace it.

Technology Used

VITAL assessed 50 parameters, including clinical trials, share price, research funding, and intellectual property ownership, but detailed information about its programming remains scarce.

9.3.1.4 Advice and Information

The technological component of an AI system performs an advisory function when it uses technology or an AI/ML model to assist decision-making and create value by offering guidance. This function aims to reveal hidden insights that clarify, enhance, or improve existing data, with the goal of providing better service to consumers. Examples of AI system components performing an advisory role are available for a range of use cases.

Here are a few examples:

Personalized Customer Service
In an AI system designed to personalize customer services, AI/ML processes aim to gain deeper insights into customer behavior and tailor advice or proposals for new offerings accordingly. This function uncovers hidden customer needs and creates new opportunities for financial institutions (Sowmya and Sathisha 2023). Mobile applications, which are transforming how customers communicate with financial institutions, enable the delivery of personalized, on-demand financial services. Additionally, analyzing data from connected devices, such as point-of-sale systems, GPS trackers, and inventory monitoring tools, enhances these services. This deeper understanding also plays an important role in detecting unusual behavior and preventing fraud.

Advice for Investors
In an AI system designed to support investors' decisions, AI/ML processes are utilized to provide investment recommendations by leveraging new, easily accessible data sources such as social media, financial reports, blogs, and news. It also incorporates sentiment analysis of securities and their market trends, along with a detailed examination of portfolio managers' behaviors. This function helps deliver tailored advice and create customized dashboards for portfolio managers, enhancing decision-making and portfolio optimization.

ESG Footprint Reporting
In an AI system designed to guide companies on reporting their ESG footprint, AI/ML processes are employed to develop strategies for disclosing ESG information effectively. These processes focus on showcasing the efforts made to improve ESG aspects, helping to shift the perception of the public and investors toward companies that might otherwise struggle to highlight their initiatives. This function fosters greater transparency regarding the measures taken, promoting accountability and enhancing the visibility of ESG improvements (Khalfaoui et al. 2022).

Case 4. Truvalue Labs

Sector: Investment.

Application: Advice and information.

Secondary Application: Optimization.

Context

ESG factors are criteria used to assess the societal impact and sustainability of investment or a company. The acronym ESG represents environmental, social, and governance factors. From an environmental perspective, concerns focus on the impact of business activities on global warming. Socially, key issues include diversity, equity, human rights, and individual protections. In terms of corporate governance, the emphasis is on responsible employee rights, ethical handling, and the management structures of private data. Truvalue Labs use this technology to analyze, monitor, and score companies in real time based on their adherence to ESG criteria.

System Description and Objectives

Truvalue Labs seeks to establish common metrics and provide consistent reports on the sustainable value created by a company's ESG performance indicators. It identifies key material issues, quantifies them, and generates a comprehensive suite of data and analyses, including trend graphs, sources, and scores. This valuable information helps investment firms process and decode large volumes of unstructured data in a very short time, empowering them to make the right investment decisions.

System Benefits

By utilizing AI, Truvalue Labs can scan millions of data points each month, offering a high level of efficiency in comparison with traditional analysis. The technology has the potential to replace several analysts while processing vast amounts of data at remarkable speed. In a large retrospective test, Truvalue Labs demonstrated that integrating data focused on a company's ESG criteria can lead to above-average returns.

Technology Used

The platform leverages two advanced AI technologies—natural language processing and large-scale data analysis—to achieve its objectives. This enables it to continuously capture unstructured data from thousands of online sources, 24/7, and generate valuable insights for investment firms.

After examining the various functions of artificial intelligence systems in finance, we will explore the role of AI in sustainable finance. This is a crucial tool for managing risk, enhancing long-term financial returns, and addressing investor demands for a more sustainable economy.

9.4 The Role of Artificial Intelligence in Sustainable Finance

In essence, ensuring the responsible and efficient use of resources, harnessing the potential of AI, and prioritizing sustainable investments are crucial steps toward creating a prosperous future for humanity and the environment. It is imperative that companies use resources responsibly and efficiently, particularly technology, to enhance the well-being of present and future generations and foster a harmonious relationship with the environment.

9.4.1 The Evolution of Artificial Intelligence in Sustainable Finance

According to the United Nations, between $5000 and $7000 billion were estimated to be needed to achieve the Sustainable Development Goals by 2020, highlighting the importance of examining investor behavior and how investments can address critical sustainability issues such as pollution, inequality, poverty, and environmental degradation.

Artificial intelligence (AI) holds immense potential for tackling these societal challenges, particularly those related to sustainability. The climate crisis and environmental deterioration demand innovative and advanced solutions, making AI a valuable tool for advancing environmental and social governance, as well as addressing pollution, poverty, and resource depletion (Chayjan et al. 2020; Alonso and Marqués 2019). The financial sector, particularly institutional investors, plays a crucial role in these ventures, as they are responsible for financing transformative investments and technologies, such as renewable energy and AI-driven solutions, required for a sustainable economy.

As a result, many countries have developed plans to increase investment in clean resources, which can stimulate economic growth and significantly contribute to sustainability goals. In the era of AI, companies are relying heavily on knowledge management, data science, and Big Data to achieve their sustainability objectives.

This rapid proliferation of intelligent systems is generating a greater volume of financial data, driving an increased demand for financial solutions and advanced accounting to address these emerging challenges. This, in turn, requires skilled accountants capable of operating AI-based financial systems.

Sustainable investing has become more prominent in recent years, attracting the attention of researchers, policymakers, and academics alike. Understanding the motivations behind sustainable investment is crucial not only for academia but also for institutional investors who frequently act on behalf of individuals, particularly in light of AI's growing role in shaping investment strategies and outcomes.

9.4.2 Practical Applications of AI in Sustainable Finance

Looking ahead, the field of financial reporting is expected to undergo significant transformations over the next two decades, driven by innovations such as artificial intelligence, blockchain technology, machine learning, and the use of Big Data. The combination of artificial intelligence and blockchain emphasizes the potential of automation to support the work of accounting professionals in the near future.

Although AI systems cannot fully replicate human intelligence, they are capable of delivering highly accurate results, reducing the need for repetitive human tasks in certain areas. In accounting, it is essential to evaluate the strengths and limitations of AI systems to address accounting and business challenges effectively.

This assessment will also identify the training and skills accountants need to effectively leverage intelligent systems. By adopting AI systems, accountants can enhance their skills and produce high-quality financial reports for investors (Duan et al. 2019). Additionally, the integration of fuzzy models, predictive models, data analytics, and algorithms into AI holds immense potential to support sustainable finance. Investors companies are increasingly demanding comprehensive financial reporting from accountants and companies.

Sustainable finance, a key concept at the intersection of the Sustainable Development Goals (SDGs) and finance, focuses on investments that promote long-term sustainability. However, according to some experts, the current definition of sustainable finance, which focuses solely on environmental, social, and governance (ESG) factors, is too narrow. Consequently, this section proposes a broader understanding of sustainable finance, encompassing activities and factors that ensure long-term financial sustainability.

Advances in robotics and artificial intelligence have undeniably propelled humanity forward. It all began after the Second World War, when researchers began exploring the capabilities of machines to replicate human processes. During the 1960s, AI research experienced significant growth, leading to the development of projects focused on robotics and chess games (Haenlein and Kaplan 2019). More recently, research has introduced neural networks and expert systems that mimic human behavior, including reasoning, problem-solving, learning, mathematical applications, and computational intelligence inspired by natural systems.

9.5 Analysis of the Economic Benefits of Synergy

Artificial intelligence (AI) in sustainable finance has significant potential to generate long-term economic value, especially as ESG considerations play a growing role in investment decisions. This integration addresses complex challenges while meeting the growing demands of stakeholders and investors.

Recent studies have highlighted several ways in which AI could improve sustainable finance practices. Portfolio managers can more effectively identify sustainable investment opportunities and assess associated risks using advanced algorithms and predictive models. For example, machine learning algorithms can analyze large datasets to identify emerging ESG trends. This capability enables investors to anticipate changes in consumer behavior or legislative developments.

Studies, such as those by Haenlein and Kaplan (2019), have shown that risk-adjusted returns and portfolio diversification improve significantly with the inclusion of AI in sustainable asset management. The authors emphasize that machine learning algorithms enhance investors' ability to recognize nonlinear relationships between ESG factors and financial performance, enabling more strategic and informed decisions.

Furthermore, the benefits of integrating AI into sustainable finance extend beyond financial gains. According to a 2019 World Bank report, AI can accelerate decision-making and reduce operating costs when applied to the selection process for sustainable investment projects. Asset managers can optimize due diligence procedures and allocate staff to more in-depth analysis by automating repetitive tasks like risk assessment and data analysis.

9.5.1 Optimizing Sustainable Investments with AI

Artificial intelligence (AI) in the context of sustainable investing opens new opportunities for portfolio optimization and balancing returns with adherence to environmental, social, and governance (ESG) standards. Beyond providing more targeted and intelligent investment options, this integration of sustainable finance and technology contributes to advancing social and environmental goals.

According to a study by Garcia (2016), machine learning algorithms are capable of identifying complex patterns and analyzing large ESG datasets, helping investors discover business opportunities that align with their sustainability objectives. AI also provides tools to effectively manage risks associated with sustainable investments.

Zhang et al. (2020) found that AI-based predictive models help investors make informed decisions and reduce potential losses by assessing ESG risks and suggesting mitigation strategies. Additionally, AI's ability to customize investment strategies according to investors' specific preferences and objectives is a significant advantage. For instance, (Li et al. 2019) demonstrated that AI systems can recommend tailored portfolios that satisfy individual ESG requirements, offering a personalized approach aligned with investors' sustainability goals.

9.5.2 Reducing Financial Risks and Enhancing Stability

Integrating environmental, social, and governance (ESG) factors into investment decisions is essential to mitigate risk and enhance stability within the financial system. Investors and asset managers are increasingly adopting this strategy, which aims to identify and mitigate risks linked to non-financial factors. At the same time, it promotes sustainable, long-term returns.

According to a recent study, companies that integrate ESG standards into their strategies generally exhibit lower risk profiles. According to a survey by Tucker and Jones (2020), companies with strong ESG performance are less likely to face lawsuits or corruption scandals. These situations often result in significant financial losses.

Consequently, investors can reduce their exposure to legal, reputational, and financial risks by incorporating ESG variables into their investment selection process. Additionally, integrating ESG criteria promotes improved corporate governance and transparency, reducing the likelihood of mismanagement and unethical business practices. Companies that adopt open and responsible governance practices generally achieve better financial outcomes and are more resilient to economic shocks, as demonstrated by Zhang et al. (2024).

Furthermore, ESG risk analysis helps investors anticipate long-term trends and regulatory developments, reducing uncertainty and improving investment resilience. By considering ESG criteria, investors are better positioned to identify emerging risks, such as climate change, social responsibility challenges, and other non-financial factors, enabling more informed decisions and effective risk management.

9.5.3 Improving the Performance of Sustainable Portfolios

In the financial sector, integrating environmental, social, and governance (ESG) standards into sustainable investment portfolio management is becoming increasingly common. This strategy fosters ethical and sustainable business practices while generating competitive financial returns.

According to recent research, incorporating ESG factors significantly improves the performance of sustainable portfolios. Wu et al. (2021) found that companies with high ESG scores in areas such as management diversity or environmental risk management deliver better returns to investors. Their conclusions were based on financial data from thousands of companies.

What's more, by including ESG criteria, investors can assess the opportunities and risks associated with their investments more effectively. According to a study, sustainable investment portfolios constructed with ESG factors are less volatile and demonstrate greater resilience to market shocks. By adopting a sustainability approach, investors can reduce their exposure to systemic risks and better predict long-term trends.

In addition, integrating ESG standards into portfolio management can enhance investment diversification. According to research by Capelle-Blancard and Monjon (2014), sustainable investment portfolios, composed of companies from diverse sectors and geographical regions, generally deliver consistent returns and maintain a balanced risk profile across various market scenarios.

9.6 Conclusion

Generative artificial intelligence (AI) has become an essential tool for financial institutions. While its adoption raises concerns about implementation and ethical considerations, it offers significant efficiency gains by automating certain administrative tasks.

Generative AI is transforming the way users interact with financial technologies. Thanks to AI, employees become more efficient and save time, allowing them to focus on strategic tasks instead of routine administrative processes. Financial institutions are discovering multiple applications for AI, particularly in enhancing customer service by reducing call volumes.

Generative AI also improves processes related to regulatory compliance, complaint handling, and fraud prevention. It can summarize lengthy documents and extract relevant information, helping identify operational risks or detect fraud (Awosika et al. 2024).

Moreover, generative AI analyzes financial incidents by scanning complex databases and identifying risks through detailed textual descriptions. This capability strengthens the ability of financial institutions to prevent financial crimes, such as money laundering, by offering a richer context compared to traditional, isolated searches.

Generative artificial intelligence (AI) has become an essential tool for financial institutions. While its adoption raises concerns about implementation and ethical challenges, it offers significant efficiency gains by automating specific administrative tasks.

"Hallucinations," or errors generated by AI models, can be problematic if not detected and addressed promptly. To mitigate these risks, experts emphasize the importance of constant human supervision and specialized employee training. Skepticism regarding "black box models" is another obstacle to the widespread adoption of AI in the financial sector. Their opacity raises concerns about the transparency and fairness of AI-driven decisions. Additionally, the environmental impact of AI, particularly regarding its carbon footprint and sustainability, is a growing concern (Cevik 2024).

Despite these reservations, the introduction of ChatGPT in 2022 sparked significant changes in the financial sector. Fintechs like Arteria AI were quickly inundated with requests from financial institutions. Many institutions still rely on legacy IT systems that are incompatible with modern technologies like generative AI, which can produce diverse media in response to queries. Financial institutions must adapt existing systems or create new ones to fully leverage generative AI.

These processes require substantial investment, with returns that are not always immediate. However, AI has already streamlined customer communication processes, accelerating interactions by at least 50%. This measurable improvement has helped justify investments. To adapt to this technological shift, financial institutions often adopt a gradual approach, transforming systems incrementally (Papathomas and Konteos 2023).

To successfully integrate generative AI, companies need to identify areas where this technology can drive substantial improvements, while recognizing the critical role of human expertise. Striking the right balance between technology and human interaction is essential, as noted by several experts (Eapen et al. 2024).

Adapting legacy systems to AI presents significant technical and operational challenges, but ongoing progress shows that transformation is achievable. In this evolving landscape, financial institutions aim to balance investments with tangible gains in productivity, risk management, and customer satisfaction. Thanks to their agility and expertise, fintechs play a pivotal role in this transformation.

References

Abdoli M, Akbari M, Shahrabi J (2022) Bagging supervised autoencoder classifier for credit scoring. Exp Syst Appl 213:118991

Ahn J, Park M, Shin H, Paek J (2019) A model for deriving trust and reputation on blockchain-based e-payment system. Appl Sci 9(24):5362

Alasmri N, Basahel S (2022) Linking artificial intelligence use to improved decision-making, individual and organizational outcomes. Int Bus Res 15(10):1

Alexander K, Karametaxas XE (2021) Digital transformation and financial inclusion. Routledge eBooks, p 273

Alonso A, Marqués JM (2019) Financial innovation for a sustainable economy. In: Banco de España Occasional Paper

Antwi BO, Adelakun BO, Eziefule AO (2024) Transforming financial reporting with AI: enhancing accuracy and timeliness. Int J Adv Econ 6(6):205

Apruzzese G, Laskov PEM, Mallouli W, Rapa LB, Grammatopoulos AV, Franco FD (2022) The role of machine learning in cybersecurity. Dig Threats Res Pract 4(1):1

Awosika T, Shukla RM, Pranggono B (2024) Transparency and privacy: the role of explainable AI and federated learning in financial fraud detection. IEEE Access 12:64551

Barber BM, Morse A, Yasuda A (2021) Impact investing. J Finan Econ 139:162–185

Burnell R, Schellaert W, Burden J, Ullman T, Martínez-Plumed F, Tenenbaum JB, Rutar D, Cheke LG, Sohl-Dickstein J, Mitchell M, Kiela D, Shanahan M, Voorhees EM, Cohn AG, Leibo JZ, Hernández-Orallo J (2023) Rethink reporting of evaluation results in AI. Science 380(6641):136

Capelle-Blancard G, Monjon S (2014) The performance of socially responsible funds: does the screening process matter? Eur Finan Manag 20(3):494–520

Cevik S (2024) Is schumpeter right? Fintech and economic growth. In: Cevik S (ed) IMF working paper, vol 2024. International Monetary Fund, p 1

Chayjan MR, Bagheri T, Kianian A, Someh NG (2020) Using data mining for prediction of retail banking customer's churn behaviour. Int J Electr Bank 2(4):303–320

Chen D, Esperança JP, Wang S (2022) The impact of artificial intelligence on firm performance: an application of the resource-based view to e-commerce firms. Front Psychol 13:5985

Chohlas-Wood A, Coots M, Goel S, Nyarko J (2023) Designing equitable algorithms. Nat Comput Sci 3(7):601

Duan Y, Edwards JS, Dwivedi YK (2019) Artificial intelligence for decision-making in the age of big data-evolution, challenges and research agenda. Int J Inform Manag 48:63–71

Eapen T, Finkenstadt DJ, Folk J, Venkataswamy L (2024) How generative AI can augment human creativity. SSRN Electr J 12:3781

Garcia M (2016) Racist in the machine: the disturbing implications of algorithmic bias. World Policy J 33(4):111–117

Gupta U (2023) GPT-InvestAR: enhancing stock investment strategies through annual report analysis with large language models. SSRN Electr J 12:548

Haenlein M, Kaplan A (2019) A brief history of artificial intelligence: on the past, present, and future of artificial intelligence. Calif Manag Rev 61(4):5–14

Hambly B, Xu R, Yang H (2023) Recent advances in reinforcement learning in finance. Math Finance 33(3):437

Hoang D, Wiegratz K (2022) Machine learning methods in finance: recent applications and prospects. Eur Finan Manag 29(5):1657

Jain R, Vanzara R (2023) Emerging trends in AI-based stock market prediction: a comprehensive and systematic review. Eng Proceed 216:254

Khalfaoui H, Guenich H, Saâda MB (2022) Does corporate financial performance promote ESG: evidence from US firms. Cogent Bus Manag 9(1):5485

Kuiper OX, Berg M, Burgt J, Leijnen S (2022) Exploring explainable AI in the financial sector: perspectives of banks and supervisory authorities. Commun Comput Inform Sci 15:105

Li Y, Yi J, Chen H, Peng D (2021) Theory and application of artificial intelligence in financial industry. Data Sci Finan Econ 1(2):96

Li Y, Li X, Li Y, Xu Y (2019) An analysis of factors influencing consumers' continued purchase intention based on online shopping pre-sale mode. In: WHICEB 2019 proceedings

Maple C, Szpruch L, Epiphaniou G, Staykova K, Singh SB, Penwarden W, Wen Y, Wang Z, Hariharan J, Avramović P (2023) The AI revolution: opportunities and challenges for the finance sector. Cornell University

Papathomas A, Konteos G (2023) Financial institutions digital transformation: the stages of the journey and business metrics to follow. J Finan Serv Market 29(2):590

Pramod A, Naicker HS, Tyagi AK (2021) Machine learning and deep learning: open issues and future research directions for the next 10 years. Computational analysis and deep learning for medical care: principles, methods, and applications. Wiley, Amsterdam, pp 463–490

Prasad GNR (2021) Evaluating student performance based on bloom's taxonomy levels. J Phys Conf Ser 1797(1):12063

Saeedi S, Fong ACM, Mohanty SP, Gupta A, Carr S (2021) Consumer artificial intelligence mishaps and mitigation strategies. IEEE Consum Electr Mag 11(3):13

Shihadeh F (2020) Online payment services and individuals' behaviour: new evidence from the MENAP. Int J Electr Bank 2(4):275–282

Sowmya GS, Sathisha HK (2023) Detecting financial fraud in the digital age: the AI and ML revolution. Int J Multidiscip Res 5(5):587

Tidjon LN, Khomh F (2022) The different faces of AI ethics across the world: a principle-implementation gap analysis. arXiv (Cornell University). Cornell University.

Tucker J, Jones S (2020) Environmental, social, and governance investing: investor demand, the great wealth transfer, and strategies for ESG investing. J Financ Serv Prof 74:56–75

Wu F, Hu HZ, Lin HY, Ren XY (2021) Digital transformation of enterprises and capital market performance: empirical evidence from stock liquidity. J Manag World 37(07):130–144

Zhang X, Li W, Ji T, Xie H (2024) The impact of ESG performance on firms' technological innovation: evidence from China. J Front Environ Sci 12:652

Zhang Y, Liao QV, Bellamy RK (2020) Effect of confidence and explanation on accuracy and trust calibration in AI-assisted decision making. In: FAccT proceedings

Fostering Sustainable Urban Development Through Digital Commerce Districts

10

Nádia Leitão

Abstract

This study explores the integration of digital commerce districts as a pathway to sustainable urban development. By leveraging emerging technologies such as the Internet of Things (IoT), robust information networks, and smart governance tools, these districts foster improved resource management, enhanced citizen engagement, and economic innovation. A focal point is the PRR Digital Commerce District, which demonstrates how embedding smart technologies into urban infrastructure can promote local commerce, drive economic growth, and minimize environmental impacts. The study emphasizes the importance of balancing technological advancements with the preservation of local culture and community well-being, offering a model for sustainable, inclusive urban transformation.

Keywords

Sustainable urban development • Digital commerce districts • Smart technologies • Digital commerce district

10.1 Introduction

In the twenty-first century, cities are growing fast, becoming more and more centers of economic, cultural, and technological development. According to the United Nations (2018), almost 68% of the world's population is expected to live in urban areas by 2050.

N. Leitão (✉)
Doctoral Researcher in Tourism, University of Algarve, Faro, Portugal
e-mail: a83185@ualg.pt

M. J. Sousa et al. (eds.), *Economic Sustainability Potentiated by Artificial Intelligence*, Contributions to Environmental Sciences & Innovative Business Technology,
https://doi.org/10.1007/978-981-96-8089-4_10

This means that it is important to rethink the way we design and manage cities, as this growth is accompanied by challenges including environmental degradation, traffic congestion, social inequality, and resource scarcity. To address these challenges, sustainable urban development has emerged as a key approach to urban planning.

The sustainable urban development seeks to balance economic growth, social equity, and environmental sustainability. Cities are seen as dynamic ecosystems where resources are managed efficiently to meet current needs without compromising the ability of future generations to succeed (Brundtland Report, United Nations 1987; Singh et al. 2024; Li 2024).

While traditional urban planning mostly emphasizes physical infrastructure, sustainable urban development integrates digital technologies and smart governance strategies to address urban issues. "*The future of smart cities lies in their ability to integrate emerging technologies to meet the evolving demands of urban life.*" (Batty et al. 2012, p. 485).

The fundamental principles of sustainable urban development include:

- Sustainable infrastructure, energy efficiency, integration of green spaces (Raghubanshi 2024; Li 2024).
- The promotion of social equity and community involvement to ensure that the needs of different populations are met, while addressing issues such as resource reduction and social inequality (Singh et al. 2024; Elgohary et al. 2024).
- Effective public transport systems and the use of renewable energy sources (Li 2024).

In other words, sustainable urban development envisions a multifaceted holistic planning approach designed to promote resilient, inclusive, and environmentally responsible cities, thus paving the way for sustainable urban futures (Raghubanshi 2024; Singh et al. 2024; Li 2024).

10.2 Digital Commerce Districts: A New Approach to Urban Development

> With the adoption of the New Urban Agenda and 2030 Agenda for Sustainable Development, the international community affirmed that urbanization is a driver of positive change with the genuine aspiration of leaving no one and no place behind. The New Urban Agenda is the framework to integrate and elevate the vital role that cities must play in decision-making and realizing development transformations. (UN-Habitat 2020, p. 40)

The concept of Digital Commercial Districts (DCDs) is an evolving paradigm that integrates digital technologies and urban planning to improve commercial activities in specific geographical areas (Shen et al. 2022). In the context of smart cities, the development of digital business districts is intertwined with the need to balance commercial interests with social benefits, as seen in the sustainable development models that aim to minimize negative externalities while promoting economic growth (Suyendikova et al. 2022).

DCDs emerge as an innovative solution for sustainable urban development, combining technological advances with economic growth and environmental sustainability. These neighborhoods integrate digital infrastructures to revitalize local economies, promote social inclusion, and minimize environmental impact (Graham and Marvin, 2001).

By adopting a holistic approach, digital business districts align with the principles of smart cities and offer a practical model for cities targeting to achieve sustainability goals (Chourabi et al., 2012). They are urban areas designed to merge the potential of digital technologies and data-based infrastructures, ecosystems where local businesses, governments, and residents collaborate to promote innovation, create jobs, and improve quality of life (Yoo and Lee, 2015).

These districts are defined by their reliance on e-commerce platforms, Internet of Things (IoT) enabling infrastructures and collaboration hubs that drive both economic activity and sustainable urban planning (Latief and Javeed 2024; Neto et al. 2024). "*(...) cities are giving increasing importance to the leading role that technological innovation plays in sustainable development, particularly in increasing productivity, employability and urban governance, which has fostered the growth of "smart city" solutions.*" (UN-Habitat, 2020, p. 212).

The integration of digital networks into urban systems significantly transforms cities into spaces where commerce, governance, and everyday life converge, with DCDs serving as a crucial link between traditional commerce and digital innovation. The development of smart cities, characterized by the use of technologies, plays a focal role in fostering innovative ecosystems and enhancing urban living standards (Rudnev et al. 2024).

This digital transformation is evident in the commercial circulation industry, where digital city competitiveness has been shown to enhance total factor productivity, particularly through investments in urban information infrastructure (Meng et al. 2023). Moreover, the digital economy's role in urban and rural integration highlights its capacity to drive economic growth and optimize industrial structures, further supporting the integration of traditional commerce with digital platforms (Wu et al. 2023).

The spatial network of urban digital financial technology also contributes to this transformation by reducing commercial bank risks and promoting balanced development across cities (Chang and Xu 2022). Additionally, the e-commerce transformation of cities has been linked to improvements in green total factor productivity, driven by industrial structure upgrades and green technology innovation (Ding and Gao 2023).

In the context of urban governance, digital technologies have reshaped resident interactions with government services, enhancing administrative efficiency and convenience (Li et al. 2024). The creation of digital twins of urban systems, as demonstrated in Spanish cities, provides a framework for exploring various scenarios, including pandemic preparedness, further illustrating the potential of digital networks to transform urban spaces (Rodríguez et al. 2024).

Finally, the impact of new digital infrastructures on urban carbon emissions underscores the dual role of digital networks in promoting economic growth while addressing environmental concerns (Wang et al. 2024).

Collectively, these developments underscore the transformative potential of digital networks in creating interconnected urban spaces that seamlessly integrate traditional commerce with digital innovation (Sassen, 2001).

So, the convergence of digital and physical realms not only enriches commerce but also redefines urban living in the digital age. The DCDs emerge in the context of Portugal's Recovery and Resilience Plan (PRR), as an innovative and strategic solution to drive economic growth and urban sustainability. These districts act as catalysts for digital transformation and urban rehabilitation, aligning with the national digital strategy. Their primary goal is to integrate digitalization into the business models of economic operators, foster online commerce, and modernize value and supply chains. Furthermore, the initiative promotes the requalification of neighborhoods, contributing to the revitalization of urban-commercial planning, optimizing accessibility for consumers to local commerce, and strengthening the connection between communities and local services. This holistic approach not only revitalizes urban and economic fabrics but also reinforces the integration of technological innovation with the sustainable development of local communities (Eurocid 2024).

10.3 Economic, Social, and Environmental Impacts

According to Tiantian et al. (2023), the proliferation of digital technologies thus reshapes urban landscapes, affecting economic, social, and environmental dynamics. Urban digital transformation is more than just a technological innovation; it is at the heart of future urban competitive strategies.

The adoption of digital platforms by local businesses and SMEs (small and medium enterprises) has been shown to significantly enhance their reach and competitiveness, leading to increased revenue and customer engagement (Ngoma et al. 2024).

Local businesses and SMEs are increasingly taking advantage of digital platforms to strengthen their market presence and competitiveness, which allows them to interact efficiently, increasing their ability to compete in the market, regardless of their size or location, hereby fostering economic growth and job creation (Besario and Encarnacion 2024), contributing to broader economic development goals (Mekaniwati et al. 2024; Mulyana and Irfan 2024).

Online business directories are emphasized as vital tools in the local economy. They help consumers discover local businesses and understand the services available (Besario and Encarnacion 2024). This change significantly impacts the social dynamics of commerce, also influencing consumer behavior. As businesses transition to digital platforms, they face the challenge of maintaining customer engagement in a digital-first environment, which is crucial for fostering brand trust and loyalty.

Cities are in this digital terrain as much as in a physical terrain—where data and information flows have their own geographies which, in turn, position some parts of cities differently in the global environment. As a result of this competitive digital geography, we must challenge our usual definition of cities in terms of location and spatial extent. Instead, we need to think of cities as simultaneously facilitators and constrains of interactions, as well as products of them. Seen in this way, the mediation of such interactions by ICTs can have profound effects Hollands (2008).

This then "*shifts the emphasis from the urbanity of the built physical form to the quality of interaction in cultural life through the exchange of information*" (Little 2000, p. 1814). It thus suggests a rescheduling of urban interaction that challenges conventional planning and governance. Consequently, rather than seeing cyberspace as separate, it focuses on "a multi-scalar mixture of electronic and physical space" of digital flows and physical flows, of virtual and real places (Page and Phillips 2003, p. 73). Daily lives do not find a great division between the offline and online worlds, but feed into each other in continuous interaction (Crang 2010).

From an environmental point of view, the integration of smart technologies significantly reduces the urban ecological footprint. Smart city technologies have been shown to reduce energy consumption by up to 25% and improve waste management by 30%, thereby decreasing carbon footprints in cities that implement these systems (Khadka et al. 2024). These technologies facilitate sustainable urban development by optimizing resource management, improve the efficiency of public service, and mitigate environmental impacts through real-time data collection and analysis (Narain 2024; Singh et al. 2024).

Smart energy networks and intelligent transport systems are pivotal in reducing pollution and traffic congestion, while smart grids enhance the management of renewable energy sources, contributing to ecological resilience (Khosla 2024). Furthermore, green building technologies increase water and energy efficiency, supporting sustainable urbanization (Khosla 2024). The strategic application of these technologies not only fosters innovation and resilience but also promotes inclusive growth and a greener future for urban areas (Narain 2024; Ruíz-Vanoye et al. 2024). However, challenges such as ICT infrastructure, regulatory frameworks, and community participation must be addressed to fully realize the potential of smart city technologies in reducing urban ecological footprints (Singh et al. 2024).

According to Umesh et al. (2024), when analyzing data from existing smart city projects around the world, there are significant reductions in energy consumption (up to 25%) and improvements in waste management (up to 30%), leading to a decrease in carbon footprint. Therefore, cities can take advantage of advanced technologies to develop their sustainable initiatives, contributing to the strategic development of smarter urban environments.

Hemanth Kumar et al. (2024) emphasize that advanced technologies play a crucial role in enabling sustainable urban development within smart cities, particularly by improving resource management and enhancing urban governance. By leveraging information and communication technology (ICT), these cities effectively address urban challenges, improve the quality of life, and support sustainability initiatives.

10.4 Challenges and Lessons Learned

The deployment of digital commerce districts is not without challenges, several challenges that must be faced to ensure their successful implementation and sustainability. A significant barrier is the digital gap, as some residents, particularly seniors, lack the skills needed to engage with digital platforms. To address this, Portugal initiative has implemented extensive digital literacy programs targeting vulnerable groups, for example EU SOU DIGITAL.

Additionally, there is the unequal availability of technological infrastructures. In many regions, particularly in developing areas, gaps in broadband connectivity and IoT networks, and digital literacy, make it difficult to fully realize these initiatives. Without robust and equitable access to technology, the digital divide risks exacerbating socioeconomic disparities. To address this situation, significant investments are needed in fundamental ICT infrastructure.

To successfully implement the PRR DCD, I also highlight the importance of community involvement. The initiative will need to align with local needs and promote a sense of ownership. Smart technologies, driven by advancements in digital infrastructure, have significantly transformed urban management by optimizing resource use and enhancing citizen participation, crucial for sustainable urban development. The integration of artificial intelligence (AI), machine learning (ML), and deep learning (DL) in urban systems has enabled real-time analytics for predictive maintenance, energy optimization, and traffic management, thereby improving the efficiency and sustainability of city operations (Rane et al. 2024).

These technologies facilitate the development of smart energy networks, intelligent transport systems, and green building technologies, which collectively enhance environmental resilience and public service efficiency (Khosla 2024). Public participation, facilitated by smart governance platforms, is increasingly recognized as essential for sustainable urban development, allowing citizens to engage in the legislative process and contribute to the formation of urban regulations (Pamungkas et al. 2023). Together, these technologies and participatory frameworks ensure that urban areas function efficiently and inclusively, addressing both immediate operational challenges and long-term strategic planning needs.

Another challenge is related to governance and regulatory barriers. Rapid technological developments often outpace the ability of regulatory frameworks to adapt, creating obstacles in areas such as data privacy, cybersecurity, and the ethical use of digital tools. These challenges are compounded by bureaucratic inertia and fragmented urban governance structures, which can delay decision-making and policy implementation. Adaptive governance models that prioritize collaboration between governments, businesses, and local communities are crucial to overcoming these barriers. Clear policies on data management and cybersecurity are essential to building trust and ensuring effective implementation.

There is a need for updated legal frameworks, with calls for technology-neutral laws and robust data governance to support sustainable e-governance (Tojiev 2024). The global nature of digital trade further complicates these issues, as cross-border data flows must navigate multiple privacy and cybersecurity laws, which can make digital trade negotiations difficult (Laidlaw 2021).

The concept of "datafication" underscores the immense governance challenge, as data serve multiple roles in the digital economy, necessitating urgent regulatory measures to safeguard democratic processes and functional markets (Ciuriak et al. 2020). Additionally, the rise of the Internet of Things (IoT) introduces further governance complexities, requiring extensions of data protection laws and modifications to competition law to protect consumer privacy and ensure accountability (Sutherland 2020). To overcome these barriers, adaptive governance models that foster collaboration between governments, businesses, and communities are essential, alongside clear policies on data management and cybersecurity to build trust and ensure effective implementation. These themes are crucial for shaping future digital trade policies and agreements.

Financial constraints also pose significant challenges. Developing DCDs demands substantial investment in digital infrastructure, training programs, and system integration.

The digital divide and skills gap in certain regions require financial investments and targeted training to improve technological skills among residents. Economic challenges, such as the high costs of digital tools and telecommunications services, hinder the adoption of digital technologies, especially for small businesses and enterprises, which often face financing constraints due to limited investor support and inadequate capital markets. Least developed countries are also particularly affected, with gaps in digital readiness highlighting the need for robust digital infrastructures and supportive policies to promote e-commerce and the digital economy. Even regions with improved ICT infrastructures need to keep pace with digital advances, emphasizing the importance of strategic policies and financial incentives to fill these gaps. Collectively, these challenges underscore the urgent need for financial and policy support to drive the successful development of digital business districts globally (Nikakhtar 2013; Liu 2023; Hou 2024).

The PRR Digital Commerce Districts initiative in Portugal serves as a prime example of such incentives, demonstrating how targeted investments and strategic policies can drive the digital transformation of commercial districts, enhance local economic resilience, and bridge the digital divide.

Community engagement and social inclusion are equally critical aspects in implementing DCD. Resistance to change, particularly among traditional businesses and older demographics, can delay adoption. Marginalized groups often face barriers in accessing digital tools, which can exacerbate existing social inequalities. Transparent communication and active community engagement from the outset can help build trust and ensure that DCDs meet specific local needs.

Globally, numerous initiatives are paving the way for greater digital inclusion and literacy, particularly targeting marginalized groups and underserved communities. These programs not only foster access to digital tools but also empower individuals to actively participate in the digital economy. The Her Digital Skills initiative, spearheaded by the EQUALS Global Partnership for Gender Equality in the Digital Age, offers free foundational IT skills training and e-mentoring programs for women and girls worldwide.

The European Digital Skills and Jobs Coalition is a European Union initiative addressing the digital skills gap across member states. This coalition brings together governments, businesses, and training providers to upskill workers, unemployed individuals, and youth in digital competencies.

The Reskilling Revolution aims to provide one billion people with better education, skills, and jobs by 2030. This initiative emphasizes digital upskilling, recognizing that digital literacy is fundamental for global economic resilience and inclusion.

These global initiatives demonstrate a collective effort to ensure that digital transformation is inclusive and equitable. They serve as exemplary models.

In Portugal, as well, several programs demonstrate commitment to fostering an inclusive digital society by equipping its citizens with the necessary skills to engage in the evolving digital landscape, such as:

- Eu Sou Digital: A national program aimed at enhancing digital literacy among adults over 45 with little or no internet experience. It utilizes a network of volunteers and training centers to provide basic internet skills, fostering digital inclusion.
- Training Voucher + Digital: Part of the Employment + Digital 2025 Program, this initiative offers financial support up to €750 for workers to develop digital skills. It targets employees, self-employed individuals, and company partners, encouraging participation in digital training courses.
- Portugal Digital Academy: A free course aggregator platform that allows individuals to assess their digital skills, receive training, and explore career opportunities. It offers multiple options to help test, improve, and certify digital proficiency, promoting digital inclusion across various demographics.

10.5 Conclusion

The integration of Digital Commerce Districts has the potential to transform cities into innovative, resilient, and inclusive ecosystems. By combining emerging technologies such as IoT and smart grids with participatory governance, cities can revitalize local economies, improve citizens' quality of life, and reduce environmental impacts. However, there are significant challenges, especially with regard to digital inclusion, infrastructure, and governance. Initiatives like the PRR Digital

Commerce Districts in Portugal illustrate that strategic investments and community collaboration can effectively overcome these barriers, promoting sustainable digital transformation.

DCDs not only drive economic development and urban modernization, but also serve as a replicable model for sustainable digital transformation, emphasizing the need to balance technological innovation with cultural preservation and social inclusion. Furthermore, the concept of "datafication" presents governance challenges, and data play a fundamental role in the digital economy. Regulatory measures are urgently needed to safeguard democratic processes and ensure functional markets. Governance models that promote collaboration between governments, companies, and communities are essential for the successful implementation of these digital initiatives.

The integration of smart technologies in urban areas also offers notable environmental benefits, significantly reducing ecological footprints. For example, smart city technologies can reduce energy consumption by up to 25% and improve waste management by 30%, leading to substantial reductions in carbon footprints. Public participation, facilitated by smart governance platforms, plays a key role in achieving sustainable urban development. It allows citizens to actively participate in legislative processes and urban regulations, ensuring that urban areas function inclusively and efficiently.

Finally, combating digital illiteracy remains a crucial priority. Financial investments and specific training are essential to fill the skills gap in disadvantaged regions and groups. The high costs of digital tools and technologies pose significant barriers, especially for small businesses, underlining the importance of economic support.

These findings highlight the importance of integrating digital commerce into urban planning. By promoting sustainable development and addressing associated challenges, cities can unlock new opportunities for growth and innovation.

References

Batty M, Axhausen KW, Giannotti F et al (2012) Smart cities of the future. Eur Phys J Spec Top 214(1):481–518. https://doi.org/10.1140/epjst/e2012-01703-3

Chourabi H, Nam T, Walker S et al (2012) Understanding smart cities: an integrative framework. In: Proceedings of the 45th Hawaii international conference on system sciences. IEEE. https://doi.org/10.1109/HICSS.2012.615

Graham S, Marvin S (2001) Splintering urbanism: networked infrastructures, technological mobilities and the urban condition. Routledge, New York

Hollands RG (2008) Will the real smart city please stand up? Intelligent, progressive or entrepreneurial? City 12(3):303–320. https://doi.org/10.1080/13604810802479126

Li L, Lin X, Yang X, Luo Z, Wang M (2024) Digital governance and urban government service spaces: understanding resident interaction and perception in Chinese cities. Land 13(9):1403. https://doi.org/10.3390/land13091403

Sassen S (2001) The global city. Princeton University Press, New York

Singh S, Rai A (2024) Harnessing innovation for sustainable urban futures. In: Advances in computational intelligence and robotics book series, pp 341–356. https://doi.org/10.4018/979-8-3693-5918-1.ch013
Singh S, Rai A, Özsungur F (2024) Nexus of AI climatology and urbanism for smart cities. In: Harnessing innovation for sustainable urban futures a deep dive into smart cities and sustainable urban development. IGI Global, pp 341–356
United Nations Human Settlements Programme (UN-Habitat) (2020) World cities report 2020: the value of sustainable urbanization. UN-Habitat, Nairobi. https://unhabitat.org/
United Nations (1987) Report of the world commission on environment and development: our common future. United Nations
United Nations (2018) World urbanization prospects: the 2018 revision. UN Department of Economic and Social Affairs
Yoo SH, Lee SY (2015) Smart city development in South Korea: Songdo as a new green city. Asian J Innov Policy 4(1):52–77

Immersive Learning: Leveraging VR for Sustainable Soft Skills Development

11

Agnes Jenei and Maria José Sousa

Abstract

This study investigates the impact of hybrid training (cooperation skills development training enhanced with VR elements), by utilizing a diverse sample of 43 individuals from various professional backgrounds and age groups. Pre-and post-experience surveys were employed to collect quantitative data on cooperation (the topic of the soft skills development training), openness to VR, prior VR experience, comfort with technology, and expectations. The results indicate a generally positive shift in attitudes towards VR, with many participants moving from negative or neutral to positive or extremely positive attitudes. Comfort with VR technology also increased, suggesting that the immersive nature of VR effectively familiarizes users with new tools. In a learning-by-doing process, cooperation levels improved across all age groups, highlighting VR's potential to enhance teamwork and collaborative skills. Participants provided constructive feedback, emphasizing the need to address technical issues and showing interest in trying more VR experiences. Recommendations for future research include conducting longitudinal studies to assess long-term impacts, including more diverse demographics, comparing the same training with and without VR elements, and investigating factors influencing user engagement. Overall, the study demonstrates VR's multifaceted potential to improve cooperation skills and positively influence attitudes towards VR technology in carefully designed, soft skills training enhanced by VR elements, providing valuable insights for refining and enhancing future VR-enhanced hybrid training programs.

A. Jenei (✉)
Ludovika University of Public Service, Budapest, Hungary
e-mail: Jenei.Agnes@uni-nke.hu

M. J. Sousa
Instituto Universitário de Lisboa, Lisbon, Portugal
e-mail: maria.jose.sousa@iscte-iul.pt

M. J. Sousa et al. (eds.), *Economic Sustainability Potentiated by Artificial Intelligence*, Contributions to Environmental Sciences & Innovative Business Technology,
https://doi.org/10.1007/978-981-96-8089-4_11

Keywords

Sustainability • VR training • Hybrid VR escape room • Soft skills training with VR elements • Quantitative methodology

11.1 Introduction

According to the EC's Directorate-General for Research and Innovation, in the document released in 2024 on Emerging Digital Technologies in the Public Sector, virtual reality is a transformative tool that aligns with long-term goals of accessibility and environmental responsibility while enabling efficient, scalable learning and operational practices (European Commission, 2024).

Virtual reality technology has rapidly evolved over the past decade, transforming from a niche entertainment medium into a powerful tool for education, training, and professional development. The global market size for VR in education has seen rapid expansion: In 2023, it is valued at approximately USD 14.55 billion. In 2024, it was expected to reach USD 17.18 billion and is projected to expand to USD 65.55 billion by 2032 (Fortune Business Insights, 2024).

VR's immersive and interactive nature offers unique opportunities to enhance learning experiences, making it an attractive option for various fields, including engineering, healthcare, environmental management, and information technology. This study explores the impact of a hybrid cooperation skills development VR training on participants' attitudes, comfort levels, and cooperation, providing valuable insights into the effectiveness of VR as a training tool.

The potential of VR to revolutionize training and education lies in its ability to create realistic, engaging, and interactive (recently also adaptive and collaborative) environments that are experienced by their user as real. Unlike traditional training methods, VR allows users to experience scenarios that are difficult or impossible to replicate in real life. This capability is particularly beneficial for fields that require hands-on practice, a learning-by-doing approach and experiential learning. For instance, engineers can simulate complex machinery operations, healthcare professionals can practice surgical procedures, and environmental managers can explore virtual ecosystems.

The introduction into the market of the first "stand-alone" VR headsets (operating without needing a PC or laptop and long cables, e.g., Oculus Rift, 2016 or HTC Vive, 2016, or the mixed reality Microsoft HoloLens, 2016) paved the way for VR technology to leadership skills/soft skills development.

The research underscores that learners study approximately four more faster in virtual reality in the case of leadership skill development training (PwC 2020). In addition, learners have approximately 4 times better concentration, and they are 3,75 more attached emotionally to the learning content. No wonder if they feel app. 3 times more confident when applying the acquired new skills. In another context, Smooth Soft shared how VR training can significantly shorten upskilling training time in a quick-service restaurant: time dropped from 40 h to 3–6 h (Smooth Soft, 2023).

Currently, in some sectors, VR is widely adopted almost all over the world (e.g., in aviation, in pilot training), and VR-based training has become a standard practice in many industries; in other sectors, the demand for immersive learning is still limited. Despite its promising potential, the introduction of VR to soft skill training is still in its early stages, even if the available content is rapidly growing. Many organizations are hesitant to invest in VR technology due to concerns about cost, technical challenges, and the effectiveness of VR in a training context compared to traditional training methods. In addition to that, there is a diffused lack of knowledge on how to use VR in general and for training purposes.

This study aims to address these concerns by analyzing the impact and reception of concrete soft skills development training with VR elements. The study seeks to demonstrate the value of the "VR training" and identify areas for improvement to maximize its benefits.

The methodology of this study involved collecting and analyzing data from a diverse sample of 43 participants, all of whom participated in the 240-min cooperation skills development training, enhanced by VR elements, between August 2024 and December 2024, in five training groups. It is essential to emphasize that participants were not selected for the research; they attended real onsite trainings. The age range of participants spanned from 7 to 64 years old, with a relatively balanced gender distribution. Regarding the participants' backgrounds, various professions were represented (law, international relations, project management, chemistry, environmental engineering, IT engineering, etc.) This diversity in the sample provided a robust dataset for analyzing the effectiveness and reception of the cooperation skills development training enhanced by VR elements across different participants.

Pre-and post-experience surveys were used to collect quantitative data on participants' cooperation, openness to VR, prior VR experience, comfort with technology, and expectations. The pre-experience survey established a baseline for each participant's attitudes and comfort levels, while the post-experience survey assessed any changes in these areas after the VR training. The data was then analyzed to identify patterns and differences in responses based on gender and age, providing insights into how different demographics responded to the VR training. The data was then analyzed to identify patterns and differences in responses based on gender and age, providing insights into how different demographics responded to the soft skill development training with VR elements.

The results indicate a generally positive shift in attitudes towards VR, with many participants moving from negative or neutral to positive or extremely positive attitudes. Comfort with VR technology also increased, suggesting that the immersive nature of VR effectively familiarizes users with new tools. In a learning-by-doing process, cooperation levels improved across all age groups, highlighting VR's potential to enhance teamwork and collaborative skills with carefully selected VR experiences (content) for a well-structured training design. Participants provided constructive feedback, emphasizing the need to address technical issues and expand VR applications. Recommendations for future research include conducting longitudinal studies to assess long-term impacts, including more diverse demographics, and investigating factors influencing user engagement. Overall, the study demonstrates VR's multifaceted potential to improve cooperation skills and positively influence attitudes towards VR technology in a carefully designed, soft skills training, providing valuable insights for refining and enhancing similar future training programs, ensuring they meet the diverse needs and expectations of participants. Organizations can leverage VR technology to create more effective and engaging soft skills training experiences by addressing the identified areas for improvement.

11.2 Literature Review

A growing number of studies state that VR can give a significant contribution to soft skills development not only in the corporate sector but also in educational settings. Recently, Hill and du Preez (2022) working with undergraduate finance students in Spain confirms the idea. Kirwan (2024) highlights the potential power of immersive storytelling in soft skills development, team building, and cooperation.

The classical approach to developing soft skills in VR is about creating realistic, safe scenarios that mimic real-life situations where learners alone can practice their skills anytime. Taupic emphasizes the positive role of role plays in VR environments (Taupiac et al. 2019). VR's immersive nature allows for the simulation of complex scenarios that require group decision-making and problem-solving, skills that are crucial for effective leadership (Regal et al. 2022).

Award-winning content providers (e.g., VirtualSpeech, Bodyswaps) integrate AI into virtual reality, giving learners the opportunity to master specific soft skills (public presentation, job interview, managing angry clients, and more) entirely alone in a virtual, free-judgmental space 24 h per day, with instant AI-generated self-assessment.

While Kwame and McAllister (2024) call this type of VR training Next-Gen Engineering, Regal et al. (2022) underlines the need of social and multi-user aspects in VR training. Govaerts and De Wever (2023) suggest that the combination of technology and regular training can foster a more engaging learning experience, particularly in the case of teamwork and communication skills development.

Friedrich et al. (2019) and Manojlovic (2022) investigate the operation of the escape room considering it an innovative method for soft skills training. As activities require participants to solve puzzles collaboratively, they immediately improve their teamwork abilities and problem-solving skills.

The first physical escape room started their activities in Japan around the new millennium, and the first multiplayer VR escape room were open around 2015, e.g., Zero Latency in Australia and The Void in the USA. Multyplayer VR escape rooms have become popular in Hungary as well (LogIQrooms). In 2024, a new edition of Omescape gives a fresh overview on this popular and well-researched topic.

Feedback from participants in interprofessional settings has highlighted the value of these activities in promoting positive team dynamics (Friedrich et al. 2019). However, Dahl (2021) states that despite VR escape room presents numerous opportunities for enhancing soft skills training, challenges such as cost, accessibility, and the need for technical expertise must be considered.

The demand of a VR headset in corporate training is growing, many organizations recognizing its potential to provide realistic and engaging learning experiences that traditional methods cannot offer (Dahl 2021). However, many of the companies find complicated to move to an external venue for an hour of multiplayer VR escape room.

Mobil equipment and VR experiences (e.g., Mindfall of IDEGO) offer unique opportunities for corporates to held a cooperation skills development training with VR elements (form of mobile hybrid VR escape room) at their company.

It is better to keep in mind that the acceptance of VR in corporate settings may vary and necessitating a tailored approach to implementation. The trainer has to carefully align the VR scenarios with training purposes and incorporate feedback mechanisms to enhance learning outcomes (Herdegen 2023) (Zhao and Ma 2020). Despite these challenges, the potential of VR to revolutionizing soft skills training is undeniable, offering immersive and engaging experiences that traditional methods don't do.

Virtual reality technology has rapidly evolved over the past decade, transforming from a niche entertainment medium into a powerful tool for education, training, and professional development. The global market size for VR in education has seen rapid expansion: In 2023, it is valued at approximately USD 14.55 billion. In 2024, it was expected to reach USD 17.18 billion and projected to expand to USD 65.55 billion by 2032 (Fortune Business Insights 2024).

VR's immersive and interactive nature offers unique opportunities to enhance learning experiences, making it an attractive option for various fields, including engineering, healthcare, environmental management, and information technology. This study explores the impact of a hybrid cooperation skills development VR training on participants' attitudes, comfort levels, and cooperation, providing valuable insights into the effectiveness of VR as a training tool.

The potential of VR to revolutionize training and education lies in its ability to create realistic, engaging, and interactive (recently also adaptive and collaborative) environments that are experienced by their user as real. Unlike traditional training methods, VR allows users to experience scenarios that are difficult or impossible to replicate in real life. This capability is particularly beneficial for fields that require hands-on practice, learning-by-doing approach, and experiential learning. For instance, engineers can simulate complex machinery operations, healthcare professionals can practice surgical procedures, and environmental managers can explore virtual ecosystems.

The introduction into the market of the first "stand-alone" VR headsets (operating without needing a PC or laptop and long cables, e.g., Oculus Rift, 2016 or

HTC Vive, 2016, or the mixed reality Microsoft HoloLens, 2016) pave the way for VR technology to leadership skills/soft skills development.

Research underscores that learners study approximately 4 more faster in virtual reality in case of leadership skill development training (PwC 2020). In addition, learners have approximately 4 times better concentration, and they are 3,75 more attached emotionally to the learning content. No wonder if they feel app. 3 times more confident when applying the acquired new skills. In another context, Smooth Soft shared how VR training can significantly shorten upskilling training time in a quick-service restaurant: time dropped from 40 h to 3–6 h (Smooth Soft 2023).

Currently, in some sectors, VR is widely adopted almost all over the world (e.g., in aviation, in pilot training), and VR-based training has become a standard practice in many industries; in other sectors, the demand for immersive learning is still limited. Despite its promising potential, the introduction of VR to soft skill training is still in its early stages, even if the available content is rapidly growing. Many organizations are hesitant to invest in VR technology due to concerns about cost, technical challenges, and the effectiveness of VR in a training context compared to traditional training methods. In addition to that, there is a diffused lack of knowledge on how to use VR in general and for training purposes.

This study aims to address these concerns by analyzing the impact and reception of concrete soft skills development training with VR elements. The study seeks to demonstrate the value of the "VR training" and identify areas for improvement to maximize its benefits.

The methodology of this study involved collecting and analyzing data from a diverse sample of 43 participants, all of whom participated in the 240-min cooperation skills development training, enhanced by VR elements, between August 2024 and December 2024, in five training groups. It is essential to emphasize that participants were not selected for the research; they attended real onsite trainings. The age range of participants spanned from 7 to 64 years old, with a relatively balanced gender distribution. Regarding the participants' backgrounds, various professions were represented (law, international relations, project management, chemistry, environmental engineering, IT engineering, etc.) This diversity in the sample provided a robust dataset for analyzing the effectiveness and reception of the cooperation skills development training enhanced by VR elements across different participants.

Pre-and post-experience surveys were used to collect quantitative data on participants' cooperation, openness to VR, prior VR experience, comfort with technology, and expectations. The pre-experience survey established a baseline for each participant's attitudes and comfort levels, while the post-experience survey assessed any changes in these areas after the VR training. The data was then analyzed to identify patterns and differences in responses based on gender and age, providing insights into how different demographics responded to the VR training. The data was then analyzed to identify patterns and differences in responses based on gender and age, providing insights into how different demographics responded to the soft skill development training with VR elements.

The results indicate a generally positive shift in attitudes towards VR, with many participants moving from negative or neutral to positive or extremely positive attitudes. Comfort with VR technology also increased, suggesting that the immersive nature of VR effectively familiarizes users with new tools. In a learning-by-doing process, cooperation levels improved across all age groups, highlighting VR's potential to enhance teamwork and collaborative skills with carefully selected VR experiences (content) for a well-structured training design. Participants provided constructive feedback, emphasizing the need to address technical issues and expand VR applications. Recommendations for future research include conducting longitudinal studies to assess long-term impacts, including more diverse demographics, and investigating factors influencing user engagement. Overall, the study demonstrates VR's multifaceted potential to improve cooperation skills and positively influence attitudes towards VR technology in a carefully designed, soft skills training, providing valuable insights for refining and enhancing similar future training programs, ensuring they meet the diverse needs and expectations of participants. Organizations can leverage VR technology to create more effective and engaging soft skills training experiences by addressing the identified areas for improvement.

11.3 Methodological Approach

11.3.1 Methods and Materials

The study employed a quantitative methodological approach to systematically collect and analyze numerical data from participants who underwent a VR training experience. This approach was chosen to objectively measure changes in participants' attitudes, comfort levels, and cooperation before and after the VR experience. The use of structured surveys allowed for the collection of standardized data, which facilitated statistical analysis and comparison across different demographic groups.

Participants were asked to complete pre-and post-experience surveys that included Likert-scale questions to quantify their levels of cooperation, openness to VR, prior VR experience, comfort with technology, and expectations. The post-experience survey included similar questions to assess any changes in these areas and additional questions about their VR experience, attitudes towards VR, comfort with VR technology, and suggestions for future improvements. This structured format ensured consistency in data collection and enabled the researchers to capture the responses.

Participants were diverse in terms of professional backgrounds. This diversity was crucial in providing a comprehensive understanding of how different demographics respond to VR training. Data was collected through pre- and post-experience surveys, which included questions on cooperation, openness to VR, prior VR experience, comfort with technology, and expectations.

Before the VR experience, participants were asked to rate their levels of cooperation, openness to VR, prior VR experience, and comfort with technology. They were also asked about their expectations for the VR training. These pre-experience surveys helped establish a baseline for each participant's attitudes and comfort levels. After the VR experience, participants completed a post-experience study with similar questions to assess any changes in their responses. The post-experience survey also included questions about their VR experience, attitudes towards VR, comfort with VR technology, and suggestions for future improvements.

The collected data was then analyzed to identify patterns and differences in responses based on gender and age. Statistical analysis was used to compare pre- and post-experience responses, highlighting significant changes in attitudes and comfort levels. The study also included comparing responses by gender and age to understand how different demographics responded to the VR training. The results provided insights into the effectiveness of the VR training, participants' experiences, and areas for future improvement. This comprehensive approach ensured that the study captured various perspectives and provided valuable recommendations for enhancing VR training programs.

11.3.2 Sample

The study's sample consisted of 43 participants who underwent cooperation skills development training enhanced by VR elements. These participants' diverse professional backgrounds included engineers, IT professionals, environmental managers, and more. This diversity was crucial in providing a comprehensive understanding of how different demographics respond to VR training. The participants' interests varied widely, ranging from gaming and sports to professional development and technology, adding richness to the data collected.

The participants' age range was broad, spanning from 17 to 64. This wide age range allowed the study to capture responses from younger and older individuals, providing insights into how different age groups perceive and interact with VR technology. Most participants were in the 30–49 age range, indicating a mid-career demographic that may be particularly interested in professional development and new technologies.

The gender distribution among the participants was relatively balanced, with 18 females and 25 males. This balance helped ensure that the study could analyze differences in gender responses, providing a more nuanced understanding of how VR training impacts different groups. Overall, the diverse sample in terms of age, gender, professional background, and interests provided a robust dataset for analyzing the effectiveness and reception of VR training.

Table 11.1 Gender

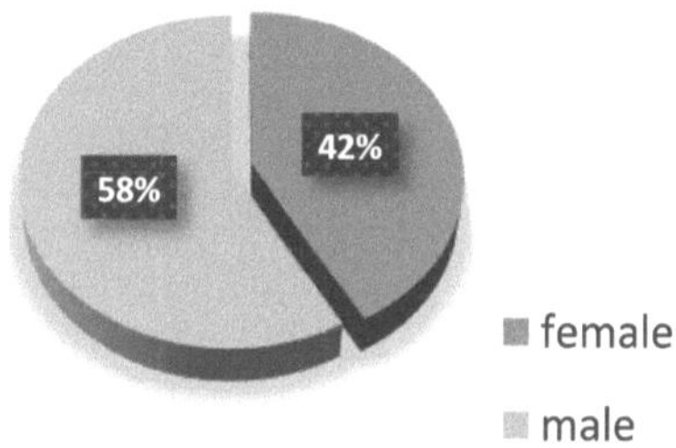

Gender	Frequency
Female	18
Male	25
Total geral	43

11.4 Data Analysis and Interpretation

Descriptive data analysis was conducted to summarize and interpret the collected data. This involved calculating frequencies, percentages, and means to provide an overview of the participants' responses. The analysis focused on identifying patterns and trends in the data, such as changes in cooperation levels, comfort with technology, and attitudes towards VR before and after the experience.

The data was also disaggregated by gender and age to explore differences in responses among various demographic groups. These insights were derived from comparing the frequencies and means of responses across different categories.

Visual representations, such as tables and charts, were used to illustrate the distribution of responses and highlight key findings. This helped in making the data more accessible and easier to interpret. The descriptive analysis provided a comprehensive understanding of the participants' experiences and perceptions, which informed the recommendations for future research and improvements in VR training programs.

11.4.1 Descriptive Analysis

11.4.1.1 Gender

The gender distribution among participants is relatively balanced (Table 11.1), with a slight male majority. This diversity can provide varied perspectives on the VR experience.

11.4.1.2 Age

The age range of participants is broad, spanning from 17 to 64 years old (Table 11.2). The majority of participants are in the 30–49 age range, indicating a mid-career demographic that may be particularly interested in professional development and new technologies.

Table 11.2 Age groups

Age groups	Frequency
17 (such age group was not expected)	1
20–24	2
25–29	1
30–34	9
35–39	7
40–44	8
45–49	7
50–54	4
55–59	1
60–64	3

Table 11.3 Cooperation before the training

Cooperation before	Frequency
–	1
Extremely	2
Moderately	21
Not at all	1
Other: very experienced	1
Slightly	4
Very much	13

11.4.1.3 Background/Interest

Participants come from a wide array of professional backgrounds, including engineers, IT professionals, and environmental managers, among others. Their interests range from gaming and sports to professional development and technology. This diversity in backgrounds and interests can enrich the VR experience by bringing in different viewpoints and expertise.

11.4.1.4 Cooperation Before the Training

The levels of cooperation before the VR experience vary, with most participants being moderately to very cooperative (Table 11.3). This suggests a generally positive attitude towards collaborative activities.

11.4.1.5 Open to VR Before the Training

Many participants were open to VR before the experience, with a significant number being extremely or very much open (Table 11.4). This indicates a high level of interest and willingness to engage with new technology.

Table 11.4 Open to VR before the training

Open to_VR_before	Frequency
Extremely	11
Moderately	6
Slightly	3
Very much	23

Table 11.5 Tried VR before the training

Tried VR_before	Frequency
Moderately	5
Not at all	22
Slightly	15
Very much	1

Table 11.6 Comfort with technology before the training

Comfort tech_before	Frequency
Extremely	9
Moderately	22
Not at all	1
Very much	11

11.4.1.6 Tried VR Before the Training

A substantial portion of participants had not tried VR before, suggesting that the experience was new for many (Table 11.5). This provides an opportunity to observe first-time reactions and learning curves.

11.4.1.7 Comfort Tech Before the Training

Participants' comfort with technology before the VR experience was generally high, with many feelings very comfortable (Table 11.6). This likely facilitated their engagement with the VR activities.

11.4.1.8 Cooperation After the Training

Post-experience, cooperation levels remained high, with many participants reporting very much cooperation (Table 11.7). This suggests that the VR experience positively influenced collaborative behaviors.

11.4.1.9 VR Experience After the Training

The majority of participants had a positive or extremely positive experience with VR (Table 11.8). This indicates that the VR training was well-received and effective in meeting participants' expectations.

Table 11.7 Cooperation after the training

Cooperation after	Frequency
–	2
Moderately	8
Unchanged	1
Other:	1
Other: it was already perfect	1
Slightly	2
Very much	28

Table 11.8 VR experience_after the training

VR experience_after	Frequency
–	2
Extremely positive	24
Positive	17

Table 11.9 VR attitude_after the training

VR attitude_after	Frequency
–	2
It was positive and remained positive	1
It was negative and became positive	6
It was positive and became extremely positive	13
It was positive and remained positive	17
It was negative, it became slightly more positive	2
Other:	1
Other: it was neutral and became positive	1

11.4.1.10 VR Attitude After the Training

Attitudes towards VR improved for many participants, with several shifting from positive to extremely positive (Table 11.9). This demonstrates the impact of the VR experience in enhancing participants' perceptions of the technology.

11.4.1.11 Comfort VR After the Training

Comfort with VR technology increased, with many participants feeling extremely or very comfortable after the experience (Table 11.10). This suggests that the training effectively familiarized participants with VR.

11.4.1.12 VR Objectives After the Training

Participants' objectives for using VR were largely met, with many feeling that the experience was extremely or very beneficial (Table 11.11). This indicates that the VR training successfully addressed participants' goals.

Table 11.10 Comfort VR_after the training

Comfort VR_after	Frequency
–	2
Extremely	11
Moderately	11
Very much	19

Table 11.11 VR objectives_after the training

VR objectives_after	Frequency
–	2
Extremely	14
Moderately	5
very much	22

11.4.1.13 Expectations

The diverse range of expectations highlights the multifaceted potential of VR training. Participants were looking for both personal and professional growth, as well as opportunities to enhance teamwork and explore new technologies. The anticipation of enjoyment and engagement also suggests that participants were open to immersive and an interactive learning experience.

Before the VR experience, participants had a wide range of expectations that can be grouped into several categories. Many participants expected to gain new skills, broaden their perspectives, and develop personally, with examples including "acquiring new techniques," "personal development," and "learning new technology." Several participants anticipated that the VR experience would enhance teamwork and cooperation, with expectations such as "team building," "enhancing cooperative collaboration," and "improving communication with peers." Participants were also excited about exploring new technologies and staying up-to-date with modern advancements, as reflected in expectations like "discovering something new," "exploring VR technology," and "keeping up with the times."

Some participants looked forward to having fun and enjoying the experience, with expectations including "it will be fun," "pleasant; interesting; development," and "it should be a fun, humorous experience." Additionally, a few participants were interested in the practical applicability of VR in their professional fields, with expectations such as "understanding the principles of operation," "practical applicability should be evident," and "development of VR technology usage." These varied expectations highlight the multifaceted potential of VR training to meet diverse needs and interests.

The data indicates that participants had high hopes for the VR experience, expecting it to be both beneficial and enjoyable. This variety in expectations underscores the importance of designing VR training programs that cater to different needs and interests, ensuring that all participants find value in the experience.

11.4.1.14 Suggestions

The suggestions indicate that participants found the VR experience valuable but see room for improvement in several areas. Addressing technical issues is a priority to ensure smooth and effective use of VR equipment. Participants are also eager to explore more diverse applications of VR, suggesting a high level of interest and engagement with the technology.

Based on the suggestions for improvement provided by participants, we can highlight some key interpretations (Table 11.12):

The carefully designed training's hybrid elements are complex and structured, which benefits participants, so it is essential to maintain this training design. The positive feedback on pushing boundaries and personal growth highlights VR's potential to challenge and develop participants meaningfully.

These suggestions provide valuable insights for refining and expanding future VR training programs, ensuring they meet participants' needs and expectations while leveraging the full potential of VR technology, as follows:

- **Technical issues** include the physical/technical site inspection, the control of the hardware (VR headsets and laptops) reliability the day before the training, charging the laptops, headsets, and controllers, updating the VR apps on all headsets, and ensuring optimal software performance and VR experience. In addition, a B-plan for each VR element, channel, and platform should be included. Finally, always have the same technical trainer (min. one) who knows the training program and prepares all headsets. In contrast, the trainer explains the following task to the participants and helps them with their technical issues without breaking the training flow, which is structured, coordinated, and catalyzed by the trainer. I have always had a strong B-plan for technical assistance (another assistant with the same qualities).

Table 11.12 Participants suggestions

Addressing Technical Issues: Some participants highlighted the need to resolve technical problems in order to utilize the VR equipment fully. For example, one suggestion was: "Next time, the technical issues should be addressed first so that the equipment can be fully utilized."	**Exploring More VR Applications**: Participants expressed interest in trying a wider range of VR applications. Suggestions included: "It would be great to try out as many possible usage as possible!" and "I hope there will be many more such opportunities, as it is very useful and very interesting."
Enhancing the Experience: Some participants appreciated the overall experience , complex or structured. For instance: "It was good and complex just the way it was," and "The training was properly structured. It was well-organized and easy to follow."	**Pushing Boundaries and Personal Growth**: Some participants appreciated the opportunity to push their boundaries and suggested continuing this approach. For example: "I found it very interesting. I was able to push my boundaries."

- **Expanded VR Experiences**: Explore and integrate a broader range of VR applications to provide diverse and comprehensive training experiences for the group. Be ready for additional training with VR elements, such as hybrid or multiplayer simulations, interactive scenarios, and collaborative tasks that cater to various learning objectives and professional fields.
- **Personalization and Adaptability**: Develop cooperation skills development training enriched with personalized VR elements that adapt to the group's specific problems, learning styles, preferences, and progress. It can enhance the training's effectiveness by providing tailored experiences that meet the group's needs.
- **Enhanced Interactivity and Immersion**: Increase interactivity and immersion to make VR training more engaging and impactful. It can incorporate advanced VR technologies, such as haptic feedback, eye-tracking, and AI-driven interactions.
- **Comprehensive Feedback Mechanisms**: Implement comprehensive feedback mechanisms to give participants detailed insights into their performance and progress. It can include real-time feedback, post-training assessments, and personalized recommendations for further improvement.
- **Continuous Improvement and Iteration**: Gather participant feedback and iterate on the VR training program to address any issues and incorporate new features. This iterative approach ensures the training remains relevant, effective, and aligned with participants' evolving needs.

The goal is to design and deliver future training programs with VR experiences even more effective, engaging, and tailored made to the diverse needs of the different participants. This will help maximize the benefits of VR technology in professional development and learning.

11.4.2 Comparative Analysis

11.4.2.1 Cooperation VR

The data on cooperation levels before and after the VR experience shows a consistent pattern, with no significant changes observed. Most participants (21) reported moderate cooperation both before and after the experience. The number of participants who felt "very much" cooperative remained the same at 12, and those who felt "extremely" cooperative also stayed constant at 2. There were a few unique responses, such as "Other: it was already perfect" and "Other: very experienced," which also remained unchanged. Overall, the data indicates that the VR experience did not significantly alter participants' self-reported levels of cooperation, suggesting that their initial attitudes towards cooperation were maintained throughout the training.

Table 11.13 Attitude_after VR training

VR_attitude_after	Frequency
–	2
It was positive and remained positive	1
It was negative and became positive	6
It was positive and became extremely positive	13
It was positive and remained positive	17
It was negative, it became slightly more positive	2
Other: it was neutral and became positive	2

11.4.2.2 Comfort with Technology

The data comparing participants' comfort with technology before the VR experience (Comfort_Tech_Before) and their comfort with VR after the experience (Comfort_VR_After) shows a consistent pattern, indicating that participants' comfort levels remained stable. The majority of participants reported moderate comfort both before and after the experience (22 participants). The number of participants who felt extremely comfortable remained the same at 9, while those who felt very much comfortable also stayed constant at 11. There were a few unique responses, such as "not at all" and "-", which also remained unchanged. Overall, the data suggests that the VR experience did not significantly alter participants' self-reported comfort levels with technology, indicating that their initial comfort levels were maintained throughout the training.

11.4.2.3 Attitude After VR Training

The data on participants' attitudes towards VR after the experience reveals a generally positive shift, as showed in Table 11.13.

Out of 43 responses, the majority (17) indicated that their attitude was positive and remained positive, while 13 participants reported that their positive attitude became extremely positive. Additionally, 6 participants who initially had a negative attitude towards VR experienced a positive change, and 2 participants noted a slight improvement from a negative attitude. There were also 2 participants whose neutral attitude became positive. Overall, the data suggests that the VR experience significantly enhanced participants' attitudes towards VR, with many moving from positive to extremely positive or from negative to positive.

11.4.2.4 Comparison by Gender

The main differences in responses by gender reveal that males were slightly more likely to report "extremely" and "very much" cooperation before the VR experience, and they were also more likely to be "very much" open to VR. Both genders had a similar distribution of prior VR experience, with many participants having not tried VR before. Males reported slightly higher comfort with technology before the VR experience. After the VR experience, cooperation levels improved for both genders, with males showing a higher increase in "very much" cooperation. Both genders had a positive experience with VR, with males slightly more likely to report an "extremely positive" experience. Attitudes towards VR improved for both genders, with males showing a slightly higher improvement. Comfort with VR technology increased for both genders, with males reporting higher levels of extreme comfort.

Both genders felt that their objectives for using VR were largely met, and they provided positive feedback and constructive suggestions for future improvements. Females focused on addressing technical issues, while males suggested exploring more VR applications. Both genders had a positive experience with VR, with males tending to report higher levels of comfort and cooperation, while females had a more balanced distribution across the categories.

11.4.2.5 Comparison by Age

The comparison of responses by age reveals several key differences: Before the VR experience, cooperation levels varied, with younger participants (17–24) showing lower levels of cooperation, while those in the 30–44 age range reported higher levels. Openness to VR was generally high across all age groups, with strong interest from younger participants and high openness from those aged 30–44. Prior VR experience varied, with younger participants having less experience and those aged 30–44 showing more varied levels.

Comfort with technology was high across all ages, particularly among younger participants and those aged 30–44. After the VR experience, cooperation levels improved for all age groups, notably among younger participants and those in the 30–44 age range. The majority of participants across all ages had a positive or extremely positive experience with VR, indicating the training was well-received and effective in meeting expectations.

11.5 Conclusions

The conclusions drawn from this study highlight the significant impact of the Cooperation skills development training with VR elements on participants' attitudes, comfort levels, and cooperation. The data indicates that participants generally had a positive experience with VR, with many reporting improvements in their attitudes towards the technology. This positive shift was particularly notable among those initially with negative or neutral attitudes. The training with VR elements also enhanced participants' comfort with VR technology, suggesting that VR's immersive and interactive nature can effectively familiarize users with new technological tools. Additionally, cooperation levels improved across all age groups, indicating that carefully designed and delivered soft skills training with VR elements can foster teamwork and collaborative skills.

The study also underscores the importance of addressing technical issues and expanding the range of VR applications to have always several B-plans to maximize the benefits of VR training. Participants provided constructive feedback, emphasizing the need for smoother technical execution and more diverse VR experiences. These insights are crucial for refining future cooperation skills development training tailored made for the participants' diverse needs and expectations. The recommendations for future research, including longitudinal studies and comparative analyses, will further enhance our understanding of VR's potential in professional development and learning. The study demonstrates the multifaceted potential of VR to foster collaboration and positively influence attitudes towards technology.

Several recommendations for future research can be highlighted based on the data and feedback from the training enhanced by VR elements. Conducting longitudinal studies can assess the long-term impact of training with VR elements on participants' skills, attitudes, and professional development, helping to determine lasting benefits and areas for improvement. Including a more diverse demographic in future studies will provide insights into how different age groups, genders, and professional backgrounds respond to the cooperation skills development training with VR elements. Comparative studies on cooperative skills development training, with or without VR elements, could help quantify the difference. Focusing on the specific needs of a training group, specific VR experiences can be optimized to tailor-make the training and enhance customer satisfaction, further increasing participant engagement and learning outcomes.

Acknowledgements This work would not have been possible without the invaluable support of Prof. Balázs Benjámin Budai, Head of the Department of Public Management and Information Technology and mentor of the XR Research Group, as well as the generous backing of Areus Zrt. Hungary. I would also like to extend my heartfelt gratitude to Péter Nagy, Dr. Noémi Császár-Nagy, András Váradi, Marcella Ficsur, Orsolya Böröcz Ágfalvyné, Prof. Gábor Keve, Béla Hevesi, Anna Páli, and Sára Seres for their invaluable contributions and unwavering support throughout this journey.

References

Dahl TL (2021) A preliminary scoping review of immersive virtual soft skills learning and training of employees. In: 2021 7th International conference of the immersive learning research network (ILRN). IEEE, pp 1–2. https://doi.org/10.23919/iLRN52045.2021.9459246

European Commission (2024) Emerging digital technologies in the public sector: the case of virtual worlds. Directorate-General for Research And Innovation. https://op.europa.eu/en/publication-detail/-/publication/eaa56c9b-44af-11ef-865a-01aa75ed71a1

Fortune Business Insights (2024) Virtual reality in education market size, share & COVID-19 impact analysis. Fortune Business Insights. https://www.fortunebusinessinsights.com/industry-reports/virtual-reality-in-education-market-101696

Friedrich CL, Teaford HR, Taubenheim A, Boland P, Sick BT (2019) Escaping the professional silo: an escape room implemented in an interprofessional education curriculum. J Interprof Care 33(5):573–575. https://doi.org/10.1080/13561820.2018.1538941

Govaerts S, De Wever B (2023) Preparing for (future) work? Supporting soft skill development in the context of higher education with a digital instrument for 360° feedback. INTED2023 Proceedings, pp 5201–5209. https://doi.org/10.21125/inted.2023.1347

Herdegen M (2023) Extended reality vocational training's ability to improve soft skills development and increase equity in the workforce. AI Comput Sci Robot Technol 2. https://doi.org/10.5772/acrt.22

Hill T, du Preez H (2022) Reflections on a virtual reality soft skills teaching intervention. In: 8th International Conference on Higher Education Advances (HEAd'22). Universitat Politècnica de València, pp 841–848. https://doi.org/10.4995/HEAd22.2022.14468

Kirwan E (2024) The power of virtual reality performance experiences for education. Virtual Creat 13(2):231–243. https://doi.org/10.1386/vcr_00086_1

Kwame S, McAllister J (2024). Next-gen engineering: virtual reality for soft skills development and enhanced work efficiency in educational contexts. In: Proceedings of the 2024 ASEE midwest section conference

Manojlovic H (2022) Escape room as a teaching method. Opus et Educatio 9(2). https://doi.org/10.3311/ope.504

PwC (2020) The effectiveness of virtual reality soft skills training in the enterprise. Price-waterhouseCoopers. https://www.pwc.co.uk/issues/technology/immersive-technologies/study-into-vr-training-effectiveness.html

Regal G, Schrom-Feiertag H, Nguyen Q, Aust M, Murtinger M, Smit D, Tscheligi M, Billinghurst M (2022) VR [we are] training: workshop on collaborative virtual training for challenging contexts. In: Extended abstracts of the 2022 CHI conference on human factors in computing systems. Association for Computing Machinery, pp 1–5. https://doi.org/10.1145/3491101.3503710

Smooth Soft (2023) How VR training saves restaurants time and money. ArborXR. https://arborxr.com/blog/customer-story-how-smooth-soft-vr-training-saves-restaurants-time-and-money/

Taupiac J-D, Rodriguez N, Strauss O, Beney P (2019) Social skills training tool in virtual reality, intended for managers and sales representatives. In: 2019 IEEE conference on virtual reality and 3D user interfaces (VR). IEEE, pp 1183–1184. https://doi.org/10.1109/VR.2019.8798317

Zhao Z, Ma X (2020) Situated learning of soft skills with an interactive agent in virtual reality via multimodal feedback. In: UIST '20 Adjunct: adjunct publication of the 33rd annual ACM symposium on user interface software and technology. Association for Computing Machinery, pp 25–27. https://doi.org/10.1145/3379350.3416183

Engineering Education in a Sustainable Context, Based on AI

12

João Alvarez Peixoto, Andreia de Bem Machado, and Marc François Richter

Abstract

Sustainability plays an essential role in training engineers to meet contemporary challenges, as they are responsible for designing products that must consider the sustainable origin of inputs, ease of transport, energy efficiency, and customer requirements and generate the lowest possible level of waste during the recycling process. The research problem is how sustainability can be effectively integrated into engineering curricula, and which technologies? In order to answer this question, the aim was to analyse the integration of sustainability into engineering education, identifying its integration into curricula and proposing technologies to aid this process of inclusion. To achieve the proposed objective, the study's methodology is a bibliographic review on the Web of Science (WOS) platform, in order to identify the recurrences of the integration of sustainability premises with engineering teaching, in a bibliographic review and bibliometric analysis on the Scopus database, focusing on innovative approaches and emerging technologies. The results indicated a significant growth in publications since 2020, especially after the COVID-19 pandemic. Between 1990 and 2024, 4053 documents were identified, of which 1671 were

J. A. Peixoto
Postgraduate Programme in Science, Technology, Engineering and Mathematics (PPGSTEM), State University of Rio Grande do Sul, Porto Alegre, Brazil
e-mail: joao-peixoto@uergs.edu.br

A. de Bem Machado
Engineering and Knowledge Management Department, Federal University of Santa Catarina (UFSC), Florianópolis, Brazil
e-mail: andreia.bem@ufsc.br

M. F. Richter (✉)
Graduate Programme in Environment and Sustainability (PPGAS), State University of Rio Grande do Sul (UERGS), Porto Alegre, Brazil
e-mail: marc-richter@uergs.edu.br

M. J. Sousa et al. (eds.), *Economic Sustainability Potentiated by Artificial Intelligence*, Contributions to Environmental Sciences & Innovative Business Technology,
https://doi.org/10.1007/978-981-96-8089-4_12

scientific articles, highlighting areas such as engineering (2396 publications) and computer science (989 publications). The most active countries were the USA and the UK. The findings show that practices such as the circular economy, eco-design and product-service systems (PSS), combined with the use of Industry 4.0 technologies (IoT, big data, blockchain), promote the development of curricula that are more aligned with sustainability. These practices offer tools for designing innovative and responsible solutions, using artificial intelligence for life cycle analysis, simulations and personalisation of teaching. In addition, the data highlights the need for curricula that train engineers to address social and environmental impacts from the earliest stages of projects, promoting a holistic and sustainable vision. This is followed by a pedagogical proposal based on project-based learning in engineering education, which involves integrating teaching with the development of real projects in all their phases, and culminating in a proposal to use artificial intelligence tools in the process of integrating sustainability into the development/learning phases of product design. Artificial intelligence comes in as a point of analysis, as well as a look at sustainability, among the TRLs, whether as a scientific search tool or tools for predicting the use, reuse and disposal of the product, even if it is still at the idea, research, technology, proof of concept, prototype, validation or production level. In this sense, a more effective engineering education, contextualised with the world and consciously sustainable, is fundamental.

Keywords

Sustainability • Artificial intelligence • Engineering education • Learning • Projects

12.1 Introduction

Despite the growing global concern about sustainability and the urgency of more conscious practices in all sectors, the level of integration of sustainable concepts in engineering education is still limited. Many engineering curricula remain focused on traditional approaches, with little attention paid to the environmental, social and economic issues surrounding sustainability (Ashford, 2004). This mismatch is reflected in the training of engineers, who often enter the labour market with an incomplete view of the importance of sustainability in their practices. To overcome this challenge, it is necessary to reformulate engineering curricula, promoting a multidisciplinary approach that explicitly integrates sustainability into all phases of product and project development (Arefin et al. 2021).

This reformulation of the curriculum cannot just be superficial; it requires a profound pedagogical transformation that includes both theoretical content and practical methodologies that connect students to real sustainability challenges. This means that the premises of sustainability—such as the efficient use of resources, minimising waste, promoting clean energy and taking social impacts into account—must be incorporated into technical subjects and also into the development of projects and applied case studies (Monteiro and Sousa 2024). Furthermore, this integration needs to be continuous and comprehensive in order to prepare future engineers to face the complex challenges of sustainability, which are central to the development of contemporary society (de Oliveira et al 2023).

What the present study aims to address is precisely how sustainability can be effectively incorporated into engineering curricula, overcoming the fragmented approach that still prevails. To answer this question, the study will investigate both pedagogical strategies and the use of innovative educational technologies—such as digital simulations, problem-based learning and prototyping labs with sustainable materials (Singha and Singha 2024). These technological tools can provide more active and participatory learning, connecting students directly with practice and allowing them to experiment with sustainable solutions in a controlled environment. The proposal thus seeks to contribute to a more complete and conscious education of future engineers, promoting a paradigm shift towards a professional practice that prioritises sustainability.

The study of engineering education with a focus on sustainability, supported by the use of artificial intelligence (AI), has the central objective of analysing and promoting the integration of sustainability concepts into engineering curricula (Staniškis and Katiliūtė 2016). Sustainability, although increasingly debated in various areas, is still not fully addressed in a practical and theoretical way in engineering education, especially at all stages of the life cycle of products and projects (Van den Beemt et al. 2020). In this context, this study seeks to map the presence of sustainability in current curricula, identify gaps and propose supporting technologies—especially AI—to help train engineers capable of considering the environmental, social and economic impacts of their creations from conception to disposal.

Artificial intelligence, in particular, has vast potential for improving the teaching of sustainable engineering, both in the pedagogical aspect and in the development of research and analysis methodologies (Nunez and Lantada 2020). AI can be used to create predictive models and detailed analyses of environmental impact that allow students to simulate and predict the effects of their design decisions and choice of materials. In addition, AI can facilitate the development of optimisation tools that help identify more efficient solutions with lower environmental impact (Kar et al. 2022). This data-driven approach provides deeper, more concrete learning, where students can visualise and understand how their choices directly affect the environment and society.

Another essential point of this study is product life cycle analysis, an approach that allows future engineers to understand the impact of a product at every stage of its development—from the extraction of materials to production, transport, use and final disposal (Moni et al. 2020). By integrating sustainability into teaching, it is possible for curricula not only to provide theoretical knowledge, but also to train students to assess and manage these impacts in a practical way, promoting the creation of products that are more sustainable and efficient throughout their life cycle (Chiu and Chai 2020). Artificial intelligence, in this sense, can be applied to carry out more accurate life cycle assessments, providing real-time data on the environmental impact of different design and manufacturing choices.

This study also sets out to develop and recommend AI-based pedagogical tools that help educators personalise learning for each student, identifying areas where knowledge can be improved and adjusting teaching to cover these gaps. These technologies can create a more dynamic and adaptable learning environment, where students receive real-time feedback on their activities and simulations. The use of AI is thus not just limited to analysing environmental impact; it can also be a powerful learning tool, providing students with a practical and interactive view of how sustainability can and should be applied in engineering projects (Dimitriadou and Lanitis 2023).

In this way, this study aims to make a significant contribution to the modernisation and adaptation of engineering education, emphasising sustainability as a central part of the training of future professionals. The integration of AI allows engineering education to evolve towards a more interactive and data-driven model, where students have the opportunity to explore sustainable scenarios while simultaneously developing the critical skills needed to deal with contemporary environmental and social challenges. Thus, the study aims to provide guidelines for more robust curricula that are aligned with global demands for sustainability, as well as suggesting technologies and methodologies that aid this transformation in a practical and effective way.

To achieve the proposed objective, the study adopts a methodology that combines a comprehensive analysis of the literature with the formulation of a pedagogical and technological proposal aimed at teaching sustainable engineering. Initially, a bibliometric review will be carried out on the Web of Science (WOS) platform, with the aim of mapping and analysing recurrences in the integration of sustainability principles with engineering education (Richter and Bem Machado 2023). The study will seek to identify the pedagogical approaches, methodologies and practices already explored in the literature, offering a detailed overview of the most effective strategies and common challenges faced in this integration. From this analysis, it will be possible to build a solid theoretical basis that will support the next stages of the study, including the identification of gaps and opportunities for improving engineering education with a focus on sustainability.

The study then proposes a pedagogical approach based on project-based learning (PBL), which aims to integrate theoretical teaching with the development of real projects. Project-based learning has been widely recognised for promoting deeper and more active learning by engaging students in practical challenges that reflect real-world problems and stages (Bressiani et al. 2020). In the context of engineering education, PBL enables students to go through all the development phases of a project—from initial research and planning to design, prototyping and final impact analysis. This allows them to directly experience the complexities and implications of their choices, especially with regard to sustainability. In addition, by working on real projects, students develop essential skills such as critical thinking, problem-solving and collaboration, which are fundamental to a conscious engineering practice aligned with sustainability principles (Singha and Singha 2024).

Finally, the methodology culminates in the proposal to use artificial intelligence tools to strengthen the integration of sustainability into the development and learning phases of engineering projects (Debrah et al. 2022). AI can play a crucial role by providing advanced analysis and simulation capabilities, allowing students to assess the environmental, economic and social impact of their projects more accurately and quickly. For example, AI tools can be used to create environmental impact simulations that show, in real time, how different choices of materials or manufacturing processes influence the product's life cycle (Wang et al. 2021). Artificial intelligence can also help optimise processes and search for more efficient solutions, offering valuable insights into sustainable alternatives that students can explore in their projects.

In addition, the use of AI in engineering education makes it possible to personalise learning, since the technology can identify the specific needs and difficulties of each student and provide tailored feedback and guidance (Bhutoria (2022). With this, the pedagogical proposal not only encourages a greater understanding of sustainability, but also improves each student's ability to apply these concepts in a practical way that is aligned with the demands of the market and society (Abulibdeh et al. 2024). By integrating AI, the study aims to promote engineering education that is both technically and socially responsible, training professionals who are prepared to face the environmental and social challenges of the twenty-first century with innovation and ethical commitment. In this way, the proposed methodology aims to contribute to building a solid educational foundation that will drive the transformation of engineering education, aligning it with the objectives of sustainable development.

12.2 The Pillars of Sustainability

Education for sustainability is essential to promote products in line with the circular economy, which value financial, social and environmental aspects. This involves creating products that use sustainable inputs, are easy to transport, meet consumer needs and encourage reuse. In addition, at the end of their useful life, these products should be recyclable with the least possible impact, minimising waste and favouring a more efficient and responsible production cycle (Wilson et al. 2024).

In recent years, sustainability has gained increasing relevance, both in global discussions and in people's daily lives. This is because sustainability addresses the conscious use of natural and social resources to meet current needs without compromising the ability of future generations to meet theirs. In a context of accelerating climate change, loss of biodiversity and social inequality, sustainability has become a fundamental guide for economic, social and environmental development. This vision, aligned with the UN Sustainable Development Goals (SDGs) (Weiland et al. 2021), emphasises the need to balance economic growth, social justice and environmental preservation, seeking alternatives that do not generate degradation and encourage fair and responsible development (Sachs 2015). Sustainability is an integrated and multifaceted concept that goes beyond the mere preservation of natural resources. It represents a balanced development model that considers the financial, social and environmental aspects holistically (Ramírez-Márquez et al. 2024). Adopting sustainable practices promotes a more robust economy, a fairer society and a healthier environment.

The financial aspect of sustainability is crucial, as it ensures that companies and organisations can maintain sustainable practices without compromising their economic viability. Adopting sustainable practices may initially involve larger investments, such as the transition to clean technologies or the use of recyclable materials. However, in the medium and long term, these practices tend to generate financial returns. Companies that invest in energy efficiency, for example, can reduce operating costs and, at the same time, stand out in a competitive market, attracting conscious consumers and investors (Celestin et al. 2024). In addition, various lines of financing and government incentives are being directed towards sustainable projects, which can facilitate the economic viability of green initiatives. In this scenario, sustainability and finance are not opposites, but complementary (Elsawy and Youssef 2023).

The social dimension of sustainability focuses on promoting well-being, equity and justice for communities and individuals impacted directly or indirectly by development practices (Talan et al. 2020). The social aspect ranges from including fair labour practices, such as equal pay and gender diversity, to ensuring decent living conditions and safety for communities. Companies that invest in the training and development of their employees, respect human rights and value diversity not only strengthen their own organisational cultures, but also contribute to building a fairer and more inclusive society. For social sustainability to be effective, there needs to be a commitment to integrating all the voices involved in the development process, promoting dialogue and cooperation for the sake of a collective future (Padilla-Rivera et al. 2020).

The environmental aspect of sustainability involves the protection of ecosystems and the conservation of natural resources essential to life on the planet (Hariram et al. 2023). It focuses on practices that minimise the negative impacts of human activities on the environment, such as excessive water use, greenhouse gas emissions and soil, air and water pollution. Environmental sustainability is particularly urgent in a scenario of climate change and declining biodiversity, which requires efforts on several fronts, such as reducing fossil fuel consumption, promoting renewable energy and encouraging the circular economy (Tang et al. 2023). Sustainable environmental practices aim to regenerate resources, ensuring that they continue to be available in the future and that the ecological balance is preserved. In addition, individuals and companies that adopt responsible environmental practices tend to have a positive impact not only on the environment, but also on people's health and quality of life (Camilleri 2022).

The circular economy is an approach that aims to redefine the way society should consume and manage resources, promoting a more efficient and sustainable system (Richter and de Bem Machado 2023). In contrast to the traditional linear model, which is based on extracting, producing, consuming and discarding, the circular economy focuses on reducing waste, reusing materials and regenerating natural resources. This vision not only reduces pressure on the environment, but also brings significant economic and social benefits (de Bem Machado et al. 2021).

In environmental terms, the circular economy reduces the extraction of natural resources and the generation of waste, preserving ecosystems and reducing pollution. For example, the use of recyclable materials in production reduces the need for virgin raw materials and the energy used in the process, minimising carbon emissions. On the economic side, this model encourages innovation by promoting new sustainable products and services and generates business opportunities such as the repair, rental and recycling market. In addition, the circular economy strengthens the resilience of supply chains and reduces dependence on finite resources, which is especially relevant in a global context of scarcity and instability of raw materials. Thus, by adopting circular economy practices, companies and consumers contribute to a more sustainable, balanced and inclusive economic system, where development does not jeopardise the future of generations to come.

12.3 Project-Based Engineering Education

Products that reach consumers are not simply assembled in a factory. There is a longer path that precedes this. The product starts from an idea or need and is developed until it becomes a commercial product that satisfies the user and encourages them to buy it. This is the path of product development, from an idea to its materialisation. This path is traced by engineering, design and marketing professionals, among others (Pazmino 2015).

Engineers turn ideas into tangible reality and fulfil the user's need for functionalities. Designing products and services is one of the competences expected to be developed in an engineering course, as stated in article 4 of resolution number 2 of the Brazilian Association of Higher Education Providers (ABMES):

> Art. 4 The undergraduate programme in Engineering must provide its graduates with the following general competences throughout their training:
>
> [...] III - conceive, design and analyse systems, products (goods and services), components or processes:
>
> a) be able to conceive and design creative, desirable and viable solutions, technically and economically, in the contexts in which they will be applied;
>
> b) design and determine the construction and operational parameters for engineering solutions;
>
> c) apply management concepts to plan, supervise, prepare and coordinate engineering projects and services. (ABMES 2019, p 2).

Developing product concepts is one of the activities that engineers will carry out in the labour market. This is why it makes sense to practise this in the teaching environment as part of their training. More than the study of methods and techniques, practical application is what guarantees the graduate profile for the engineering professional, suited to the realities of the market.

There are teaching methodologies that focus on developing project-based learning, starting from a need and arriving at a product or service, which requires knowledge of fundamentals, concepts and techniques. The content becomes a tool for a greater purpose. Project-based learning (ABProj) is an active teaching methodology that proposes practical activity as a tool. Instead of explaining all the details of an activity, the student is invited to take part in real actions to develop the competence being worked on. Project-based learning is applied to modalities in which there is a tangible product as a result, which makes the learner's interpretation of the problem more tangible to their prior knowledge (Ausubel 2003).

The process of creating a product takes place through the application of design methods, which lead the creation process in phases. Of the various methods available, two stand out: six-phase product development and TRL.

Product development in six phases considers the process to take place in the following phases: ideation, product definition, prototyping, initial design, validation and testing and commercialisation (Raeburn 2024). Figure 12.1 shows this model.

In phase 1, product ideas are conceived, based on customer needs, feasibility, the target market, minimum functionalities and value offer. Tools such as SWOT analysis (Kuazaqui 2016) and the Scamper method (Carvalho et al. 2018) can help at this stage. Branstorming a product concept to idea generation.

In phase 2, the product is defined, taking into account the commercial analysis, value proposition, success metrics and marketing strategy. This is when the assembly of the minimally viable product (MVP) begins, with initial prototyping. Scoping and refining the product definition.

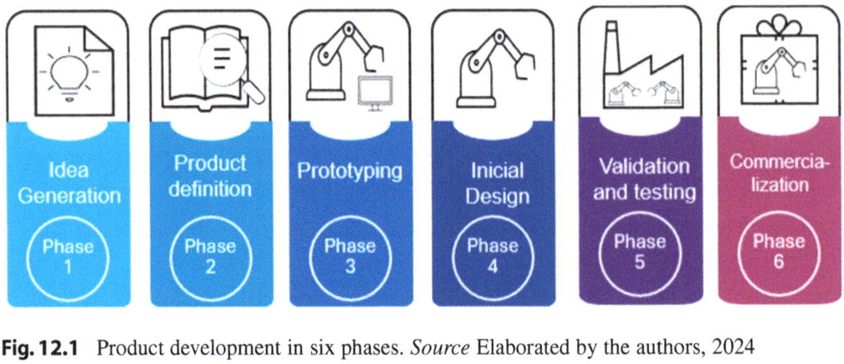

Fig. 12.1 Product development in six phases. *Source* Elaborated by the authors, 2024

Phase 3 sees the creation of the project plan, in which topics such as market risk research, development strategy, feasibility analysis and minimally viable product (MVP) are explored. The use of methods such as critical path analysis CPM—Critical Path Method (Nigam 2023) and Gantt chart (Clark 2019) help to make this process more visual. Constructing a visual representation same as a prototyping.

In phase 4, the basic model of the product is created, taking into account the target audience and the minimum functions it should contain. Analysing materials and the manufacturing process are useful at this stage. So is good project management (Gray and Larson 2010). Producing an initial design.

In phase 5, the product is validated in both, simulated and real, environments to check that the planned functionalities are being met, and inconsistencies are tested. Validating and testing the development strategy.

In phase 6, the product is offered to the market, checking marketing strategies and a good after-sales strategy. Developing and implementing the product for commercialisation.

Another approach to managing product development, which is not mutually exclusive but complementary, is to analyse its technological maturity. Technological maturity indicates the degree to which a given technology is ready to be considered finalised. This maturity can be assessed using different methodologies. The Technology Readiness Level (TRL) is one of these methodologies. Currently, the TRL consists of a set of scalable metrics, ranging from level 1 to level 9 (Andrade et al. 2019) (Fig. 12.2).

In Brazil, the application of TRL is regulated by ABNT-NBR ISO-16290 (ABNT, 2015). The assessment of technological maturity, through the TRL, plays a crucial role in indicating the effort required for the development and research of each technology, providing valuable information to researchers for conducting their projects, as well as opening up new opportunities for studies and the need for further research.

At TRL 0, the idea is presented. An idea that arises from a market opportunity, the improvement of an existing product or a functionality desired by the user.

At TRL 1, basic principles are observed and reported in a basic research. Initial scientific research begins, from studies into the basic properties of materials, physical, chemical and mathematical principles are postulated and observed qualitatively. The focus is on the fundamental understanding of a material or process, checking whether there is a theoretical basis for what is being proposed as a product.

At TRL 2, a technology concept or application is formulated. Once the basic principles are observed, practical applications can be identified. The applications are speculative, and there may be no evidence or detailed analyses to support the assumptions. It is an identification of technology pertinent to the product proposal. The transition from TRL 1 to TRL 2 moves ideas from basic research to applied research. The experimental work is designed to corroborate the basic scientific observations made during the work in TRL 1.

In an applied research study, a critical function or distinctive proof of concept is developed at TRL 3 using analytical and experimental methods. The predictions for

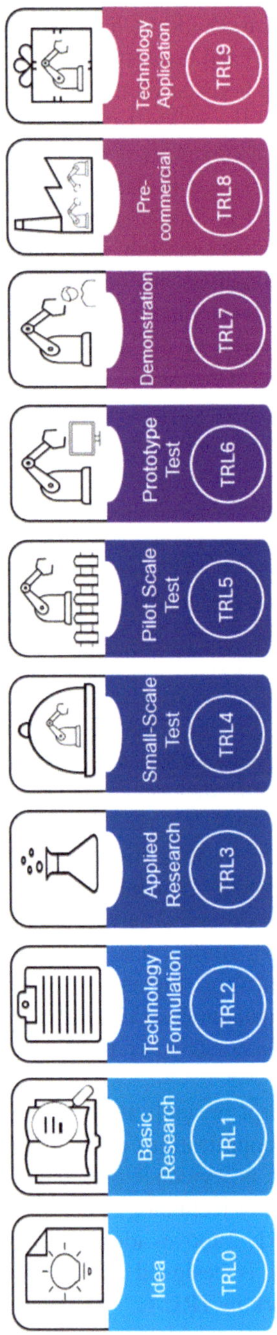

Fig. 12.2 Learning levels within the context of sustainable engineering education. *Source* Elaborated by the authors, 2024

distinct technological components are intended to be physically validated by analytical and laboratory-scale research. The outcomes of laboratory tests conducted to evaluate relevant parameters and comparison with analytical forecasts for crucial components are examples of supporting data. Verifying that the notion functions as intended is the goal of the experimental work in TRL 3. Although the technology's constituent parts have been verified, no significant effort has been made to combine them into a whole system. Physical experiments can be supplemented with modelling and simulation.

At TRL 4, a small-scale test is conducted in a lab setting to validate the technology. To ensure that the elements will function together, the fundamental technological components are integrated. Comparing this to the finished system, the fidelity is comparatively low. The results of the integrated experiments and estimations of the differences between the experimental components and test results and the system's expected performance targets are included as supporting material. The shift from scientific inquiry to engineering, from development to demonstration, is represented by TRLs 4 through 6. The first stage in figuring out whether the various parts will function as a system is TRL 4. Reducing the number of alternatives for the entire system should be the aim of TRL 4.

At TRL 5, the technology is validated in a relevant environment, as a pilot-scale test. The basic technological components are integrated so that the system configuration is similar to the final application in almost every respect. Supporting information includes the results of laboratory-scale tests, an analysis of the differences between the system and the laboratory environment and the future operating environment, and an analysis of what the experimental results mean for the eventual operating environment. The main difference between TRLs 4 and 5 is the increased fidelity of the system and environment to the real application. The system tested is almost prototypical.

At TRL 6, a technology model or prototype test is demonstrated in a relevant environment. Engineering-scale prototypes are tested in a relevant environment. This represents a major advance in the demonstrated readiness of the technology. The main difference between TRLs 5 and 6 is the transition from laboratory scale to engineering scale and the determination of the scaling factors that will enable the design of the final system. The engineering pilot-scale demonstration must be able to perform all the functions that will be required of a complete manufacturing system.

At TRL 7, a technology prototype is demonstrated in an operational environment. It is requiring the demonstration of a real prototype of the system in a relevant environment. The final design is practically complete. The objective of this phase is to eliminate engineering and manufacturing risks. To credibly achieve this objective and exit TRL 7, a scaled production proposal is necessary.

At TRL 8, the current technology is completed and qualified through testing and demonstration, for a pre-commercial phase. The technology has been proven to be functional in its final form and under expected conditions. In almost all cases, this TRL represents the end of true system development. Actual manufacturing costs

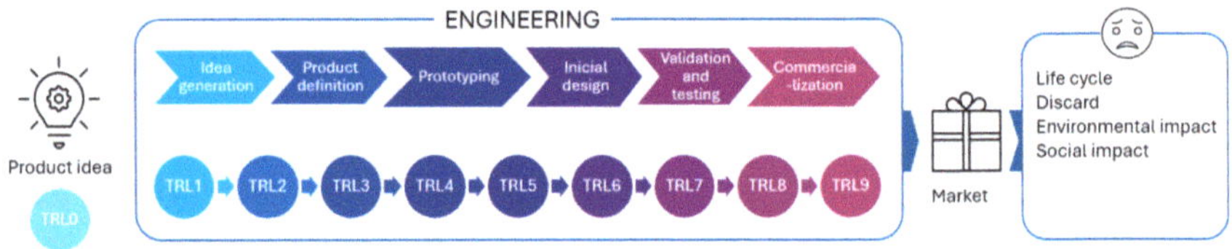

Fig. 12.3 Conceptual model for integrating sustainability into engineering education. *Source* Elaborated by the authors, 2024

will be determined. The difference in product performance relative to plan needs to be highlighted, and plans to reduce this difference will need to be developed.

At TRL 9, the actual technology application is qualified through successful mission operation. The technology is in its final form and operates under the full range of operating conditions. The emphasis shifts to statistical process control.

Figure 12.3 shows these two complementary approaches, with the path between the idea having emerged and the product conceived.

This figure also points to what happens after the product reaches the market. Once the functionalities desired by the user have been achieved, within the reality of the market, it means that the process of creating the product has worked. And in fact it has, although there might be concerns that begin to emerge, such as:

(a) Life cycle: which corresponds to the product's journey from its introduction to the market, its growth, its maturity, its decline, to its withdrawal from the market, when there is concern about its reuse or disposal (Barros 2020);
(b) Disposal: which implies a cost for the segregation and correct designation of the waste generated by the product in disuse (Instituto Butantan 2014).
(c) Environmental impact: even during the product's life cycle, there is concern about its impact on the environment, and this analysis extends to its disposal (Piatti and Rodrigues 2005).
(d) Social impact: the way in which a product is produced, the technologies used, the need for skilled labour, the degree of unhealthiness that the production process emanates, as well as the social environment in which the production process is located, brings problems to society that are not seen when the product has already reached the user, but are latent during its production (Chaves and Mezzari 2016).

Any of these concerns occur when the product is already in the customer's hands. And actions begin to be proposed to modify the product or even change the design of similar new products. But it doesn't change the fact that the product has already been designed and is already on the market, having an impact on the environment and society.

But what if product design would consider environmental and social issues at every stage of its conception? Couldn't the product's negative sustainability impacts be mitigated? Would these impacts appear between project phases or only when the product is finished? These questions can be answered during engineering training, when the need for a more holistic vision arises, and where the fundamentals of product design are studied and, above all, practised. Training engineers with this vision of a sustainable product from the earliest stages of creation is imperative if we want a society that favours minimally sustainable products.

12.4 Artificial Intelligence Tools

Artificial intelligence (AI) is a concept that the average person, who has no technical knowledge of the subject, ends up understanding as a computer that thinks. And this thinking is sometimes better than that of a human being. In fact, this interpretation is not entirely wrong. The point of the term artificial intelligence is not that it is digital, or that it is computational, but that it is intelligence. Whereas the best understanding is that it is artificial.

When performing a mathematical addition operation, humans use a technique to add two numbers and get a result. If this same addition is carried out by a digital system, there will be an algorithm that provides this sum of values, displaying the result digitally. And because it's not something that is native to human beings, it becomes an artificial sum.

Without going into the merits of the concepts of digital machines and intelligence, this study deals with the method of artificial intelligence, the application of which helps the study of engineering, when it requires analysing the sustainability of the product under development.

There are various classifications of artificial intelligence, of which the three mentioned in this paper are:

(a) Machine learning (ML): it allows systems to learn from data, without being explicitly programmed to do so. Instead of following strict instructions, ML systems use algorithms to identify patterns and make predictions based on historical data;
(b) Artificial neural networks (ANN): these are a subset of machine learning that try to imitate the way human neurons work. They are made up of layers of interconnected processing units, which process information in a hierarchical manner;
(c) Natural language processing (NLP): this is a field of AI dedicated to teaching machines to understand, interpret and generate human language in a way that is useful and relevant. This involves tasks such as sentiment analysis, machine translation, text summarisation and voice interaction.

Artificial intelligence applied to the search and analysis of scientific articles on the internet uses a combination of machine learning, artificial neural networks, and natural language processing techniques. These technologies make it possible to improve the search and recommendation of articles, organise content based on topics and categories, generate automatic summaries and even predict research trends. With this, AI makes it easier to navigate the vast and growing volume of academic articles available on the internet, making research more efficient and personalised for users.

In a common algorithm for searching articles and websites with scientific information, keywords are used in a logical chain that can relate words in a logical way. Like: contain the word house, or contain the word flat (house||flat); contain the word life and at the same time contain the word death (life & death). But even so, what occurs is a search for the appearance of the word as it was typed. It's up to the user to define the correct words that will bring the appropriate scientific information.

On the other hand, an algorithm with artificial intelligence will search for the word and all its context, as described by the user.

And it is this research that will have to take place in order to investigate the sustainable nature of the product design, at each level of maturity it reaches. Because scientific research is what gives the seal of approval to what is being investigated.

As a suggestion, Table 12.1 presents ten AI tools suitable for scientific research into topics of interest, which in this proposal will be themes relating to the sustainability of the product being designed.

There are also the ChatGPT (https://chat.openai.com/) and Copilot (https://www.microsoft.com/pt-br/bing) tools, which are AI tools that respond to user questions, but their data sources are not segmented, and they access information from suppliers, websites in the field, companies, among others. It's up to the user to ask the right questions to get the scientific answer they want.

12.5 Engineering Education Based on Sustainability

Engineering education must take into account the concepts of sustainability between each phase of a project, from TRL0 to TRL9 maturity, in order to have something more than a sustainable product, but an idea of a sustainable product.

Of course, this involves an adjustment to the course's pedagogical plan, as the teaching methodology will have to comply with project-based learning (PrjBL). This is an evolution in terms of teaching, as it guarantees the development of competences in line with the market reality of each engineering.

But this starts with a vision of a production environment with sustainable adherence. When producing a product that is intended to be as sustainable as possible, the finished item should be analysed. Checking items that can be indicative of the degree of sustainability that the product has, even in comparison to other products on the market, can add value to it.

This work proposes an analysis based on ten steps: raw materials; production process; design; social impacts; packaging; transport and logistics; recyclability; certifications; consumer; and environmental footprint. Table 12.2 shows the ten-step sustainability analysis, with its internally subdivided topics and their respective descriptions.

Table 12.1 AI tools for researching scientific references

Artificial intelligence tool	Description
Heuristica https://www.heuristi.ca/	Information is presented in the form of a conceptual map, including information from articles, videos, news, and more
OpenRead https://www.openread.academy/	Search platform with interaction through questions and answers (chat), based on published scientific articles
Paperbrain https://www.paperbrain.org/	A platform to help understand scientific topics, through interaction, with answers based on relevant publications
Explainpaper https://www.explainpaper.com/dashboard	A platform that helps in better understanding an article, by loading up its PDF version
GraphMaker https://www.graphmaker.ai/dashboard	Scientific responses are generated from graphs, CSV files, excel spreadsheets or by search request
Tavily https://app.tavily.com/sign-in	The programme performs searches based on scientific questions entered by the user, with answers based on scientific articles
PowerDrill https://powerdrill.ai/	Produces data analysis based on the user's scientific questions
Scispace https://typeset.io/	Tool for extracting information from texts, from their PDF version, interacting with the user via chat
Perplexity https://www.perplexity.ai/	Platform that answers scientific questions, based on published articles, through chat interaction
NextNet https://getnextnet.com/	AI-powered database that provides scientific articles that answer user-asked questions in the form of relational search

Source Elaborated by the authors, 2024

The proposal to create products with environmental and social concerns suggests that, at each transition in the maturity stage of the product under construction, an assessment of environmental and social aspects should be carried out. What has already been built and its potential impacts should be checked. If there is a possibility of any impact, the project is reviewed, making the necessary adjustments or even redoing the project. The aim is not to go ahead without eliminating or minimising the potential impact. Figure 12.4 shows this project proposal for the creation of products with environmental and social concerns.

In the engineering education process, at each stage of the project where there is a change in the maturity level of the product being built, a ten-step sustainability analysis is carried out at the stage of maturity it has reached. Perhaps some of the steps don't apply due to the early maturity, but the aim is to apply as many

Table 12.2 Sustainability analysis in ten steps: topics and subtopics for a product sustainability analysis

Topics	Subtopic	Description
Raw materials	Sources of raw materials	Is the source of inputs and raw materials renewable?
	Environmental certifications	Are raw materials certified by organisations that guarantee responsible production practices?
	Impact on biodiversity	Does the extraction or production of raw materials affect local ecosystems or endangered species?
Production process	Energy efficiency and emissions	Does the product's manufacturing process use renewable energy or generate high greenhouse gas emissions?
	Use of water and natural resources	Does the product's production consume large amounts of water or other scarce natural resources?
	Pollution and waste	Does the manufacturing process generate pollution or waste?
Design	Durability and useful life	Does the product allow its useful life to be measured?
	Maintainability	Can the product be easily recycled when it reaches the end of its useful life?
	Modular design	Are the product's components easily separated?
Social impacts	Working conditions	Is the product manufactured under fair and safe working conditions?
	Support for local communities	Does the product support local economies?
Packaging	Recyclable or biodegradable materials	Is the product packaging made from environmentally friendly materials?
	Reduction of packaging	Does the packaging aim to minimise impact?
Transportation	Emissions associated with transportation	What mode of transportation does the product require?
	Sustainable supply chain	Does the production require suppliers who also adopt sustainable practices?
Recyclability	Possibility of recycling	Can the product be recycled after use?
	Environmental impact of disposal	Does the product contain toxic substances or substances that are difficult to dispose of in an environmentally safe manner?
Certifications	Sustainability seal	Is it possible to have seals indicating responsible and sustainable business practices?
	Eco-labels	Are there specific eco-labels?
Consumer	Conscious consumption	Does it promote the reuse, repair or exchange of items instead of simply discarding them?

(continued)

Table 12.2 (continued)

Topics	Subtopic	Description
	Educating the consumer	Is there information on how the consumer can use or dispose of the product in a more sustainable way?
Environmental footprint	Carbon emissions	Are the carbon emissions generated throughout its life cycle known?
	Water footprint	Is water consumption in its production and use measurable?

Source Elaborated by the authors, 2024

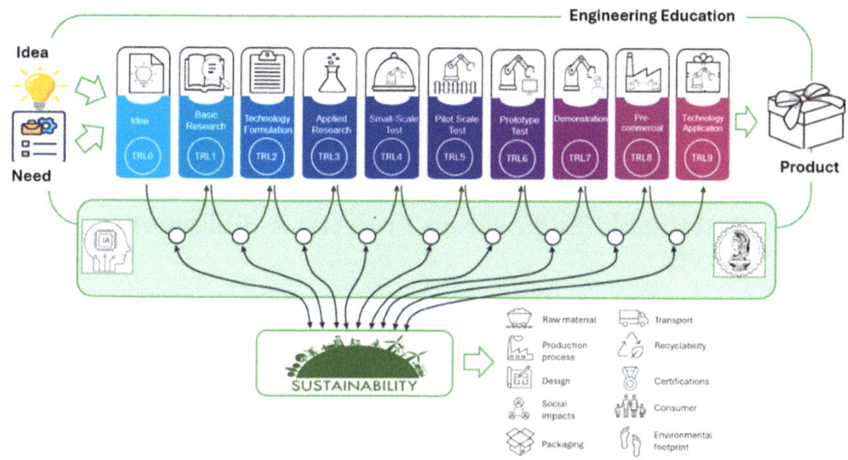

Fig. 12.4 Proposed framework for designing products with integrated environmental and social concerns. *Source* Elaborated by the authors, 2024

analyses as possible to ensure that the next stage of the project is carried out with a sustainable background. And this is carried on until the final product is obtained.

At this point, artificial intelligence makes a difference. The use of reference search tools for the sustainability topics that the product under construction must address at this stage of maturity helps to obtain correct and relevant references for product design. Figure 12.5 shows an example of a ten-step sustainability analysis for a product that is at TRL2 and intends to move on to TRL3.

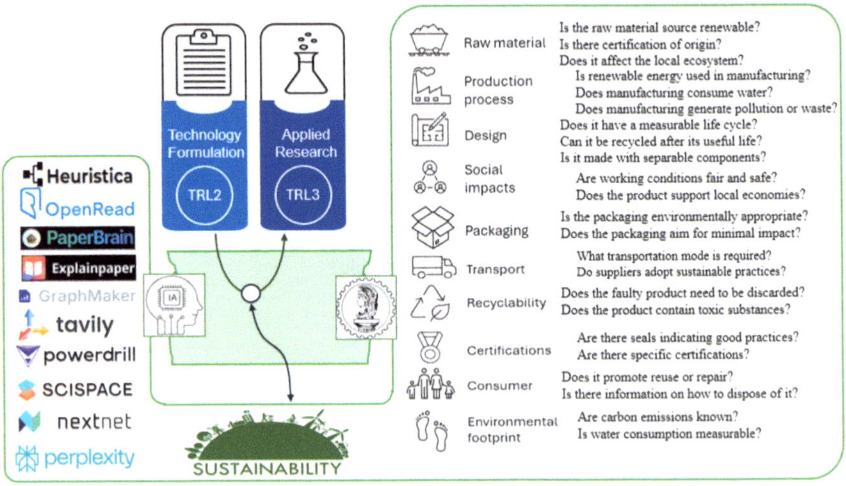

Fig. 12.5 Using AI to analyse sustainability in ten steps. *Source* Elaborated by the authors, 2024

12.6 Sustainable Engineering Projects: A Bibliometric Survey

An integrative review (IR) is a specific method that summarises past empirical or theoretical literature to provide a comprehensive understanding of a particular phenomenon. According to Whittemore and Knafl (2005), the integrative review has the potential to build science by informing research, practice and policy initiatives through the analysis of diverse data sources, which allows for an increased holistic understanding of a given topic. The present study carried out an IR followed by a bibliometric analysis. Bibliometrics makes it possible to organise and quantitatively analyse relevant data such as: the timing of publications; research by area of knowledge, among others. This analysis makes it possible to systematise the results of a study and minimise the occurrence of possible biases when looking at a particular topic (Machado and Santos 2020).

The aim of the integrative literature review, followed by a bibliometric analysis of this study, was to analyse how different leadership styles can influence digital transformation in the health sector, with a focus on the ethical implications associated with data privacy, equity in access to technology and the patient-health professional relationship. The research problem is: how can sustainability be effectively integrated into engineering curricula and which technologies? To plan this review, the method proposed by Kitchenham et al. (2009), as shown in Table 12.3, was adapted.

For the systematic search, the terms detailed in Table 12.4, were used:

From this stage, data were analysed using bibliometrics, where the data collection obtained a total of 4053 documents from Scopus database. The eligible

Table 12.3 Integrative literature review protocol

Item	Subject
1	Data
2	Research context
3	Rcscarch question
4	Databases consulted
5	Inclusion and exclusion criteria
6	Search strategies
7	Quality criteria for article selection
8	Data extraction strategies
9	Data analysis strategies
10	Knowledge dissemination strategies
11	Activity schedule

Source Adapted from Kitchenham et al. (2009)

Table 12.4 Search strategy applied in the databases without considering inclusion criteria

Database	Expression	Amount
Scopus	("product design*" and "sustainability")	4

Source Elaborated by the authors, 2024

articles were from 1990 to 2024. It can be seen that in the period of the COVID-19 pandemic, publications in this area have intensified. In 2021, there were 333 publications in the area, and in 2022, there was an increase in the number of publications to 372. In 2023, there was a significant increase in publications with 446 publications in the area. In 2024, 367 publications in the area were identified up to the date of the survey (November 2024), as shown in Fig. 12.6.

In the set of 1671 articles obtained based on the bibliometric search via Scopus, the seven authors with the most publications in the area were identified. The researcher Michele Germani from the Università Politecnica delle Marche, Ancona, Italy, stands out, with 40 citations in the area, as shown in Fig. 12.7.

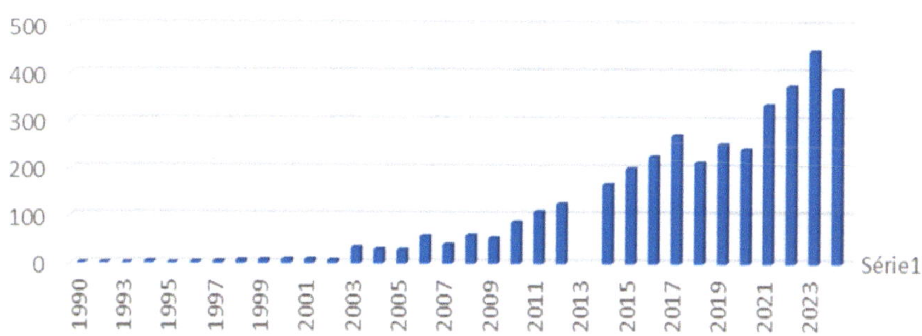

Fig. 12.6 Time distribution of papers. *Source* Elaborated by the authors, 2024

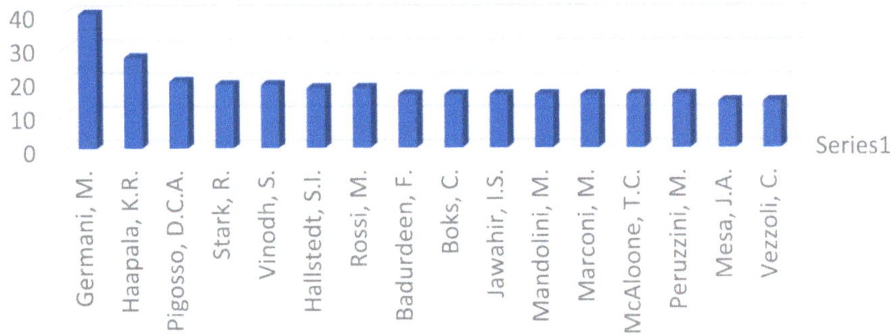

Fig. 12.7 Authors with the highest citation in the area. *Source* Elaborated by the authors, 2024

When analysing the most cited articles, by country, the USA stands out with 696 citations, followed by the UK with 403 citations in the area, as shown in Fig. 12.8.

Based on the general survey, it was also possible to analyse the areas of concentration of the articles obtained by the bibliometric search, in which the research area "engineering" stands out with 2396 and, in second place, "environmental science" with 989, as shown in Fig. 12.9.

Based on the articles retrieved from Scopus, it was possible to identify a total of 1523 different keywords. Of these, the keywords: "Life", "product", "cycle", "design", "sustainability", "sustainable", and "environmental", are the most frequent, as shown in Fig. 12.10.

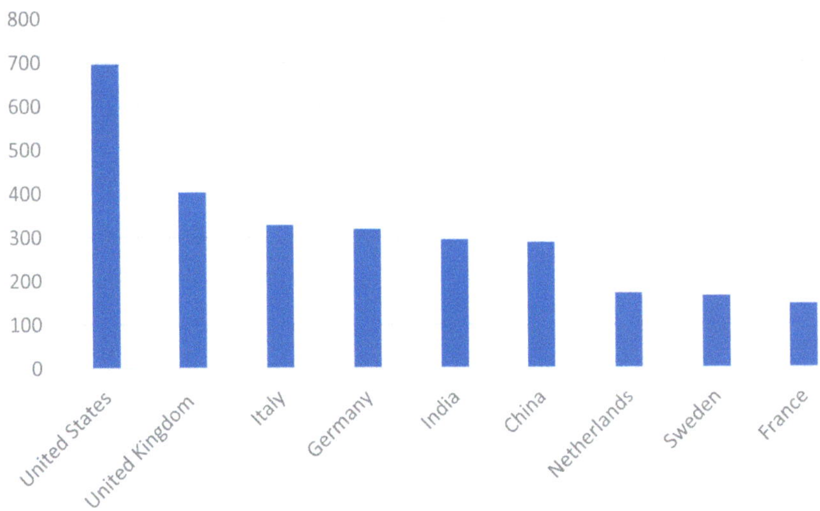

Fig. 12.8 Most cited countries in the area. *Source* Elaborated by the authors, 2024

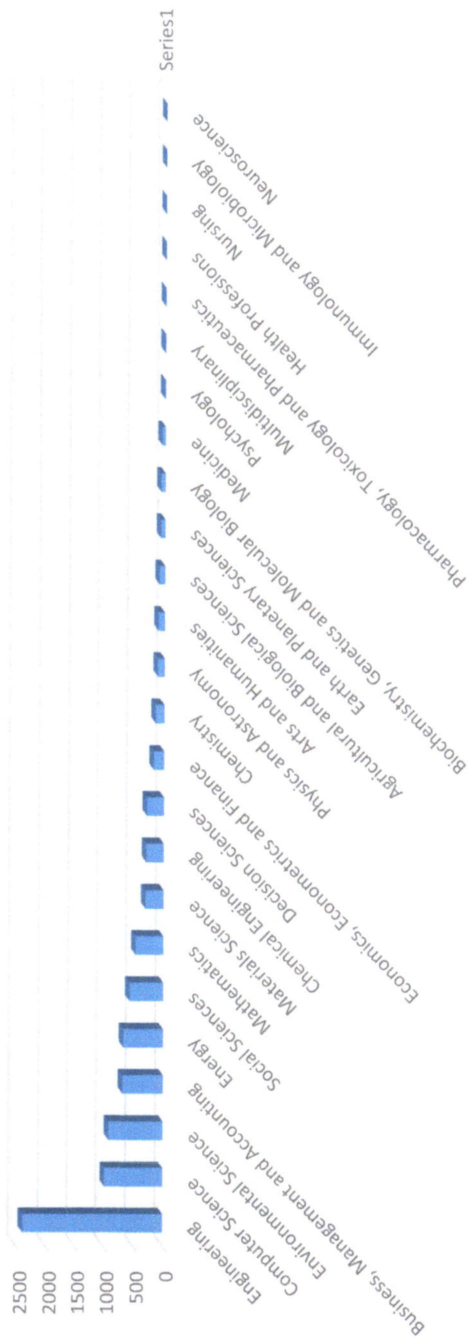

Fig. 12.9 Research concentration area. *Source* Elaborated by the authors, 2024

Fig. 12.10 Tag cloud. *Source* Elaborated by the authors, 2024

Planning of the bibliometric search began on 01 November and ended on 11 November 2024. The articles were selected on 9 November 2024 and the analysis and discussion of the articles was completed on 11 November 2024. A total of 4053 documents were found (conference paper, article, book chapter, review, book, conference review, note, short survey, erratum, editorial, data paper, retracted and letter). Of the total number of documents found, 1671 were articles, and of these, the ten most cited articles in the Scopus database were selected.

In order to answer the question: how can sustainability be effectively integrated into engineering curricula and which technologies? the Scopus database was chosen. The inclusion criteria determined for the research were: studies in the format of full open access articles, the ten most cited articles in the Scopus database and in the English language. The exclusion criteria used, were: theses, dissertations, books and book chapters.

In the second stage of the survey, titles and abstracts of the articles were read in order to detect those that answered the research problem: how can sustainability be effectively integrated into engineering curricula and what technologies? The ten most cited scientific articles selected are listed in Table 12.5.

The first article discusses design strategies and business models to promote a circular economy (Bocken et al. 2016), which can serve as a basis for engineering curricula focused on sustainability. It suggests introducing concepts such as resource cycles (closing, slowing down and narrowing), designing durable and reusable products, and integrating circular business models. Technologies such as remanufacturing systems, design for disassembly and recycling of materials and industrial symbiosis practices can be used to teach applied sustainability in engineering.

The second manuscript explores the integration of innovative technologies in design, manufacturing and service processes, focusing on the concept of "digital twins". It presents a model based on the convergence of physical and virtual data, enabling real-time interaction. This approach, applied to the product life cycle, promotes greater efficiency, intelligence and sustainability. In the context of engineering curricula (Tao et al. 2018), sustainability can be integrated through the use of technologies such as Internet of Things (IoT), big data, virtual and augmented

Table 12.5 Articles selected for the integrative review

No	Scientific articles	Number of citations
1	Bocken et al. (2016). Product design and business model strategies for a circular economy. *Journal of industrial and production engineering, 33*(5), 308–320	2114
2	Tao et al. (2018). Digital twin-driven product design, manufacturing and service with big data. *The International Journal of Advanced Manufacturing Technology*, 94, 3563–3576	2027
3	Davis et al. (2015). Smart manufacturing. Annual review of chemical and biomolecular engineering, 6(1), 141–160	929
4	Reim et al. (2015). Product–Service Systems (PSS) business models and tactics–a systematic literature review. *Journal of Cleaner Production, 97*, 61–75	613
5	Hong and Guo (2019). Green product supply chain contracts considering environmental responsibilities. *Omega, 83*, 155–166	555
6	Ceschin and Gaziulusoy (2016). Evolution of design for sustainability: from product design to design for system innovations and transitions. *Design Studies, 47*, 118–163	542
7	Esmaeilian et al. (2020). Blockchain for the future of sustainable supply chain management in Industry 4.0. *Resources, Conservation and Recycling*, 163, 105,064	533
8	Luttropp and Lagerstedt (2006). EcoDesign and: generic advice for merging environmental aspects into product development. *Journal of Cleaner Production, 14*(15–16), 1396–1408	467
9	Aguilera-Caracuel and Ortiz-de-Mandojana (2013). Green innovation and financial performance: an institutional approach. *Organization & Environment, 26*(4), 365–385	421
10	Maxwell and Van der Vorst (2003). Developing sustainable products and services. *Journal of Cleaner Production, 11*(8), 883–895	420

Source Elaborated by the authors, 2024

reality and artificial intelligence. These tools help with modelling, simulation and data analysis to optimise resources, reduce waste and predict failures, preparing future engineers to face challenges in a sustainable and innovative way.

The third article explores how sustainability can be effectively integrated into engineering curricula through the concept of smart manufacturing. It highlights the use of technologies such as advanced sensors, the Internet of Things (IoT), digital platforms and real-time data analysis to promote energy efficiency, reduce environmental impacts and improve operational performance. Integrating these themes into curricula can be done by aligning teaching practices with the reality of modern manufacturing, using computational modelling tools, automated control and data-based decision-making systems that enable future engineers to adopt sustainable strategies in the development of products and industrial processes (Davis et al. 2015).

The fourth article presents Product-Service Systems (PSS) as a strategic approach to integrating sustainability into engineering curricula. To this end, it proposes sustainable business models, categorised as product-oriented, use-oriented or outcome-oriented, and associated operational tactics, such as product and service design, contracts and marketing (Reim et al. 2015). These strategies promote efficiency in the use of resources, encourage reuse and recycling practices, and consider the full life cycle of products. Technologies such as IoT, digital platforms and data management systems can be incorporated into teaching, enabling future engineers to implement sustainable solutions in line with the circular economy.

In the fifth manuscript, the authors discuss how sustainability can be integrated into engineering curricula, emphasising the role of green supply chains and cooperation contracts. It proposes that concepts of sustainable product design, green marketing and supply chain coordination be incorporated as core elements. In addition, it highlights the use of technologies such as data analysis, IoT and simulations to optimise sustainable practices and minimise environmental impacts (Hong and Guo 2019). The inclusion of practical case studies and the application of advanced technologies in curricula can enable engineers to adopt sustainability strategies effectively.

The sixth article explains the evolution of design for sustainability, highlighting how to integrate sustainability into engineering curricula. It proposes a systemic approach that goes beyond product design to encompass product-service systems, spatial-social innovations and socio-technical transitions. It suggests incorporating concepts such as eco-design, emotionally durable design and nature-inspired design, as well as tools such as life cycle assessment and digital simulations. Technologies such as IoT, data analysis and modelling systems are recommended to facilitate the implementation of these approaches and enable engineers to promote sustainable solutions at different scales (Ceschin and Gaziulusoy 2016).

The seventh article explores how sustainability can be integrated into engineering curricula with a focus on adopting emerging Industry 4.0 and Blockchain technologies. It suggests that the introduction of concepts such as IoT-enabled energy management, smart logistics, green business models and the use of Blockchain to monitor life cycles and encourage sustainable behaviour are essential. Technologies such as IoT sensors, big data analysis, smart contracts and distributed ledger systems are seen as tools to increase efficiency, transparency and process traceability, promoting a circular economy and more sustainable industrial practices. These approaches, incorporated into curricula, can prepare future engineers to solve environmental and social challenges (Esmaeilian et al. 2020).

The eighth article discusses how sustainability can be effectively integrated into engineering curricula through EcoDesign and the "Ten Golden Rules". These guidelines provide a practical basis for incorporating environmental aspects into product development, emphasising actions such as minimising energy and resource consumption, avoiding toxic substances, promoting durability, repair and recycling, and optimising processes throughout the product's life cycle. Technologies such as life cycle analysis (LCA) tools, modular design, recyclable materials and efficient disassembly systems are essential for implementing these strategies. Integrating these practices into teaching prepares engineers to design more sustainable and environmentally conscious solutions (Luttropp and Lagerstedt 2006).

Article number 9 analyses how sustainability can be integrated into engineering curricula using the green innovation approach. It emphasises the importance of technologies and practices that promote energy conservation, pollution prevention, recycling and corporate environmental management. Effective integration of sustainability can be achieved by teaching concepts such as green product design and environmental management practices based on regulatory and cultural norms. Technologies such as IoT, energy efficiency monitoring systems and digital platforms are recommended tools to prepare engineers to address environmental challenges and create innovative and sustainable solutions (Aguilera-Caracuel and Ortiz-de-Mandojana 2013).

The tenth article presents the concept of Sustainable Product and Service Development (SPSD) as an approach to integrating sustainability into engineering curricula. The proposal is based on analysing the complete life cycle of products and services, from conception to disposal, incorporating environmental, social and economic aspects (triple bottom line). Technologies such as life cycle assessment (LCA), modular design, supply chain optimisation and product-service systems (PSS) are highlighted as tools to reduce environmental impacts, promote resource efficiency and increase product durability. Integrating these practices into teaching prepares engineers to create sustainable solutions in line with industrial demands and global sustainability challenges (Maxwell and Van der Vorst 2003).

The articles analysed converge in the proposal to integrate sustainability into engineering curricula using innovative approaches and emerging technologies, with a focus on the life cycle of products, resource efficiency and social responsibility, as shown in Table 12.6.

In response to this study's research problem, the articles explain that sustainability can be integrated into engineering curricula through the inclusion of concepts such as circular economy, eco-design and PSS, as well as the use of Industry 4.0 technologies (IoT, big data, blockchain), life cycle analysis (LCA) tools and modular systems. These practices enable engineers to design innovative, ethical and environmentally responsible solutions, addressing global challenges in a practical and strategic way.

Table 12.6 Proposals to integrate sustainability into engineering curricula

Circular economy and EcoDesign	Strategies such as the circular economy, product-service systems (PSS) and eco-design are fundamental tools. These practices emphasise reducing material consumption, promoting recycling and creating more durable products, using technologies such as IoT, sensors and life cycle analysis (LCA)
Smart manufacturing and IoT	Industry 4.0 technologies such as IoT, big data and digital simulations are presented as solutions for optimising industrial processes, reducing waste and improving energy efficiency. Incorporating these technologies prepares engineers to implement sustainable manufacturing strategies
Blockchain and green management	The application of blockchain in green supply chains is suggested to increase traceability and transparency, promoting ethical and sustainable practices. Data analysis and smart contracts complement this approach
Product-service systems (PSS)	PSS, which combine products and services to fulfil a functionality, are indicated to reduce the production of goods and encourage reuse and maintenance practices. Practical examples include equipment leasing and modular design
Holistic approach	The integration of sustainability requires the inclusion of themes such as social impact and functionality in curricula. The strategy suggests aligning environmental, economic and social criteria with traditional engineering objectives, such as quality and technical feasibility
Pragmatic methods	To facilitate adoption in teaching, a practical approach is recommended, using simple tools such as environmental impact checklists and qualitative simulations. This makes sustainability integration easier

Source Elaborated by the authors, 2024

12.7 Conclusion

The bibliometric analysis carried out in this study provided a broad and quantitative view of academic trends and contributions related to the integration of sustainability into engineering curricula. There has been a significant increase in publications since 2020, driven by the relevance of the topic in a context of global crisis and technological transition. Survey data indicates a concentration of authors, countries and areas of knowledge, highlighting the need for greater geographical diversity and interdisciplinarity in future studies.

In response to the research problem, it was found that sustainability can be effectively integrated into engineering curricula through approaches such as the circular economy, eco-design and product-service systems, combined with the use of emerging Industry 4.0 technologies, including IoT, big data and blockchain.

These tools provide engineers with practical and strategic skills to tackle environmental and social challenges in an innovative and ethical way, in line with global sustainability goals.

As well as highlighting the existing gaps in the integration of these concepts, the study points to viable paths for the practical implementation of methodologies and technologies, suggesting that higher education in engineering should align itself with the global demands of sustainable development. The AI tools highlighted not only offer support for analysing and simulating impacts, but also promote personalisation and efficiency in the learning process. This approach facilitates the creation of more dynamic educational scenarios in which students can apply and validate knowledge in real contexts.

Research into the impact of these approaches on the professional development of engineers and organisations can deepen understanding of the effectiveness of the proposed integrations, promoting progress in both the educational and industrial fields. The modernisation of curricula, combined with investments in technology and teacher training, can boost the training of professionals who are not only technically competent, but also committed to the sustainable future of the planet. In addition, future research should explore the application of these proposals in different socio-cultural contexts, as well as assessing the impact of these initiatives on the employability of graduates and the evolution of industrial practices.

References

ABMES (2019) Resolução N° 2/2019, de 24 de abril de 2019. Institui as Diretrizes Curriculares Nacionais do Curso de Graduação em Engenharia. Retrieved from: https://abmes.org.br/arquivos/legislacoes/Resolucao-CNE-CES-002-2019-04-24.pdf

Abulibdeh A, Zaidan E, Abulibdeh R (2024) Navigating the confluence of artificial intelligence and education for sustainable development in the era of industry 4.0: challenges, opportunities, and ethical dimensions. J Clean Prod 140527. https://doi.org/10.1016/j.jclepro.2023.140527

Aguilera-Caracuel J, Ortiz-de-Mandojana N (2013) Green innovation and financial performance: an institutional approach. Organ Environ 26(4):365–385. https://doi.org/10.1177/1086026613507931

Andrade HS, Chagas Junior MF, Silva MB (2019) Avaliação da maturidade tecnológica: Conceitos e aplicações. Jundiaí-SP: Edições Brasil/Editora Fibra. Retrieved from: https://www.edicoesbrasil.com.br/index.php?modulo=usuarios&acao=login

Arefin MA, Nabi MN, Sadeque S, Gudimetla P (2021) Incorporating sustainability in engineering curriculum: a study of the Australian universities. Int J Sustain High Educ 22(3):576–598. https://doi.org/10.1108/IJSHE-07-2020-0271

Ashford NA (2004) Major challenges to engineering education for sustainable development: what has to change to make it creative, effective, and acceptable to the established disciplines? Int J Sustain High Educ 5(3):239–250. https://doi.org/10.1108/14676370410546394

Associação Brasileira de Normas Técnicas (2015). NBR ISO 16290/2015: Sistemas espaciais—Definição dos níveis de maturidade da tecnologia (TRL) e de seus critérios de avaliação. ABNT

Ausubel D (2003) Aquisição e retenção de conhecimentos: uma perspectiva cognitiva. Plátano Edições Técnicas

Barros KS (Org.) (2020) A vida dos produtos: 12 ciclos de vida de produtos-referência ilustrados para orientar projetos de produtos sustentáveis. Editora UFPB, p 173

Bhutoria A (2022) Personalized education and artificial intelligence in the United States, China, and India: a systematic review using a human-in-the-loop model. Comput Educ: Artif Intell 3:100068. https://doi.org/10.1016/j.caeai.2022.100068

Bocken NM, De Pauw I, Bakker C, Van Der Grinten B (2016) Product design and business model strategies for a circular economy. J Ind Prod Eng 33(5):308–320. https://doi.org/10.1080/21681015.2016.1172124

Bressiani TS, Oliveira MD, Rainha KP, Santana IL, Barros JRPM, Lelis MDFF, Moura PR (2020) Aprendizagem Baseada em Projetos na disciplina tratamento de resíduos e meio ambiente: um estudo de caso. Revista Virtual de Química 12(2):356–367. https://doi.org/10.21577/1984-6835.20200028

Camilleri MA (2022) Strategic attributions of corporate social responsibility and environmental management: the business case for doing well by doing good! Sustain Dev 30(3):409–422. https://doi.org/10.1002/sd.2256

de Carvalho DN, Figueira FL, Lezana AGR, de Souza JA (2018) Revisão estruturada de literatura: Scamper método de geração de ideias. Revista Brasileira Multidisciplinar 21(3):6–29. https://doi.org/10.25061/2527-2675/rebram/2018.v21i3.561

Celestin M, Vasuki M, Sujatha S, Kumar AD (2024) Implementing green technologies to reduce environmental impact: economic and competitive advantages of eco-friendly practices. International Journal of Scientific Research and Modern Education 9(2):33–39. https://doi.org/10.5281/zenodo.13879990

Ceschin F, Gaziulusoy I (2016) Evolution of design for sustainability: from product design to design for system innovations and transitions. Des Stud 47:118–163. https://doi.org/10.1016/j.destud.2016.09.002

Chaves TJ, Mezzari L (2016) *Guia prático para negócios de impacto social* (1a edn). Editora Palco, p 172

Chiu TK, Chai CS (2020) Sustainable curriculum planning for artificial intelligence education: a self-determination theory perspective. Sustainability 12(14):5568. https://doi.org/10.3390/su12145568

Clark W (2019). *The gantt chart: a working tool of management*, vol 1(1), S.l. Alpha Edition

Davis J, Edgar T, Graybill R, Korambath P, Schott B, Swink D, Wetzel J (2015) Smart manufacturing. Annu Rev Chem Biomol Eng 6(1):141–160. https://doi.org/10.1146/annurev-chembioeng-061114-123255

de Bem Machado A, Richter MF (2020) Sustentabilidade Em Tempos De Pandemia (COVID-19). RECIMA21-Revista Científica Multidisciplinar 1(2):264–279. https://doi.org/10.47820/recima21.v1i2.25

de Bem Machado A, Richter MF, Figueras LVDF (2021) Sustentabilidade e economia circular: implicações para retomada verde. Revista Panorâmica online 34. Retrieved from https://periodicoscientificos.ufmt.br/revistapanoramica/index.php/revistapanoramica/article/view/1424

de Oliveira RA, Hipólito GMB, Pontes RDFF, Ferreira PHN, Moreira RS, Plácido J, Tovar LP (2023) Transdisciplinary competency-based development in the process engineering subjects: a case study in Brazil. Educ Chem Eng 44:133–154. https://doi.org/10.1016/j.ece.2023.05.007

Debrah C, Chan AP, Darko A (2022) Artificial intelligence in green building. Autom Constr 137:104192. https://doi.org/10.1016/j.autcon.2022.104192

Dimitriadou E, Lanitis A (2023) A critical evaluation, challenges, and future perspectives of using artificial intelligence and emerging technologies in smart classrooms. Smart Learn Environ 10(1):12. https://doi.org/10.1186/s40561-023-00231-3

Elsawy M, Youssef M (2023) Economic sustainability: meeting needs without compromising future generations. Int J Econ Financ 15(10):1–23. https://doi.org/10.5539/ijef.v15n10p23

Esmaeilian B, Sarkis J, Lewis K, Behdad S (2020) Blockchain for the future of sustainable supply chain management in Industry 4.0. Resour Conserv Recycl 163:105064. https://doi.org/10.1016/j.resconrec.2020.105064

Gray CF, Larson EW (2010) Gerenciamento de projetos: o processo gerencial, 4th edn. McGraw-Hill, p 601

Hariram NP, Mekha KB, Suganthan V, Sudhakar K (2023) Sustainalism: an integrated socio-economic-environmental model to address sustainable development and sustainability. Sustainability 15(13):10682. https://doi.org/10.3390/su151310682
Hong Z, Guo X (2019) Green product supply chain contracts considering environmental responsibilities. Omega 83:155–166. https://doi.org/10.1016/j.omega.2018.02.010
Instituto Butantan (2014) Guia prático de descarte de resíduos. Editora IB, p 48. Retrieved from: https://gestaoderesiduos.ufsc.br/files/2016/02/Butantan_guia_pratico.pdf
Kar AK, Choudhary SK, Singh VK (2022) How can artificial intelligence impact sustainability: a systematic literature review. J Clean Prod 376:134120. https://doi.org/10.1016/j.jclepro.2022.134120
Kitchenham B, Brereton OP, Budgen D, Turner M, Bailey J, Linkman S (2009) Systematic literature reviews in software engineering–a systematic literature review. Inf Softw Technol 51(1):7–15. https://doi.org/10.1016/j.infsof.2008.09.009
Kröplin J, Huber T, Geis C, Braun B, Fritz T (2022) ESurgery—digital transformation in surgery, surgical education and training: survey analysis of the status quo in Germany. Eur Surg 54(5):249–258. https://doi.org/10.1007/s10353-022-00747-x
Kuazaqui E (2016) Planejamento estratégico. Cengage Learning, p 90
Luttropp C, Lagerstedt J (2006) EcoDesign and the ten golden rules: generic advice for merging environmental aspects into product development. J Clean Prod 14(15–16):1396–1408. https://doi.org/10.1016/j.jclepro.2005.11.022
Machado ADB, Santos JVAD (2020) Revisão integrativa: intersectando concepções de educação ambiental com referencial teórico marxiano. Educação UFSM 45
Maxwell D, Van der Vorst R (2003) Developing sustainable products and services. J Clean Prod 11(8):883–895. https://doi.org/10.1016/S0959-6526(02)00164-6
Moni SM, Mahmud R, High K, Carbajales-Dale M (2020) Life cycle assessment of emerging technologies: a review. J Ind Ecol 24(1):52–63. https://doi.org/10.1111/jiec.12965
Monteiro F, Sousa A (2024) An educational board game to promote the engagement of electric engineering students in ethical building of a sustainable and fair future. J Environ Educ 55(2):138–152. https://doi.org/10.1080/00958964.2023.2259832
Nigam S (2023). CPM e PERT, vol 1(1). Edições Nosso Conhecimento. https://doi.org/10.1016/j.jclepro.2022.134120
Nunez JM, Lantada AD (2020) Artificial intelligence aided engineering education: state of the art, potentials and challenges. Int J Eng Educ 36(6):1740–1751. Retrieved from: https://www.ijee.ie/1atestissues/Vol36-6/03_ijee3984.pdf
Padilla-Rivera A, Russo-Garrido S, Merveille N (2020) Addressing the social aspects of a circular economy: a systematic literature review. Sustainability 12(19):7912. https://doi.org/10.3390/su12197912
Pazmino AV (2015) Como se cria: 40 métodos para design de produtos, 1st edn. S/l: Editora Blucher, p 278
Piatti TM, Rodrigues RAF (2005) Plásticos: características, usos, produção e impactos ambientais. Ed. UFAL, p 51. Retrieved from: https://usinaciencia.ufal.br/multimidia/livros-digitais-cadernos-tematicos/plasticos_caracteristicas_usos_producao_e_impactos_ambientais.pdf
Raeburn A (2024) Processo de desenvolvimento de produtos: os 6 estágios. Retrieved from: https://asana.com/pt/resources/product-development-process
Ramírez-Márquez C, Posadas-Paredes T, Raya-Tapia AY, Ponce-Ortega JM (2024) Natural resource optimization and sustainability in society 5.0: a comprehensive review. Resources 13(2):19. https://doi.org/10.3390/resources13020019
Reim W, Parida V, Örtqvist D (2015) Product-Service Systems (PSS) business models and tactics–a systematic literature review. J Clean Prod 97:61–75. https://doi.org/10.1016/j.jclepro.2014.07.003
Richter MF, de Bem Machado A (2023) Circular economy and sustainable consumption behavior for green recovery. Multitemas 183–201. https://doi.org/10.20435/multi.v28i70.3948
Sachs J (2015) The age of sustainable development. Columbia University Press

Singha R, Singha S (2024) Application of experiential, inquiry-based, problem-based, and project-based learning in sustainable education. In: Teaching and learning for a sustainable future: innovative strategies and best practices. IGI Global, pp 109–128. https://doi.org/10.4018/978-1-6684-9859-0.ch006
Staniškis JK, Katiliūtė E (2016) Complex evaluation of sustainability in engineering education: case & analysis. J Clean Prod 120:13–20. https://doi.org/10.1016/j.jclepro.2015.09.086
Talan A, Tyagi RD, Surampalli RY (2020) Social dimensions of sustainability. Sustain Fundam Appl 183–206. https://doi.org/10.1002/9781119434016.ch9
Tang W, Mai L, Li M (2023) Green innovation and resource efficiency to meet net-zero emission. Resour Policy 86:104231. https://doi.org/10.1016/j.resourpol.2023.104231
Tao F, Cheng J, Qi Q, Zhang M, Zhang H, Sui F (2018) Digital twin-driven product design, manufacturing and service with big data. Int J Adv Manuf Technol 94:3563–3576. https://doi.org/10.1007/s00170-017-0233-1
Van den Beemt A, MacLeod M, Van der Veen J, Van de Ven A, Van Baalen S, Klaassen R, Boon M (2020) Interdisciplinary engineering education: a review of vision, teaching, and support. J Eng Educ 109(3):508–555. https://doi.org/10.1002/jee.20347
Wang L, Liu Z, Liu A, Tao F (2021) Artificial intelligence in product lifecycle management. Int J Adv Manuf Technol 114:771–796. https://doi.org/10.1007/s00170-021-06882-1
Weiland S, Hickmann T, Lederer M, Marquardt J, Schwindenhammer S (2021) The 2030 agenda for sustainable development: transformative change through the sustainable development goals? Politics and Governance 9(1):90–95. https://doi.org/10.17645/pag.v9i1.4191
Whittemore R, Knafl K (2005) The integrative review: updated methodology. J Adv Nurs 52(5):546–553. https://doi.org/10.1111/j.1365-2648.2005.03621.x
Wilson A, Kask R, Ming LW (2024) Exploring circular digital economy strategies for sustainable environmental, economic, and educational technology. Int Trans Educ Technol (ITEE) 2(2):129–139. https://doi.org/10.33050/itee.v2i2.579

Economic Sustainability Potentiated by Artificial Intelligence: Transformative Impacts of Intelligent Participatory Budgeting on Policy, Digital Inclusion, and the Economy

13

Johnny Camello

Abstract

Intelligent Participatory Budgeting (i-PB) is an innovative approach integrating artificial intelligence (AI) technologies to enhance citizen participation in formulating and managing public budgets. This chapter explores the cross-cutting impact of AI on i-PB, addressing its interface with the economy, sustainable digital inclusion, and the necessary policies and regulations. The implementation of i-PB allows for an in-depth analysis of economic data, promoting more informed and efficient decisions in resource allocation. Additionally, using AI digital tools facilitates the inclusion of a broader diversity of citizens in the participatory process, ensuring that previously marginalized voices are heard and considered. The chapter also discusses the political and regulatory implications, emphasizing the need for a legal framework that supports transparency and accountability in using AI in the i-PB. By holistically integrating these technologies, i-PB improves efficiency and equity in public resource management and strengthens citizens' trust in the government, promoting a more sustainable and inclusive economy. This study provides a comprehensive overview of the opportunities and challenges in implementing i-PB, highlighting its transformative potential for modern economies.

Keywords

Artificial intelligence • Intelligent participatory budgeting • Citizen participation • Public budgets

J. Camello (✉)
Iscte Instituto Universitário de Lisboa, Lisbon, Portugal
e-mail: johnnycamello@gmail.com

M. J. Sousa et al. (eds.), *Economic Sustainability Potentiated by Artificial Intelligence*, Contributions to Environmental Sciences & Innovative Business Technology, https://doi.org/10.1007/978-981-96-8089-4_13

13.1 Introduction

Economics, as a science that organizes the production, distribution, and consumption of goods and services, has long been a central pillar of human civilization. From the earliest barter systems to modern global economies, its role has been to ensure the survival and continuity of human life. However, despite notable successes, the current economic model faces criticism for social inequality and environmental destruction (Stiglitz 2003). The unsustainability of unlimited growth and the unbridled exploitation of natural resources indicate the urgent need for a transition to a more balanced economy, which integrates economic growth with environmental sustainability (Camello 2024a; Diamond 2006; Klein 2014).

The public budget plays a crucial role in this transition. It sustains state infrastructure and can serve as an engine for sustainable economic development. Historical examples, such as the New Deal and the Marshall Plan, demonstrate how public investments can be decisive for economic recovery and social stability in times of crisis (De Long and Eichengreen 1991; Kennedy and Woodward 2005). However, the misuse of public resources, often associated with corruption and inefficient allocation, compromises the effectiveness of these policies (George W. Bush Presidential Center 2024).

In addition, contemporary global challenges, such as climate crises and economic inequalities, require more integrated and efficient approaches. The use of technologies such as artificial intelligence, in initiatives such as the i-PB, offers a new perspective. i-PB can make public management more transparent, inclusive, and adaptive, overcoming the limitations of the traditional system and promoting multilevel and multicultural governance that responds more effectively to global and local needs (Sousa et al. 2024).

This chapter proposes that implementing i-PB is key to transforming the current economic model into a more sustainable, inclusive, and solidary system, facing global challenges in an integrated way and promoting a more efficient public administration.

13.1.1 The Evolution of the Economy: Challenges and Achievements in the Current Model

The classic definition of economics is that it is the science that studies how societies organize the production, distribution, and consumption of goods and services. Economics is a human invention created to ensure the survival and sustenance of humans on Earth. From the earliest barter and subsistence systems to modern global economies, it organizes the production and distribution of essential resources to meet human needs. Thus, the economy is a central pillar for the continuity of human life on a planet with finite resources (Samuelson and Nordhaus et al. 1998). On a macrolevel, it has fulfilled its role so far. However, on a microlevel, there is significant global collective frustration regarding its outcomes

(Stiglitz 2003). A common criticism of the current economic system is that it neither promotes financial and social equality nor protects the environment. This is a valid and pertinent criticism. However, when we analyze our process of civilizational evolution, these criticisms may seem premature. They imply that the system should already be perfect and complete, which is unrealistic because time does not stand still, and we are in a continuous process of social and economic evolution (Schneider 2019).

It was the Industrial Revolution of the eighteenth century that brought technological innovations which transformed production and social organization. Before this period, people relied on hunting, gathering, and subsistence farming to survive, behaving like grasshoppers by depleting resources and causing destruction to everything and everyone to survive. With the Industrial Revolution, the emergence of wage labor replaced the need for direct resource acquisition, allowing individuals to exchange their labor power for wages. This made it possible to purchase goods and services in a structured market, promoting a more dynamic economic cycle that was less dependent on the direct exploitation of the natural environment (Polanyi 2021; Smith 2021; Thompson 2021).

The current economic system is far from perfect, marked by environmental devastation and profound global economic and social inequalities. The unbridled exploitation of natural resources and the concentration of wealth have caused significant damage to both ecosystems and societies. Yet, despite these glaring flaws, the system has managed to remarkably sustain and support nearly eight billion people. It has driven advances in health, infrastructure, and quality of life in various parts of the world. In this sense, as human civilization, it is undeniably a success (Diamond 2006; Klein 2014; Piketty 2017; Raworth 2022; Stiglitz 2003).

13.1.2 The Transition to a Sustainable Economy: The Crucial Role of the Public Budget and the Challenges of Corruption

However, as an ever-evolving system, the economy must adapt by integrating economic growth with environmental sustainability and global social equity. Thus, the transition to a sustainable economy has become indispensable, and we already possess the necessary technologies to facilitate this change. The current economic model, which relies on unlimited growth and the unbridled exploitation of resources, is unsustainable given the planet's limits. This approach is leading to the collapse of ecosystems and contributing to the extinction of countless species. A sustainable economy, by contrast, seeks to balance development with environmental preservation, promoting the conscious and renewable use of natural resources. This shift is urgent, especially in light of the critical challenges facing humanity in this century (Camello 2024b; Diamond 2006; Klein 2014; Raworth 2022).

The transition to a sustainable economy is a challenge that requires the active participation of all sectors of society. Governments and citizens must work together to build a sustainable present and future. The public budget is fundamental and strategic in this effort. Using the public budget as a driver of economic development plays a crucial role in ensuring the sustainability and progress of nations, especially in times of crisis. The public budget not only supports the state's infrastructure but also acts as a catalyst for investment and economic growth, promoting the well-being of the population (Daly 1991; Hawken 2017; Jackson and Charles 2017; Klein 2014; Raworth 2022; Stiglitz 2013).

Throughout history, emblematic examples illustrate how public intervention through the budget has been decisive in addressing macroeconomic and social challenges. The New Deal in the USA, implemented during the Great Depression in the 1930s, is a paradigmatic case of how the injection of public resources can revitalize a deeply fragile economy. This program not only created jobs but also fostered the recovery of strategic sectors, expanding the American social safety net. Similarly, the Marshall Plan, aimed at the reconstruction of Europe after the devastation of the Second World War, represented a coordinated and massive allocation of financial resources, which was decisive for the resurgence of European economies and the political stabilization of the continent (De Long and Eichengreen 1991; Kennedy and Woodward 2005).

In more recent crises, the economic stimulus package adopted in 2008 to deal with the global financial collapse and, more recently, NextGenerationEU, the European Union's post-COVID-19 pandemic recovery plan, are contemporary examples of the vital importance of the public budget in preserving jobs, boosting growth, and fostering innovation. The allocation of public resources on these occasions was not only strategic in avoiding prolonged recessions but also fundamental in protecting the population from the devastating impacts of the crises, sustaining essential sectors of the economy, and promoting social cohesion (*NextGeneration EU—European Union* 2024; Paulson 2010).

In addition to its role in crisis recovery, the public budget finances the essential public services and equipment that sustain the social fabric, such as health, education, transportation, security, and social security. These areas are fundamental to ensure a dignified quality of life for citizens, promoting human development and basic equality of opportunities. Access to quality public services, often invisible in the most immediate economic analyses, is one of the pillars of any modern and fair society. Therefore, the public budget is not just a financial tool; it is an essential lever for social and economic development. It enables not only the maintenance of jobs and economic recovery in times of crisis but also sustains the basic infrastructure that ensures the well-being of society. Its strategic and efficient application is, therefore, fundamental for citizens to enjoy quality public services and for nations to build sustainable and inclusive development (Atkinson 2014; Musgrave and Musgrave 1980; Samuelson and Nordhaus 2010; Sen 2011; Stiglitz and Rosengard 2015).

However, despite the undeniable importance of the public budget as an essential tool for investment, economic growth, and the maintenance of public services and infrastructure for the population, its use is not exempt from scathing criticism. One of the most recurrent objections refers to the misuse of public resources, often related to corruption and the inefficient allocation of funds. The global financial impact of these practices is alarming, with estimates indicating that corruption consumes between $1.5 trillion and $2 trillion annually, representing about 2% of the world's GDP. In developing countries, the amounts lost to corruption and illicit financial activities total approximately $1.26 trillion annually, which could significantly improve the quality of life for millions of people living in extreme poverty. The impact of corruption is not limited to economic losses. It also undermines governance, eroding public trust in institutions and exacerbating social inequalities. Moreover, illicit practices such as money laundering move between $800 billion and $2 trillion globally, with the USA being one of the main entry points for this 'dirty money' (George W. Bush Presidential Center 2024; World Economic Forum 2019).

These activities compromise the effectiveness of public policies and drain resources that could be used for sustainable development. The negative impact of corruption goes beyond economic and social factors, also having serious environmental implications. Corruption, often linked to the awarding of irregular contracts and non-compliance with environmental regulations, facilitates environmental degradation and the unsustainable use of natural resources. In many countries, the illegal exploitation of natural resources, such as logging and mining, is facilitated by corrupt systems. This causes irreversible damage to the environment and compromises future sustainability, especially in the face of the unprecedented and simultaneous challenges that humanity is currently facing (Camello 2024b; Williams and Le Billon 2017; Yakubu 2024).

13.1.3 Contemporary Global Challenges and the Need for New Approaches to Overcome Them

We are facing the greatest civilizational challenges ever imposed on humanity. Climate, economic, social, technological, and political issues are arising simultaneously, often antagonistically, and on a global scale. To combat them, it will be necessary to eliminate waste and corruption in public budgets.

According to Camello (2024a, b):

> *The contemporary challenges in the environmental, economic, technological, social, and political spheres are unprecedented. This occurs, firstly, because of its simultaneous occurrence. Secondly, and more worryingly, is the fact that they occur in a scenario of a global population of 8 billion inhabitants. In other words, these challenges are not unique to some, as addressing them is not restricted to national borders; They have a global nature. Thus, to overcome these challenges, it is imperative to develop global and transnational political solutions (p. 01).*

These challenges include climate crises (such as global warming), economic inequalities (the unequal global distribution of wealth), political tensions (the reconfiguration of geopolitical power structures), technological revolutions (driven by artificial intelligence and quantum computing), and social crises (such as forced migration and involuntary immigration). The yellow light is on because, even though there is incontrovertible evidence of numerous actions undertaken by international organizations, governments, government blocs, non-profit organizations, companies, and civil society to address global challenges, large-scale initiatives, such as the 2030 Agenda, composed of 17 Sustainable Development Goals (SDGs), established by the United Nations (UN) in 2015, exemplify both the collective commitment to dealing with these issues and the limitations that persist in their implementation. Although these initiatives represent a milestone in the global mobilization of public policies for sustainable, solidary, and inclusive development, they also reveal the dichotomy between the desire to act and the inertia that often prevents the achievement of the proposed goals (Aaronson 2013; Kissinger 2014; Klein 2014; Li 2021; Piketty 2017; Sachs 2015; Sassen 2014).

In principle, this demonstrates that the complexity and scale of these challenges call into question the capacity and interest of the current economic and political model in facing them. Current models, with their bureaucracy, structural limitations, and cynical, situational rhetoric, have proven inadequate to address the depth and global nature of these crises. The impression one gets is that each model focuses solely on its interests and immediate problems, forgetting that the pollution from chimneys spreads with the wind and the sewage discharged into rivers and seas is always in constant movement, affecting everyone. Economic inequality and poor income distribution have a profound global impact, contributing to increased poverty, social exclusion, and political instability (Beck 2009; Harvey 2011; Klein 2014; Piketty 2017; Stiglitz 2003).

In many regions, these inequalities limit access to education, health, and employment opportunities, perpetuating cycles of poverty and marginalization. In addition, the excessive concentration of wealth in a few hands generates social and economic tensions, making it difficult to achieve sustainable growth and reduce disparities between and within countries. These factors together affect social cohesion and balanced global development. This highlights the need for new approaches that integrate multiple levels of governance and promote more effective global coordination (Atkinson 2015; Milanović 2018; Piketty 2017; Sen 2011; Stiglitz 2013). As Kostakopoulou (2008: 2) rightly noted: ‘It is not easy to reconcile the challenges and problems of the twenty-first century with the resources of the twentieth century and the models of the nineteenth century’.

13.1.4 The Potential of i-PB: Artificial Intelligence for Sustainable and Inclusive Governance

In this context, AI-based solutions like i-PB offer the potential to promote more transparent, inclusive, and efficient decision-making processes. By leveraging data-driven insights, these systems can allocate resources more accurately, ensuring that responses to economic, environmental, and social challenges are sustainable and adaptable to the ever-changing global landscape. This approach is crucial for addressing contemporary challenges, as it promotes sustainable economic development while balancing local, regional, national, and global needs. i-PB offers a new approach, enabling the construction of public budgets that operate in a multilevel manner—covering local, regional, national, and international spheres—and are multicultural, considering not only economic, but also political, technological, social, cultural, and environmental aspects. This flexibility is possible due to the participatory nature and digital mechanics of i-PB, which uses artificial intelligence to integrate different voices. By analyzing large amounts of data, i-PB can simulate and articulate solutions that respond more efficiently to multilevel demands. In addition, i-PB allows for a more transparent, inclusive, and adaptable governance model, overcoming the limitations of the traditional system and offering an integrated and coordinated response to the management and administration of collective demands by public and private organizations (Sousa et al. 2024).

13.2 Participatory Budgeting

The Participatory Budget (PB) emerged in the city of Porto Alegre-RS, Brazil, in 1989, one year after the promulgation of the new Federal Constitution, which reestablished democracy in the country after 21 years of military dictatorship. The new Constitution allowed states and municipalities to create mechanisms for popular participation in public management, especially PB, which allowed citizens to actively engage in the preparation and execution phases of the public budget, complementing traditional political representation. The PB model in Porto Alegre stood out for its pioneering spirit, longevity, and innovation, becoming a reference both in Brazil and internationally. Recognized by organizations such as the United Nations (UN) and the World Bank, PB was considered an example of good governance and popular participation. Since its inception, the PB model has expanded and is now used as a budget management tool in thousands of cities worldwide, as well as being implemented in schools and other institutions. Over more than three decades, PB has adapted to diverse social, economic, cultural, and political contexts, demonstrating its ability to strengthen communities and promote citizenship by allowing citizens to directly influence the use of public resources (Fedozzi 2009).

The PB of Porto Alegre stood out as an innovative model of public management, capable of promoting social transformation. By granting the population the power to decide on the allocation of public resources, the PB transformed the reality of several communities. The active participation of citizens in defining priorities—such as paving streets, expanding basic sanitation, and building schools—resulted in projects that significantly improved infrastructure and public services in historically neglected areas. This experience demonstrates that participatory democracy is an effective tool for reducing inequalities and building fairer and more inclusive cities (Almeida 1997; Baiocchi 2005; Fedozzi 2009; World Bank 2003).

Over the years, the PB of Porto Alegre has become a global reference as a model for defining, managing, and overseeing the public budget, promoting decentralization of decisions and the direct participation of the population. By including various social actors in the process of defining priorities and applying public resources, the PB contributed to reducing practices such as patrimonialism and corruption. Its resilience, even in the face of challenges and crises, demonstrates its importance as an effective mechanism for strengthening democracy and ensuring that public resources are used more fairly and transparently (Camello 2019; Fedozzi 2002; Participatory Budgeting Project 2016).

13.2.1 Intelligent Participatory Budgeting

In this context, the i-PB is an innovative proposal. i-PB has the potential to transform the traditional economy by promoting fairer, more sustainable, and supportive economic ecosystems. This is because decisions are based on the real needs of communities, ensuring a more balanced, transparent, and intelligent distribution of resources. In addition, the use of new technologies—such as artificial intelligence and quantum computers—will make it possible to simulate different scenarios in the short, medium, and long term, optimizing the allocation of resources and making public policies more efficient, inclusive, and integrated in addressing global challenges. The i-PB enables the creation of a new economic system that sustains the current population, while seeking equity and sustainability for future generations (Sousa et al. 2024).

13.2.2 Active Citizen Participation

Today's technologies, such as social networks, digital platforms, and mobile telephony, offer unprecedented informational and communicational power. Through electronic devices—such as computers, tablets, and cell phones—it is possible to create information and communication channels that reach individuals, groups, or entire populations without geographical, linguistic, or cultural barriers. i-PB leverages these technologies to ensure the inclusion of all individuals and groups in active participation in each of its phases. Thus, it creates a bidirectional flow of data, which, when processed by the intelligent core of i-PB, makes it possible to suggest solutions, allowing decisions to be decentralized while ensuring that each member of the community is held accountable. Citizens are therefore authors, users, and inspectors of their budget decisions at local, regional, national, and international levels. This decentralization is essential to face the challenges humanity faces in this century. So far, it is clear that both economic and political groups have their interests and demonstrate a worrying inertia in the face of economic, political, social, and environmental issues that require urgent attention. In this context, the representative and participatory structure of i-PB decentralizes decision-making and distributes responsibilities, which are currently restricted to economic and political groups (Sousa et al. 2024).

13.2.3 Territory, Population and Representativeness

In the context of i-PB, a territory is the platform where Participatory Budgeting processes take place. The term refers to both the geographic space and the population inhabiting that space. The territory can be divided into different levels—local, regional, national, and even international—and plays a crucial role in producing data that will be processed by artificial intelligence at i-PB. The territory, together with the population, generates collective demands that are essential to the i-PB process, allowing AI to identify patterns and interconnections between these demands across various geographical scales and dimensions (such as social, economic, cultural, etc.). Representativeness refers to the inclusion and participation of citizens in all stages of the Participatory Budgeting process. The decentralization of decisions allows local demands to be taken into account, ensuring that the needs of various social, economic, and cultural groups are met equitably and inclusively (Sousa et al. 2024).

13.2.4 Processing of Collective Claims and Results

Once territories, populations, and local, regional, national, and international representatives have been defined, the i-PB is carried out to survey collective demands. However, what are collective demands?

According to Camello (2019):

> *It is the sum of the individual demands of citizens for public services and devices, financed through the apportionment of their costs (taxes), among the individuals who make up this social nucleus. The demands can be current, strategic, or emergency. Short, medium, or long term. Local, regional, national or international. Integrated or isolated at the geopolitical level. In the structural aspect, they are always multidimensional and processed by organizations (State, NGOs, and private companies), responsible for producing and delivering to their funders, the citizens, the requested public services and equipment (pp. 40–41).*

The processing of demands follows a logical and pragmatic sequence, which is detailed in Table 13.1:

Table 13.1 Dynamics of multilevel strategic planning

Multidimensional collective demands				
Location	Local	Regional	National	International
Priority	1	2	3	4
Term	Short	Short	Short	Short
	Medium	Medium	Medium	Medium
	Long	Long	Long	Long
Integration	Global multilevel integration			

Source https://repositorio.iscte-iul.pt/bitstream/10071/19338/1/master_johnny_camello.pdf

In this phase, i-PB integrates several data sources to generate a comprehensive and integrated view of each demand. The reports generated provide a systemic view of the population's needs, facilitating the identification of optimization opportunities and strategic decision-making aligned with the multilevel and multidimensional objectives of the population (Sousa et al. 2024).

13.2.5 Definition of Projects and Programs for Execution

It is at this stage that i-PB demonstrates its full potential, providing indispensable support to the government in defining the most effective strategies for executing approved projects and programs. Based on the information processed by i-PB, it becomes possible to prioritize demands and allocate resources more efficiently, ensuring the technical, financial, and environmental feasibility of each initiative. It is important to highlight the potential for integration among various projects and programs. The i-PB algorithm can identify synergies between different initiatives, facilitating the search for broader funding sources. For example, an asphalting project in a rural area, which may not be feasible with local resources, could receive support from regional, national, or international programs. Similarly, i-PB can identify opportunities for technical cooperation between projects and programs, which can accelerate the transfer of technologies and best practices (Sousa et al. 2024).

13.2.6 Enforcement Management, Supervision, and Transparency

Regarding the management, execution, inspection, and accountability of the projects and programs chosen by the population in i-PB, artificial intelligence greatly facilitates these tasks. The algorithms can monitor variables such as schedules, expenses, resources, risks, deviations, and budgets with high precision, making the process more efficient and transparent. In addition, AI enables the active participation of the population at all stages. Through multiple channels of communication, both analog and digital, i-PB can interact with community members individually and collectively at all levels. This allows the population to monitor the execution of i-PB, closely tracking the performance of those involved and ensuring their commitment to delivering the requested public services and infrastructure (Sousa et al. 2024).

13.2.6.1 Economy and i-PB: Working Together for Sustainability

The current public budget model allows public institutions to plan budgets to cover their operating costs and finance projects and programs aimed at providing public services and equipment to the population. However, many of these projects and programs end up being included in budgets to serve the interests of economic and political groups that influence these budgets due to their participation in periodic elections. This means that, in many cases, the public budget is already skewed to serve the interests of a restricted group rather than the real needs of the population, carrying with it all the sordid and harmful tricks of corruption and inefficiency (Besley and Persson 2011; Olson 2003; Pereira 1996; Piketty 2017; Ricardo 2018; Stiglitz 2013).

i-PB eliminates this distortion and dysfunction of public budgets, ensuring the delivery of budgeted public services and equipment. Unlike traditional methods, it does not pass through the 'filter of illusions' about what management and public administration are, removing all superficiality from the process. i-PB treats the management of collective demands clearly and objectively, understanding it as the identification of the needs of a group of individuals in each territory. In this simple and direct way, i-PB focuses on the real demands, the resources needed, the environmental impacts in the short, medium, and long term, and how to make these demands sustainable and solidaristic. The ultimate goal is not only to meet the needs of current generations but also to ensure the survival and well-being of future generations. And how is this possible? Simple: by using science, technology, and active, responsible citizen participation (Sousa et al. 2024).

13.2.6.2 Making It Rain in the Desert

Considering the functioning of i-PB, we can say that it allows for the identification and measurement of the geographical, financial, logistical, material, and environmental potentialities and limitations of each collective demand. i-PB processes these demands precisely, considering the unique characteristics of the territories and populations involved. This means that we can understand the capabilities and challenges of each region and population group, adapting resources according to their needs. Throughout its history, human society has developed the ability to control fundamental variables for its survival, such as food production, the organization of habitable environments, and access to health services and transportation. This control has made it possible to face climatic, demographic, and technological challenges, ensuring the continuity of life in different contexts (Camello 2024a; Sousa et al. 2024).

Currently, technological and biotechnological advances make it possible to produce strawberries in the Sahara Desert. The desalination of seawater, along with the use of nanotechnology, has made it possible to create large food farms in some of the most arid countries in the world, such as Jordan and the United Arab Emirates (Agricultura 2017; Peerzada 2018). In China, the 'green miracle' occurred thanks to the direct action of humans, who reforested thousands of hectares of desert land and created natural barriers against sandstorms, improving the quality of life for millions of people (Vidal 2023). In this sense, i-PB cannot only improve public budget planning to ensure a sustainable economic ecosystem in the present and future but also enable the regeneration and transformation of already degraded and inhospitable territories, making them suitable for population occupation and sustainable economic development (Sousa et al. 2024).

The ability to simulate scenarios, linking the public budget to sustainable economic development, offered by i-PB—initially through artificial intelligence and in the future through quantum computing—allows, in an unprecedented way, for the balancing of economic, environmental, and social development, using statistical data as a foundation. In this context, its multilevel and multidimensional implementation would make it possible to predict budget revenues that would ensure a sustainable economy, both in normal scenarios and in crises (economic, climatic, environmental, social, etc.) (Sousa et al. 2024).

13.2.6.3 Digital Inclusion at i-PB

Undoubtedly, the Achilles' heel of traditional Participatory Budgeting (PB) is the low participation of the general population in the processes. A clear example of this is the famous PB of Porto Alegre-RS, in Brazil, widely replicated internationally. At the height of the process, in the 1990s and early 2000s, Porto Alegre managed to mobilize thousands of citizens, becoming a global model for participatory democracy. In some years, the process involved about 20,000 to 40,000 people, in a municipality with more than 1 million inhabitants, which corresponds to approximately 2–4% of the population (Baiocchi and Baiocchi 2005; Wampler 2009). One reason for this may be the delay in integrating new technologies into PB processes. Resistance to the use of these technologies, especially in contexts where face-to-face methods still predominate, maybe a key cause of low popular participation. This resistance, often linked to cultural and social factors, limits the expansion of engagement channels offered by digital tools, such as greater accessibility and flexibility. Without the adoption of complementary technologies, such as e-participation platforms, many citizens—especially younger people and those who prefer digital interactions—may feel discouraged from participating (Baiocchi and Baiocchi 2005; Wampler 2009).

One of the reasons may be the delay in integrating new technologies into PB processes. Resistance to using these technologies, especially in contexts where face-to-face methods still predominate, may be a cause of low popular participation. This resistance, often linked to cultural and social factors, limits the expansion of engagement channels that digital tools offer, such as greater accessibility and flexibility. Without the adoption of complementary technologies, such as e-participation platforms, many citizens—especially younger people and those who prefer digital interactions—may feel discouraged from participating (Sousa et al. 2024).

The COVID-19 pandemic has shown how digital tools can complement, rather than replace, traditional methods, opening multiple channels of information, communication, and interaction, and significantly increasing popular participation in PB processes. This has allowed for the involvement of a larger portion of the population, even during periods of physical restrictions, expanding participation and making the process more accessible and inclusive. The pandemic accelerated the digitization of PB processes, demonstrating that digital tools can enhance traditional methods by opening new communication channels and expanding participation, particularly during times of physical restrictions. Many municipalities quickly adopted online platforms and mobile apps to enable remote participation, as seen in several cities in Brazil and Europe. Therefore, the pandemic forced the adoption of new technologies and paved the way for the development of hybrid forms of participation, combining face-to-face and digital methods, accelerating existing trends such as e-voting and e-participation. However, this digitalization also revealed inequalities in access to technology, with less-connected regions and populations unfamiliar with digital tools facing disadvantages, highlighting the need for public policies focused on digital inclusion (Dias & Júlio, 2019).

i-PB reflects the transformation of society in a hyper-technological era, where digital tools are an integral part of everyday life. Mobile apps and online platforms facilitate access and allow more citizens to participate remotely, overcoming physical barriers that limit traditional processes. This evolution is a natural consequence of technological advancement and the need for public organizations to adapt to modern realities, where technology permeates all aspects of life. In addition, i-PB promotes digital inclusion by integrating technologies into its implementation phases. Through web platforms, mobile apps, and social networks, citizens can participate without the traditional obstacles to access. The use of AI and big data ensures that contributions from a wide range of participants are processed and considered, giving marginalized groups a voice as well. With multiple digital channels, i-PB offers greater accessibility and transparency in the democratic budget decision-making process. For this inclusion to be effective, however, there must be a broadband and mobile Internet infrastructure that universalizes access to digital technologies, ensuring that all citizens can engage equitably in the democratic process, regardless of their location or socioeconomic status (Sousa et al. 2024).

Countries with the highest public Internet coverage per capita, such as Iceland, Norway, Sweden, Denmark, Finland, and Estonia, have heavily invested in digital infrastructure, resulting in near-universal Internet penetration. These nations offer high-quality broadband coverage and mobile services, even in remote areas, to ensure accessibility and digital inclusion. Estonia is another notable example, being one of the most advanced countries in the world in terms of public connectivity. With a solid digitalization strategy, Estonia has been providing largely digital public and administrative services, including an Internet-accessible e-government system for all citizens, which positions it as an international model of excellence in terms of access to the public Internet (Gil García 2013; ITU 2024; RankingRoyals 2024) In the 2022/23 Participatory Budget of the city of Tallinn, Estonia, 94% of the 9433 citizens who participated in the vote chose to vote electronically, corresponding to 8873 people, while only 560 voted on paper. The use of digital platforms has not only facilitated the process but has also demonstrated progress in digital inclusion, with the majority of the population preferring electronic voting to decide on the 114 ideas submitted to the referendum (*Summary of the Participatory Budget 2022/2023|Tallinn* 2024).

In conclusion, the digital transformation of PB processes, as exemplified by i-PB, represents a significant advance in democratization and social inclusion by allowing greater access and citizen participation. The COVID-19 pandemic has accelerated this transition, highlighting the importance of digital tools in complementing traditional methods of participation, thereby increasing accessibility and flexibility. However, digitalization has also revealed disparities in access to technology, underscoring the need for public policies focused on digital inclusion to ensure that all citizens, regardless of their socioeconomic status or location, can fully participate in these processes. The example of countries like Estonia, which invest in digital infrastructure, demonstrates that with an effective approach to digital inclusion, it is possible to create broader and more equitable participation, thereby strengthening democracy and decision-making (INCT Observatório das Metrópoles 2024).

13.3 Public Policies and Regulation: A Long Road for i-PB

Paraphrasing the Brazilian poet Carlos Drummond de Andrade: 'In the middle of the road there was a stone. There was a stone in the way' (*No Meio Do Caminho—Carlos Drummond de Andrade*, sem data 2024). The stumbling block in the path of i-PB is, without a doubt, the dependence on the will and actions of politicians to create and approve laws that make i-PB mandatory as a democratic process for defining public budgets. However, this does not mean that it is an insurmountable challenge. The legal adoption of Participatory Budgeting (PB) in different countries has been an important milestone in the institutionalization of participatory democracy, with global implications for governance and citizenship. One pioneering example is Peru, which, in 2004, implemented a national law making PB mandatory at various levels of government. This legislation paved the way for active citizen participation in the allocation of public resources, highlighting the country's commitment to transparency and accountability in the public sector. Other Latin American countries, such as the Dominican Republic and Ecuador, have also followed this trend, making PB mandatory in their municipalities and reinforcing popular inclusion in decision-making and budgeting processes (INCT Observatório das Metrópoles 2024).

In Asia, countries such as Indonesia and South Korea have made progress in adopting legal frameworks that promote PB as a tool for democratic management. Indonesia has implemented legislation that promotes popular participation in public policymaking, while South Korea has ensured citizen participation in budget decision-making processes, emphasizing the importance of collective engagement in public governance. The introduction of laws that institutionalize PB in countries with different political and social contexts reflects the flexibility of this mechanism, which is adaptable to different cultures and systems of government (INCT Observatório das Metrópoles 2024).

In Europe, Portugal also stands out for its innovative legislation by making PB mandatory in public schools, ensuring the participation of students in the management of school resources. This example demonstrates how PB can be applied not only at broad government levels but also at local and specific institutions, such as the education system. Additionally, Poland and Kazakhstan are among the countries that have adopted laws to encourage PB at local and national levels, promoting a governance model that values the active participation of civil society in the definition and execution of public budgets (INCT Observatório das Metrópoles 2024). In more than 30 years of existence, PB has demonstrated resilience in the face of political sabotage, and today it is mandatory in 10 countries, which together represent 65% of active PBs globally. While the adoption of innovative solutions, such as i-PB, may take time, the incorporation of new technologies (AI and quantum computing) into public democratic and administrative processes is irreversible (Camello 2024a; INCT Observatório das Metrópoles 2024; *Lei da UE sobre IA* 2023; *Regulamento IA entra em vigor—Comissão Europeia* 2024).

This inevitability has recent precedents. Just as the private sector has had to adapt to technological transformations since the late 1990s, the public sector now faces the challenge of regulating the use of new technologies, especially artificial intelligence (AI). Governments around the world are establishing guidelines to ensure that AI development is safe and ethical. In this scenario, the European Union (EU) stands out as a pioneer, implementing the first comprehensive regulatory framework for AI to mitigate risks and promote the responsible use of technology for the benefit of society. The European Artificial Intelligence Regulation came into force on 1 August 2024, to promote the responsible development of AI in the EU. It establishes a risk-based approach, classifying AI systems into different categories, from minimal to unacceptable risks. The regulation aims to protect the fundamental rights of citizens by ensuring greater transparency, human oversight, and quality standards for high-risk AI applications while promoting innovation and competitiveness in the sector (Camello 2024a; *Lei da UE sobre IA* 2023; *Regulamento IA entra em vigor—Comissão Europeia* 2024).

13.4 Conclusion

Given that the world's population currently stands at 8 billion people, the solutions created by humanity to ensure the survival and growth of the species have been successful so far. We have managed to control and manipulate all the variables that put our lives at risk—from food and environments to diseases. However, the current environmental, technological, and socioeconomic scenario reveals the exhaustion of this model. This exhaustion occurs at a crucial moment, considering the simultaneous and unprecedented challenges imposed on the human race in this century (Camello 2019, 2024b).

The planet's cyclical climate changes, intensified by human action, are already showing signs that difficult times are ahead. In recent years, extreme weather events have demonstrated the devastating impacts of global warming. Rising temperatures and changing weather patterns have intensified disasters such as wildfires, floods, cyclones, and droughts. Examples of this include the record fires in Australia from 2019 to 2020, which devastated 50 million hectares, and the accelerated melting of ice in the Arctic, which could result in the loss of all its ice cover by 2035. The side effects of global warming will be extremely serious and will cause profound transformations. The most significant impact, without a doubt, will be the forced migration of millions of people, affecting social, political, economic, and environmental spheres across the planet (*10 Eventos Climáticos Extremos Causados pela Mudança Climática*, sem data 2024; Klein 2019).

By the end of the twenty-first century, millions of people will face the loss of their homes due to rising sea levels, resulting in massive displacement. According to a study by Climate Central, if global temperatures rise by 3 °C compared to pre-industrial levels, about 800 million people could live in areas that will be below the high tide level by 2100. Involuntary migration caused by rising sea levels could reshape the global demographic map by the end of the twenty-first century. The forced exodus of millions of people will transform cities, nations, and borders, generating new challenges for global governance and the social and political cohesion of nations. Decisions made in the coming decades will be key to determining whether migrations caused by climate change can be managed in a way that minimizes human suffering and promotes just and inclusive adaptation, or whether they will result in catastrophic consequences (Keefe 2021; World Bank 2021).

The development of new technologies and biotechnology will have a major impact on the labor market, significantly changing both the types and the number of jobs available. With the aging of the population and the reduction of jobs due to automation and robotization, the increase in the social security deficit will be a challenge we will need to face. In other words, employment, which sustains corporate profits, generates taxes for the public sector, and keeps the economy running, will undergo a structural and market revolution whose exact form we still do not know (Ford 2016; Harari 2018; Schwab 2016; Susskind 2020).

Global power dynamics are changing, provoking violent attacks against representative democracies by extremists. The resurgence of populist fascism poses a real threat to current democracies and, indeed, to the future of the democratic rule of law. The 2024 US election demonstrated that hate speech and the deliberate dissemination of false information have a strong influence on voters, who elected Donald Trump, even though he incited and led the invasion of the Capitol in 2021 in a clear coup attempt. At the same time, the Chinese and Russian dictatorships have financed and promoted attacks on democracies and their institutions (Levitsky and Ziblatt 2018; Snyder 2017; Stanley 2018; Williams and Peniston 2021).

In this context, i-PB plays a crucial role both in promoting a sustainable economy and in defending democratic institutions and addressing climate challenges. By integrating the collective demands of different levels and dimensions and using AI to optimize the use of public resources, i-PB can simulate and propose economically sustainable solutions for the short, medium, and long term in an integrated manner. Another central aspect of its functioning is decentralization and democratic participation, which allow communities to directly define budget priorities. This decentralization reduces the concentration of power and promotes a fairer distribution of resources, which is essential for mitigating threats such as extremist populism and social inequality—factors that erode trust in democratic institutions. By ensuring that decisions are made more equitably, i-PB strengthens social cohesion and increases the resilience of local economies, mitigating the effects of political polarization. In addition, i-PB is designed to be inclusive, ensuring that all layers of society, especially the most vulnerable, participate in the allocation of resources. At a time when democracies are facing attacks, i-PB acts as a tool for institutional strengthening, promoting transparency and accountability. The direct participation of the population in the choice of budget projects reduces opportunities for corruption and reinforces trust in public institutions. By allowing communities to choose projects that prioritize inclusion, i-PB contributes to an economy with broader and more equitable opportunities (Sousa et al. 2024).

The optimization of public resources through AI represents a significant leap in the management of taxpayer money. i-PB uses AI algorithms to process large volumes of data and identify patterns in collective demands, offering a comprehensive view of how to allocate resources efficiently, reduce waste, and promote sustainable development. This approach is essential in scenarios of climate change and economic crises, as it allows for the rapid and effective implementation of solutions. For example, AI can suggest projects that meet both local and global needs, such as building infrastructure resilient to climate disasters. In addition, i-PB enables communities to set their priorities according to environmental threats, such as rising sea levels and forced migration. By integrating environmental data, the system can prioritize investments in sustainable infrastructure, including clean energy solutions and improvements in disaster resilience. This is crucial for mitigating the consequences of global warming, which is transforming economies and forcing the displacement of populations (Sousa et al. 2024).

i-PB allows for integrated and sustainable management of collective demands, ranging from local to international needs. In doing so, it facilitates cooperation between different levels of government, promoting the equitable distribution of resources and the implementation of policies that respect cultural and environmental diversity. This strengthening is essential to resist external threats, such as interference from authoritarian regimes, while ensuring that democratic governance remains robust. By allowing global data to be processed to meet local needs, i-PB creates solutions tailored to each context, promoting sustainable development on a large scale. These i-PB mechanisms help to build a sustainable economy, not only in ecological terms but also in social and political terms, directly addressing the challenges posed by climate change, socioeconomic inequalities, and threats to democracy (Sousa et al. 2024).

References

10 Eventos Climáticos Extremos Causados pela Mudança Climática (sem data). https://www.benandjerry.com.br. Obtido 14 de novembro de 2024, de https://www.benandjerry.com.br/novidades/2020/11/extreme-weather-events

Aaronson S (2013) Quantum computing since Democritus. Cambridge University Press

Agricultura (2017) Fazenda usa luz solar e água do mar para cultivo no deserto. Gazeta do Povo. https://www.gazetadopovo.com.br/agronegocio/agricultura/com-agua-do-mar-e-luz-solar-fazendas-cultivam-alimentos-no-deserto-6mgc5my0o4ouz8idylpfpbt93/

Almeida MA (1997) Orçamento Participativo. Instituto Pólis. https://polis.org.br/publicacoes/orcamento-participativo/

Atkinson AB (2015) Inequality: what can be done? Harvard University Press

Atkinson T (2014) Public economics in an age of Austerity. Taylor and Francis

Baiocchi G (2005) Militants and citizens: the politics of participatory democracy in Porto Alegre. Stanford University Press

Baiocchi G, Baiocchi G (2005) Militants and citizens: the politics of participatory democracy in Porto Alegre. Stanford University Press

Beck U (2009) Risk society: towards a new modernity (repr). Sage

Besley T, Persson T (2011) Pillars of prosperity: the political economics of development clusters. Princeton University Press

Camello J (2019) CAPAS: Célula Administrativa Pública Autossustentável: Contributo para a gestão da administração pública: Uma visão [Dissertações de mestrado, ISCTE—Instituto Universitário de Lisboa]. http://hdl.handle.net/10071/19338

Camello J (2024a) A utilização das novas tecnologias nos processos democráticos da União Europeia [doctoralThesis]. https://repositorio.iscte-iul.pt/handle/10071/31867

Camello J (2024b) A utilização das novas tecnologias nos processos democráticos da União Europeia [doctoralThesis]. https://repositorio.iscte-iul.pt/handle/10071/31867

Daly HE (1991) Steady-state economics, 2nd ed with new essays. Island Press

De Long JB, Eichengreen B (1991) The Marshall plan: history's most successful structural adjustment program. National Bureau of Economic Research, pp w3899 https://doi.org/10.3386/w3899

Diamond JM (2006) Collapse: how societies choose to fail or succeed. Penguin Books

Dias N, Júlio S (2019) Trinta anos de Orçamentos Participativos no Mundo. Entre o passado, as tendências mais recentes e os desafios futuros. Epopeia/Oficina. https://www.oficina.org.pt/30anosopmundo.html

Fedozzi L (2009) Orçamento Participativo de Porto Alegre: Gênese, avanços a limites de uma idéia que se globaliza. CIDADES, Comunidades e Territórios, 18, Artigo 18. https://revistas.rcaap.pt/cct/article/view/9288

Fedozzi LJ (2002) O eu e os outros: a construção da consciência social no orçamento participativo de Porto Alegre. https://lume.ufrgs.br/handle/10183/85057

Ford M (2016) Rise of the robots: technology and the threat of a jobless future (First paperback edition). Basic books

George W. Bush Presidential Center (2024) The U.S. is taking steps to tackle corruption in 2024. George W. Bush Presidential Center. https://www.bushcenter.org/publications/the-u-s-is-taking-steps-to-tackle-corruption-in-2024/

Gil García JR (Ed) (2013) E-government success around the world: cases, empirical studies, and practical recommendations. Information Science Reference

Harari YN (2018) 21 Lições para o século 21. Companhia das Letras

Harvey D (2011) The enigma of capital: and the crises of capitalism (Pbk. ed). Oxford University Press

Hawken P (2017) Drawdown: the most comprehensive plan ever proposed to reverse global warming. Penguin Publishing Group

INCT Observatório das Metrópoles (Diretor) (2024) Os Orçamentos Participativos no mundo: Difusão, tipologia e perspectivas [Gravação de vídeo]. https://www.youtube.com/watch?v=pFuTel8VWkE

ITU (2024) Statistics. https://www.itu.int/en/ITU-D/Statistics/pages/stat/default.aspx

Jackson T, Charles (2017) Prosperity without growth: foundations for the economy of tomorrow, 2nd edn. Routledge

Keefe RR, John (2021) Futuro debaixo d'água: Como será a elevação do nível do mar no mundo todo. CNN Brasil. https://www.cnnbrasil.com.br/internacional/futuro-debaixo-dagua-como-sera-a-elevacao-do-nivel-do-mar-no-mundo-todo/

Kennedy DM, Woodward CV (2005) Freedom from fear: the American people in depression and war, 1929–1945. Oxford Univ Press

Kissinger H (2014) World order. Penguin Press
Klein N (ed) (2014) This changes everything: capitalism versus the climate (First Simon & Schuster export edition). Simon & Schuster
Klein N (2019) On fire: the (burning) case for a green new deal (First Simon & Schuster hardcover edition). Simon & Schuster
Lei da UE sobre IA: Primeira regulamentação de inteligência artificial|Atualidade|Parlamento Europeu (2023). https://www.europarl.europa.eu/news/pt/headlines/society/20230601STO93804/lei-da-ue-sobre-ia-primeira-regulamentacao-de-inteligencia-artificial
Levitsky S, Ziblatt D (2018) How democracies die, 1st edn. Crown
Li K (2021) AI superpowers: China, Silicon Valley, and the new world order. Houghton Mifflin Harcourt
Milanović B (2018) Global inequality: a new approach for the age of globalization (First Harvard University Press paperback edition). The Belknap Press of Harvard University Press
Musgrave RA, Musgrave PB (1980) Public finance in theory and practice, 3rd edn. McGraw-Hill
NextGeneration EU—União Europeia. (sem data). Obtido 24 de setembro de 2024, de https://next-generation-eu.europa.eu/index_pt
No Meio do Caminho—Carlos Drummond de Andrade. (sem data). Escritas.Org. Obtido 13 de novembro de 2024, de. https://www.escritas.org/pt/t/9468/no-meio-do-caminho
Olson M (2003) The logic of collective action: public goods and the theory of groups (21. printing). Harvard Univ. Press
Participatory Budgeting Project (2016) Next generation democracy. https://www.participatorybudgeting.org/white-paper/
Samuelson PA, Nordhaus WD (1998) Economics: sixteenth edition (Di 1 ban, Ying yin ban)
Paulson HM (2010) On the brink: inside the race to stop the collapse of the global financial system, 1st edn. Business Plus
Peerzada AR (2018) A inovadora técnica capaz de transformar areia do deserto em terra fértil. BBC news Brasil. https://www.bbc.com/portuguese/internacional-44034760
Pereira LCB (1996) A Crise do Estado. Nobel
Piketty T (2017) Capital in the twenty-first century (A. Goldhammer, Trad.). The Belknap Press of Harvard University Press
Polanyi K (2021) A grande transformação: As origens políticas e econômicas de nossa época. Contraponto
RankingRoyals (2024) Highest internet penetration rates by country 2024—RankingRoyals. https://rankingroyals.com/tech/highest-internet-penetration-rates-by-country-2024/
Raworth K (2022) Doughnut economics: seven ways to think like a 21st-century economist. Penguin Books
Regulamento IA entra em vigor—Comissão Europeia (2024). https://commission.europa.eu/news/ai-act-enters-force-2024-08-01_pt
Ricardo D (2018) Princípios de economia política e tributação. Lebook Editora e Livraria
Sachs J (2015) The age of sustainable development. Columbia University Press
Samuelson PA, Nordhaus WD (2010) Economics (19th edn). McGraw-Hill/Irwin
Sassen S (2014) Expulsions: brutality and complexity in the global economy. The Belknap press of Harvard university press
Schneider GE (2019) The evolution of economic ideas and systems: a pluralist introduction. Routledge, Taylor & Francis Group
Schwab K (2016) The fourth industrial revolution. World Economic Forum
Sen A (2011) Development as freedom. Knopf Doubleday Publishing Group
Smith A (com Pessôa, V., & Brito, R.) (2021) A riqueza das nações: Investigação sobre a natureza e as causas da riqueza das nações. Roncel L, ed. D. M. Miranda, Trad. Edipro
Snyder T (2017) On tyranny: twenty lessons from the twentieth century, 1st edn. Tim Duggan Books
Sousa MJ, Pani SK, Dal Mas F, Sousa S (Eds) (2024) Incorporating AI technology in the service sector: innovations in creating knowledge, improving efficiency, and elevating quality of life, 1st edn. Apple Academic Press, CRC Press

Stanley J (2018) How fascism works: the politics of us and them (1st edn). Random House
Stiglitz JE (2003) Globalization and its discontents. W. W. Norton
Stiglitz JE (2013) The price of inequality: how today's divided society endangers our future, 1st edn. W. W. Norton & Company, Incorporated
Stiglitz JE, Rosengard JK (2015) Economics of the public sector (Fourth edition). W. W. Norton & Company, Inc.
Summary of the participatory budget 2022/2023|Tallinn (2024). https://www.tallinn.ee/en/participatorybudget/summary-participatory-budget-20222023
Susskind D (2020) A world without work: technology, automation, and how we should respond. Metropolitan Books (Macmillan)
Thompson EP (2021) A formação da classe operária inglesa. A árvore da liberdade, vol 1. Editora Paz e Terra
Vidal I (2023) Milagre verde no maior deserto da China. Revista fórum. https://revistaforum.com.br/global/chinaemfoco/2023/9/26/milagre-verde-no-maior-deserto-da-china-144766.html
Wampler B (2009) Participatory budgeting in Brazil: contestation, cooperation, and accountability. Pennsylvania State University Press
Williams A, Le Billon P (Eds) (2017) Corruption, natural resources and development: from resource curse to political ecology. Edward Elgar Publishing. https://doi.org/10.4337/9781785361203
Williams K, Peniston B (2021) Right-wing extremists storm US capitol. Defense one. https://www.defenseone.com/threats/2021/01/right-wing-extremists-storm-capitol-building/171216/
World Bank (2003) Case study 2—Porto Alegre, Brazil: participatory approaches in budgeting and public expenditure management. https://hdl.handle.net/10986/11309
World Bank (2021) Groundswell: acting on internal climate migration. World bank. https://www.worldbank.org/en/news/feature/2021/09/13/millions-on-the-move-in-their-own-countries-the-human-face-of-climate-change
World Economic Forum (2019) Corruption costs developing countries $1.26 trillion every year—yet half of EMEA think it's acceptable. World economic forum. https://www.weforum.org/agenda/2019/12/corruption-global-problem-statistics-cost/
Yakubu IN (Ed) (2024) Examining corruption and the sustainable development goals. IGI Global

14 Unpacking the Impacts of Digital Leadership: How Digitalization is Reshaping HRM in Public Administration Governance?

David Ferraz

Abstract

In the contemporary era, digital technology is profoundly transforming the landscape of public administration and its human resource management (HRM). This research explores the multifaceted impacts of digitalization on HRM, focussing on its influence on work models and leadership profiles in public administration (PA). The objective is to delve into the effects of digital innovations on work models, HRM, and leadership within the context of PA. The theoretical framework considers the transition from Bureaucracy models to New Public Management and then to New Public Governance, evaluating how current models are aligned with digital transformation and digital leadership concepts. Therefore, a qualitative approach was adopted, integrating data from literature review and content analysis of relevant literature and public policy documents. This study captured insights from HRM public policies, managers, HR professionals, and other stakeholders involved in digital transformation initiatives in public administration. It was possible to identify how digital technology, particularly digital transformation, is impacting HRM and leadership profiles in the public sector. The findings highlight themes such as leadership in multicultural teams, data-driven analytics and fostering engagement of digital citizens in networks, as well as the interconnectedness of public managers' leadership with internal and external environments. As public administration evolves in the digital age, it becomes imperative to understand the dynamics between digital technology, PA models, and HRM and leadership practices. This article seeks to fill this knowledge gap, offering valuable insights for future research and practice.

D. Ferraz (✉)
ISCTE—Instituto Universitário de Lisboa, Lisboa, Portugal
e-mail: david.ferraz@iscte-iul.pt

CIES, Centre for Research and Studies in Sociology, Lisbon, Portugal

M. J. Sousa et al. (eds.), *Economic Sustainability Potentiated by Artificial Intelligence*, Contributions to Environmental Sciences & Innovative Business Technology,
https://doi.org/10.1007/978-981-96-8089-4_14

Keywords

Digital transformation • Human resource management • Leadership • Public management • Governance adaptive leadership

14.1 Introduction

The second decade of the twenty-first century has witnessed an exponential acceleration in the adoption of technological innovations, profoundly reshaping organizational operations to better meet the needs of their stakeholders. Public Administration is no exception to this transformative wave. Digital technologies are fundamentally altering the way public administration interacts with society to fulfil the public interest. These changes encompass not only the delivery of services within the framework of public policy implementation but also the digital environments in which these services are designed and executed. This transformation impacts both the outcomes of services and the mechanisms through which public policies are operationalized. As governments and public sector organizations strive for greater efficiency and responsiveness, the integration of digital tools has become indispensable. This shift represents more than a mere technological advancement; it signifies a paradigm shift in the delivery of public services and the management of public sector workforces.

In this evolving landscape, the role of human resource management (HRM) in Public Administration gains prominence. Leadership within a digital environment necessitates the development of new competencies and capabilities, emphasising e-skills that facilitate agile leadership, data-driven decision-making, and a democratising digital literacy. These capacities are essential for navigating the complexities of digital transformation while ensuring that public servants are equipped to meet contemporary challenges.

Despite the rapid adoption of digital technologies, the impact of these changes on HRM work models and leadership within public administration remains underexplored. The intricate ways in which digital tools are reshaping organizational structures, decision-making processes, and employee management practices warrant further investigation, highlighting a significant gap in both academic research and the practical formulation of HRM policies.

This study, conducted as part of an exploratory research project grounded in a systematic literature review, seeks to categorize and evaluate the multifaceted impacts of digitalization on HRM within public administration. The analysis is framed by a conceptual model developed using bibliographic data extracted from the Scopus repository, encompassing literature published since 2003. This endeavour employed the Scopus AI tool, using integrated search expressions such as 'digital leadership', 'HRM', 'Public Administration Governance', and 'Models of Administration' to uncover key insights and trends in the field.

14.2 Methodology

This study adopts an exploratory research design to examine the impacts of digitalization on human resource management (HRM) within public administration. Grounded in a systematic literature review, it employs a qualitative approach to analyse and synthesize existing knowledge, aiming to uncover the multifaceted roles of digital leadership in reshaping HRM practices and governance models. The research follows a structured methodology encompassing the selection of key analytical dimensions, conceptual framing, and the use of advanced bibliometric tools to extract and analyse relevant data.

The research framework integrates two key analytical dimensions identified through a review of existing literature on digital leadership and HRM in the public sector, as stated on Fig. 14.1:

1. **Digital Leadership for HRM Transformation**: This dimension focuses on the leadership required to drive change within an organization's workforce. It explores the evolving competencies and strategies necessary to manage digital transitions, emphasising workforce adaptability, capacity building, and the alignment of HR practices with technological innovations.
2. **Digital Leadership for Data-Driven Service Provision**: This dimension examines leadership roles in managing digital work processes, particularly the use of data to inform and support public decision-making. It highlights the importance of fostering a data-driven culture and ensuring that leaders can leverage digital tools for enhanced service delivery and policy implementation.

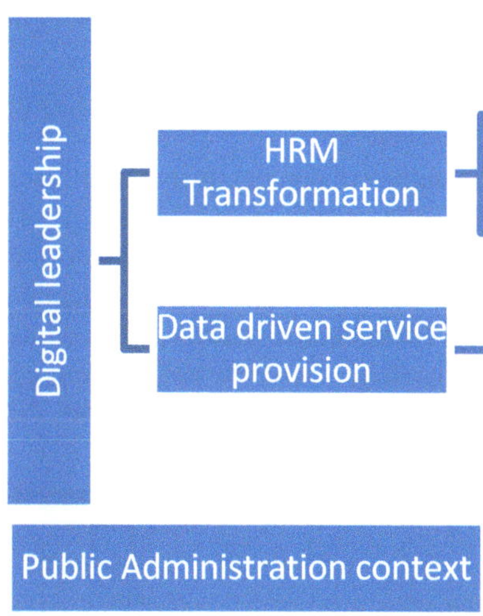

Fig. 14.1 Conceptual model of digital leadership impact in Public Administration

While recognising that additional dimensions could enrich the analysis, the study prioritizes these two to ensure a focussed and in-depth exploration of the subject matter. Both dimensions are contextualized within the broader theoretical framework of public administration, specifically the paradigm shift from traditional bureaucratic models to New Public Governance.

The study employs bibliometric and qualitative content analysis techniques to develop a conceptual model for understanding the impacts of digitalization on HRM. Bibliographic data were sourced from the Scopus repository,[1] encompassing articles and academic publications.

The study began with the establishment of clear objectives, focussing on understanding the influence of digital leadership on the digital transformation of human resource management (HRM) within the public sector. The research scope was deliberately set from 2003 onwards to align with the period marking the broader and more systematic adoption of digital technologies in public administration. Nonetheless, the study emphasized more recent publications to capture the ongoing and dynamic nature of digital transformation processes. To ensure the comprehensiveness and relevance of the literature, the Scopus database was selected as the primary source for identifying relevant studies. This choice was driven by Scopus's recognized authority in indexing high-quality publications across social sciences and applied research fields. Specific search terms such as 'digital leadership', 'HRM', 'Public Administration governance', and 'administrative models' were employed in Boolean combinations to retrieve articles directly addressing the intersection of digitalization, leadership, and HRM.

The inclusion criteria ensured that only publications indexed in Scopus were considered, focussing on empirical, theoretical, or review studies explicitly discussing digital transformation within public sector contexts. Exclusion criteria were equally important to refine the dataset. Articles that were not specific to public administration contexts or those lacking empirical evidence or robust theoretical frameworks were excluded. This rigorous selection process ensured that the final set of literature was both relevant and of high quality, forming a solid foundation for subsequent analysis.

The selected articles underwent a qualitative content analysis, following a structured and systematic approach. The first stage involved initial coding, where recurring themes, concepts, and patterns were identified. Particular attention was given to digital leadership competencies and the impacts of digitalization on HRM practices. This process allowed for the emergence of meaningful insights that were further refined in the thematic grouping stage. Here, the data were organized into key analytical categories, including 'internal process transformation', 'data-driven

[1] The Scopus AI tool was utilised to facilitate systematic searches, combining expressions such as 'digital leadership', 'HRM', 'Public Administration Governance', and 'Models of Administration'. These search terms were carefully selected to capture the intersection of digitalization, leadership, and HRM within public governance contexts.

decision-making', and 'development of digital competencies'. This step was crucial for structuring the findings in a way that could be synthesized into a coherent framework.

Based on these findings, a conceptual model was developed to link the dimensions identified in the literature with the competencies necessary for effective digital leadership in HRM. This model aimed to provide a structured representation of how digital leadership supports the broader goals of digital transformation in the public sector. To ensure its robustness, the model was iteratively validated through feedback from subject-matter experts and critical cross-referencing with existing literature. This iterative process enhanced the model's relevance and applicability.

While bibliometric indicators, such as citation counts, were considered during the literature selection process, they were not given exclusive priority. The aim was to balance the inclusion of widely cited works with those that, although less cited, provided significant contextual insights into the public sector. This approach allowed the study to capture not only the dominant narratives in the field but also the nuances and gaps that are often overlooked in highly cited literature. By maintaining this balance, the study achieved a nuanced understanding of digital leadership's role in transforming HRM within Public Administration. The extracted data were analysed to identify recurring themes, patterns, and conceptual frameworks related to digital leadership and its implications for HRM. The study integrates these findings into a conceptual model designed to categorize and evaluate the identified impacts systematically.

Theoretical Context

Central to this research is the concept of digital leadership, defined as the ability to lead and manage organizational change in digital environments. The study situates digital leadership within the evolution of public administration theory, highlighting the transition from bureaucratic paradigms to more collaborative and networked governance models characteristic of New Public Governance. This theoretical contextualization provides a lens for understanding the role of digital leadership in transforming organizational structures, decision-making processes, and HRM practices.

14.3 Digital Leadership

It is worth noting that the concept of digital leadership remains vaguely defined (Imhof & Grivas, 2022), although it has garnered enthusiasm and prominence in recent years. Zhang and Chen (2024) explored how HRM can be transformed through the digitalization of economies. Another study, conducted by Tigre et al. (2024), employed a systematic literature review and the Delphi method, identifying that digital leadership requires leaders to focus on organizational mission strategies and interpersonal skills.

In the specific context of Public Administration, studies have highlighted the crucial role of leaders in digital environments in aligning human resources with digital transformation processes (Yildirim, 2023). The same author emphasizes effective digital competencies in these contexts, such as communication skills, innovative thinking, and digital literacy.

From a broader perspective related to organizational management processes and their respective models, other authors stress the significance of algorithms in decision-making processes (Conte, 2023) and the need to incorporate investments in human capital development within public policies to ensure the success of digital transformation (Febiri et al. 2024; Holdorff et al. 2023).

As Gutkowski (2022) points out, digital leadership necessitates changes that impact organizational strategy, workplace mindsets, and skill sets. Digitalization is increasingly affecting human resource management in public administration governance, making it imperative to estimate its impacts.

The data collected allowed for the design of a conceptual model linking the concept of digital leadership with the most relevant competencies.

14.3.1 Digital Leadership for HRM Transformation: eSkills and e-Culture Management

Digital transformation processes have driven leadership towards adapting new styles in digital environments, particularly those capable of guiding people through the digital transformation processes organizations and societies are undergoing.

The Fourth Industrial Revolution has necessitated progress in human resource management, resulting in the creation of digital models for HRM activities (Shava & Ndebele, 2024).

Abbu et al. (2023) argue that leaders in digital environments play a vital role in fostering collaborative settings that facilitate the digital transformation process, which requires a new set of competencies, including a growth mindset, ethical AI awareness, and a focus on data.

Kusanke et al. (2023) highlight that changing competency requirements for public sector leaders emphasize the importance of social skills, such as communication, in driving digital transformation in public administration.

Other authors (Ahmed et al. 2024; Malik et al. 2024; Wang et al. 2023a, b; Wang et al. 2023a, b; Wang et al. 2022; Živković, 2022) emphasize various competencies for digital leadership, including vision, innovation, flexibility, understanding of digital technologies, empowerment, collaboration, multiple intelligences, experimentation, continuous learning, software proficiency, strategic thinking, leadership and teamwork, adaptability, technological education and training, communication, knowledge of construction processes, standards and regulations, technological know-how, innovativeness, the ability to lead and steer, honesty, integrity, transformative vision, and digital adaptability and resilience.

Leaders must also be capable of managing multicultural environments where teams are composed of individuals from diverse cultural backgrounds, requiring additional competencies such as respect and tolerance. In digital environments where supranational spaces are increasingly prominent, different needs are likely to arise, including managing various time zones and reconciling work and non-work schedules across distinct contexts.

Furthermore, significant legal and regulatory issues surrounding work—such as working conditions and tax regimes—impact leaders' ability to effectively manage teams, adding complexity to leadership in digital environments.

14.3.2 Digital Leadership for Data Driven Service Provision: EDecision-Making and Evidence-Based Policy

Digital leadership is expected to revolutionize HRM processes in public sector organizations, resulting in improved employee performance, enhanced decision-making, and better public services (Shava & Ndebele, 2024).

Studies have found that digital leadership positively impacts exploratory innovation, mediated by digital entrepreneurial orientation and digital organizational culture, and moderated by big data analytics capabilities (Wang et al. 2023a, b; Wang et al. 2023a, b; Wang et al. 2022; Xiao & Wang, 2022).

The integration of digital technologies is crucial for the development of countries and economies (Nazmetdinova et al. 2023), enabling better relationships between public entities and society. Big data management enhances service quality by incorporating a broader range of variables into decision-making models, thereby increasing the efficacy and transparency of processes (Mihai et al. 2022).

Data-Driven Decision-Making (DDDM) enables public leaders to make more informed decisions. Leaders who effectively use data can navigate systems and drive meaningful change. The use of data in public administration promotes accountability by providing a clear foundation for decisions and actions. For example, data-driven tools can enhance transparency by legitimizing public spending and improving the allocation of resources (Ghosh, 2024).

Within the dimension of analysis related to digital leadership in service provision, Abbu et al. (2023) highlight the necessity for digital leaders to focus on understanding and leveraging data.

Leaders must balance the rational use of data with the political realities of their environments. This dual approach allows them to use data to support objectives while navigating the complexities of public administration (Voort et al. 2019).

Developing data literacy and analytical skills among public leaders is essential. Advances in data literacy are critical to ensuring that leaders can interpret and utilize data effectively, enabling evidence-based decision-making and the implementation of policies grounded in robust data analysis (Slemp, 2023).

However, key concerns remain regarding the digital divide and equal access in the digital transformation of public services (Sani & Mandina, 2024).

14.4 Digital Leadership: From Bureaucracy to New Public Governance

The concept of bureaucracy, as formulated by Max Weber, has historically underpinned public administration. This model is characterized by its structured hierarchy, formalized rules, and impersonal relationships, designed to ensure operational efficiency and predictability (Bouckaert, 2023; Pollitt & Bouckaert, 2011; Weber et al. 1961). While the bureaucratic approach has facilitated standardized administration, it often lacks flexibility and responsiveness, traits increasingly critical in dynamic environments (Peters, 2001).

By the late twentieth century, critiques of bureaucratic inefficiency gave rise to new public management (NPM), which introduced private sector practices into public administration to enhance responsiveness and results orientation. NPM emphasizes efficiency, effectiveness, and performance measurement, advocating for managerial autonomy, accountability, and customer-oriented service (Hood, 1991). These principles set a foundation for incorporating digital technologies to streamline processes and improve service delivery (Zuccarini, 2007).

New public governance (NPG) emerged as a response to the limitations of NPM, particularly its inadequacy in addressing complex social challenges that require collaboration across boundaries (Osborne et al. 2012). NPG emphasizes networked governance, engaging diverse stakeholders in the policy-making process to leverage their collective expertise and resources (Ansell & Gash, 2008). This model fosters greater transparency and accountability, with digital platforms playing a crucial role in facilitating communication and collaboration among stakeholders (Loeffler, 2023; McMullin, 2021; OECD, 2020; Scott & Gong, 2021).

Digitalization intersects with these theoretical shifts by enhancing bureaucratic efficiency, enabling NPM's performance management, and supporting NPG's collaborative networks. For instance, digital tools can streamline bureaucratic processes, such as permit approvals and record management, thereby reducing procedural delays (Fountain, 2001). In the context of NPM, digital technologies facilitate real-time performance tracking and customer feedback mechanisms, enhancing decision-making and service quality (Margetts & Dunleavy, 2013). Within NPG frameworks, digital platforms like social media and collaborative software enhance stakeholder engagement and participatory governance, contributing to more informed and inclusive decision-making processes under the co-creation of policies and governance (Torfing et al. 2021).

The digital transformation in HRM within public administration is not merely a technological upgrade but a strategic enabler that aligns with and accelerates the shift from traditional bureaucracies to more dynamic governance models. Understanding these theoretical contexts is imperative for developing and implementing effective digital HRM strategies that address contemporary challenges and opportunities in services of public administration.

14.4.1 Digital Transformation and Public Administration HRM

Digital transformation in public administration refers to the integration of digital technologies into all aspects of public service delivery, fundamentally reshaping how services are provided and managed. This transformation is driven by the need for greater efficiency, improved service quality, and increased transparency in government operations. In the domain of human resource management (HRM), digital transformation involves adopting technologies such as digital databases, online recruitment systems, automated workflow management, and advanced analytics for workforce planning and management.

To frame the discussion, this section will introduce relevant theories that explain the adoption and impact of digital technologies in organizations. The technology acceptance model (TAM) posits that perceived usefulness and ease of use are primary predictors of technology adoption (Davis, 1989). Another relevant theory is the diffusion of innovations theory, which explores how, why, and at what rate new ideas and technology spread through cultures (Rogers et al. 2014; Rogers, 2004), particularly within public organizations.

Digital technologies have revolutionized traditional HRM practices by automating routine tasks, facilitating strategic decision-making, and enhancing employee engagement and development. That is the case of human resource information systems (HRIS) enable better data management and accessibility, improving decision-making processes in recruitment, training, and performance evaluations (Ruël et al. 2004). Additionally, digital platforms have transformed communication between HR departments and employees, making it more frequent, transparent, and interactive (Mergel, 2016). Despite its benefits, the digital transformation in HRM is facing, as with any structural change, several challenges, among which three key issues stand out:

1. ***Resistance to change*** Both employees and managers may resist the adoption of new technologies due to discomfort with change or concerns about job displacement (Kotter, 2007).
2. ***Privacy and security concerns*** The management of sensitive employee data requires robust cybersecurity measures to prevent data breaches and ensure privacy (Bhave et al. 2020).
3. ***Skills gaps*** There is often a significant skills deficit that must be addressed to enable the effective use and benefits of new HR technologies (Capgemini, 2017).

14.5 Strategies for Effective Digital Transformation on HRM

The integration of digital technologies into public administration HRM is a transformative process that requires careful consideration of theoretical inputs, practical challenges, and strategic implementations. By understanding and addressing these factors, public sector organizations can leverage digital transformation to enhance their HRM practices, thereby improving overall organizational effectiveness and responsiveness. To address these goals and overcome the challenges previously settled, public administration may incorporate on them HR policies and practices the following pillars on it' strategies:

1. *Digital Readiness—prepare to eSkills and e-Culture*: Assess the organization's digital maturity and develop a comprehensive change management plan to address the cultural and technological shifts required for successful digital transformation. This includes providing training, resources, and support to help employees adapt to new digital tools and processes. Provide digital platforms and technologies to enhance the employee experience, streamline HR service delivery, and foster greater employee engagement and collaboration. This may involve self-service portals, mobile apps, and social collaboration tools.

Reskilling and upskilling: Digital learning management systems can facilitate continuous skill development and career growth for public sector employees. Develop strategies to attract, develop, and retain employees with the digital skills and competencies required to drive innovation and transformation within the public sector. This includes upskilling and reskilling existing employees as well as incorporating digital aptitude into recruitment and succession planning.

2. *Data-Driven—prepare for e-Decision-Making:* prepare and implement robust data management and analytics capabilities to enable evidence-based decision-making. This includes leveraging digital HR information systems, workforce analytics, and predictive modelling to optimize workforce planning, talent management, and performance evaluation.

Advanced analytics on employee data can provide deeper insights to inform strategic HR decisions and align workforce planning with organizational goals (Hamoud et al. 2019; McGinty & Lylova, 2020).

3. Fostering Difficulties:
   1. *Resistance to change:* Employees and managers may resist adopting recent technologies due to discomfort with change or fear of job displacement (Kotter, 2007).
   2. *Privacy and security concerns:* Handling sensitive employee data requires robust cybersecurity measures to prevent data breaches and ensure privacy (Bhave et al. 2020).
   3. *Skill gaps:* There is often a significant skill gap that needs to be addressed to effectively use and benefit from new HR technologies (Capgemini, 2017).

By incorporating these strategic pillars, public administration organizations can navigate the digital transformation of their HRM practices more effectively, aligning their people, processes, and technologies to enhance organizational performance and serve the evolving needs of citizens. Web-based portals, mobile apps, and social media platforms can promote greater not only citizen participation but also civil servants' participation, feedback, and transparency in public sector governance (Speer, 2012).

14.6 Conclusion

The digital transformation of human resource management (HRM) within public administration represents a pivotal shift, driven by a complex interplay of internal and external factors. These include the evolving expectations of employees for more flexible, transparent, and digitally enabled workplaces; rapid innovations within industry that set new benchmarks for efficiency and effectiveness; competitive pressures for public institutions to match or exceed private sector standards; shifting governance structures prioritizing accountability and collaboration; and overarching societal demands for greater inclusivity, responsiveness, and transparency in public service delivery.

Although the literature provides significant insights into the impact of digitalization on human resource management (HRM), it also presents notable limitations that warrant attention. A considerable portion of existing studies focuses predominantly on private sector contexts, which restricts the direct applicability of their findings to the public sector. This emphasis on the private sector often overlooks the unique structural, cultural, and operational characteristics of public administration, where priorities such as accountability, transparency, and inclusivity take precedence over purely market-driven objectives.

Moreover, research on digital leadership frequently addresses general competencies without adequately exploring the specific demands of public governance environments. For instance, while general leadership frameworks highlight skills such as adaptability and digital literacy, they often fail to account for the complexity of navigating multilayered accountability structures, engaging diverse stakeholder groups, or integrating ethical considerations into digital transformation processes. These gaps limit the utility of existing models in addressing the nuanced challenges faced by HRM in public administration.

With this study, we aimed to address these limitations by proposing a conceptual model that contextualizes digital leadership within the unique framework of HRM in public administration. The model emphasizes competencies that align with the sector's distinctive characteristics. By tailoring the discussion to the public sector's governance dynamics, the study contributes to bridging the gap between generic theories of digital leadership and the practical realities of HRM in public administration. This approach not only enriches the academic discourse but also offers actionable insights for practitioners seeking to implement effective digital transformation strategies in public sector contexts.Understanding and addressing these multifaceted drivers is essential for public administrators to design and implement digital technologies that not only modernize HRM practices but also align them with the broader goals of public administration. Digital transformation is not merely a technological upgrade; it is a strategic endeavour that reshapes how public sector organizations operate, how they engage with employees, and how they fulfil their mandate to serve citizens effectively. Failure to navigate these changes risks perpetuating inefficiencies, disengaging public sector workforces, and eroding public trust in government institutions.

Future research plays a critical role in advancing this agenda by examining the dynamic interactions of these factors across diverse public sector contexts. Such studies can uncover context-specific challenges and opportunities, refine theoretical frameworks, and provide actionable insights for policymakers and practitioners. Furthermore, research should delve into the long-term impacts of digitalization on organizational culture, leadership models, and workforce dynamics to ensure that digital HRM strategies are sustainable, inclusive, and adaptable to evolving societal needs.

In conclusion, the digital transformation of HRM in public administration is both an opportunity and a necessity. By embracing this transformation strategically and informed by evidence-based insights, public sector organizations can enhance their operational effectiveness, foster an engaged and skilled workforce, and deliver on their commitment to serve the public with greater efficiency, equity, and accountability. This transformative journey requires vision, adaptability, and a relentless focus on aligning digital innovation with the core principles of public service.

References

Abbu H, Mugge P, Gudergan G, Hoeborn G (2023) Assessing Digital leadership: towards developing a scale to measure the human dimensions of digital leaders. In: Proceedings of the 29th international conference on engineering, technology, and innovation: shaping the future, ICE

Ahmed S, Rahman RA, Lee YS, Shahruddin S (2024) Effective digital leadership among construction industry professionals: analysis of core competencies. Lecture Notes in Civil Engineering

Ansell C, Gash A (2008) Collaborative governance in theory and practice. J Public Adm Res Theory 18(4):543–571

Bhave DP, Teo LH, Dalal RS (2020) Privacy at work: a review and a research agenda for a contested terrain. J Manag 46(1):127–164. https://doi.org/10.1177/0149206319878254

Bouckaert G (2023) The neo-Weberian state: from ideal type model to reality? Max Weber Stud 23(1):13–59

Capgemini (2017) Digital Transformation Institute, State of AI surve https://www.capgemini.com/nl-nl/wp-content/uploads/sites/19/2022/05/dti-ai-report_final1-5.pdf

Conte F (2023) The digital transformation of public administration: the transition process towards algorithmic government. Federalismi.it (11):54–75. https://www.scopus.com/inward/record.uri?eid=2-s2.0-85160868309&partnerID=40&md5=7ec3ac482704ea79e59ae003aebb232e

Davis FD (1989) Perceived usefulness, perceived ease of use, and user acceptance of information technology. MIS Q 319–340

Febiri F, Gariba MI, Hub M, Provaznikova R (2024) The synergy between human factors, public digitalization and public administration in the European context. J Open Innov: Technol, Mark, Complex 10(4):100424. https://doi.org/10.1016/j.joitmc.2024.100424

Fountain JE (2001) The virtual state: transforming American government? Nat'l Civic Rev 90:241

Ghosh R (2024) Data-driven governance and performances of accountability: critical reflections from US agri-environmental policy. Sci Cult 33(1):70–96. https://doi.org/10.1080/09505431.2023.2175654

Gutkowski G (2022) Digital leadership: how to thrive in the 21st century. In: Leadership strategies for the hybrid workforce: best practices for fostering employee safety and significance, pp 132–144. https://doi.org/10.4018/978-1-6684-3453-6.ch010

Hamoud B, Laszlo V (2019) Will artificial intelligence take over human resources recruitment and selection? Netw Intell Stud 7(13):21–30

Holdorff AS, Kaesmayr J, Schorn M, Steidle A (2023) Facilitating digital work in public agencies through job design and digital communication management: validating empirically and theoretically derived implications on digital leadership using expert interviews. ACM Int Conf Proceeding Ser

Hood C (1991) A public management for all seasons? Public Adm 69(1):3–19. https://doi.org/10.1111/j.1467-9299.1991.tb00779.x

Imhof D, Grivas SG (2022) Holistic digital leadership and 20 factors relevant for its understanding and implementation. In: Proceedings of the European conference on management, leadership and governance

Kotter JP (2007) Leading change: why transformation efforts fail. In Museum management and marketing, Routledge, pp. 20–29

Kusanke K, Pilgenröder S, Kendziorra J, Winkler TJ (2023) Digital leadership in the public sector: towards a public sector digital leadership competency model. In: 29th annual Americas conference on information systems, AMCIS

Loeffler E (2023) Co-production of public services and outcomes. In Public management and governance routledge, pp 299–314

Malik M, Raziq MM, Sarwar N, Gohar M (2024) Navigating the change: a case study of the textile industry on digital leadership, digital transformation and innovative business models. Benchmarking. https://doi.org/10.1108/BIJ-06-2023-0376

Margetts H, Dunleavy P (2013) The second wave of digital-era governance: a quasi-paradigm for government on the Web. Philos Trans R Soc a: Math, Phys Eng Sci 371(1987):20120382

McGinty N, Lylova E (2020) Transformation of the HR management in modern organizations. https://doi.org/10.2991/aebmr.k.200201.004

McMullin C (2021) Challenging the necessity of New Public Governance: co-production by third sector organizations under different models of public management. Public Adm 99(1):5–22

Mergel I (2016) Agile innovation management in government: a research agenda. Gov Inf Q 33(3):516–523

Mergel I, Rethemeyer RK, Isett K (2016) Big data in public affairs. Public Adm Rev 76(6):928–937

Mihai PD, Liviu T, Maria N (2022) E-Government and the general population's digital skills in the European Union and OECD Member States. Springer Proc Bus Econ

Nazmetdinova K, Kalmykova S (2023) E-Government in Russia: developing and Improving the quality of implementation of the e-Government program. In: Ilin I, Petrova MM, Kudryavtseva T (eds) Digital transformation on manufacturing, Infrastructure & Service. DTMIS 2022

OECD (2020) Policy framework on sound public governance. https://doi.org/10.1787/c03e01b3-en

Osborne SP, Radnor Z, Nasi G (2012) A new theory for public service management? toward a (Public) service-dominant approach. Am Rev Public Adm 43(2):135–158. https://doi.org/10.1177/0275074012466935

Peters BG (ed.) (2001) The politics of bureaucracy (5 ed.) Routledge

Pollitt C, Bouckaert G (2011) Public management reform: a comparative analysis-new public management, governance, and the Neo-Weberian state. Oxford University Press

Rogers EM, Singhal A, Quinlan MM (2014) Diffusion of innovations. In: An integrated approach to communication theory and research, Routledge, pp 432–448

Rogers R (2004) Information politics on the Web. mit Press

Ruël H, Bondarouk T, Looise JK (2004) E-HRM: innovation or irritation. An explorative empirical study in five large companies on web-based HRM. Management revue, pp 364–380

Sani S, Mandina SP (2024) Examining the challenges of adopting modern technologies in public sector human resource management. In: Digital transformation in public sector human resource management, pp 60–89 https://doi.org/10.4018/979-8-3693-2889-7.ch004
Scott I, Gong T (2021) Coordinating government silos: challenges and opportunities. Glob Public Policy GovAnce 1(1):20–38
Shava E, Ndebele N (2024) Digital transformation in public sector human resource management. https://doi.org/10.4018/979-8-3693-2889-7
Slemp CC (2023) Using data to guide public health action and societal change. In: Building strategic skills for better health: a primer for public health professionals, pp 150–167. https://doi.org/10.1093/oso/9780197744604.003.0012
Speer J (2012) Participatory governance reform: a good strategy for increasing government responsiveness and improving public services? World Dev 40(12):2379–2398
Tigre FB, Henriques PL, Curado C (2024) The digital leadership emerging construct: a multi-method approach. Manag Rev Q. https://doi.org/10.1007/s11301-023-00395-9
Torfing J, Ferlie E, Jukić T, Ongaro E (2021) A theoretical framework for studying the co-creation of innovative solutions and public value. Policy Polit 49. https://doi.org/10.1332/030557321X16108172803520
van der Voort HG, Klievink AJ, Arnaboldi M, Meijer AJ (2019) Rationality and politics of algorithms. Will the promise of big data survive the dynamics of public decision making? Gov Inf Q 36(1):27–38. https://doi.org/10.1016/j.giq.2018.10.011
Wang K, Li B, Tian T, Zakuan N, Rani P (2023) Evaluate the drivers for digital transformation in higher education institutions in the era of industry 4.0 based on decision-making method. J Innov Knowledg 8(3):100364. https://doi.org/10.1016/j.jik.2023.100364
Wang MH, Chen CC, Chen KY, Lo HW (2023) Leadership competencies in the financial industry during digital transformation: an evaluation framework using the Z-DEMATEL Technique. Axioms 12(9):855. https://doi.org/10.3390/axioms12090855
Wang T, Lin X, Sheng F (2022) Digital leadership and exploratory innovation: from the dual perspectives of strategic orientation and organizational culture. Front Psychol 13:902693. https://doi.org/10.3389/fpsyg.2022.902693
Weber M, Gerth H, Mills CW (1961) From max weber: essays in sociology. Routledge. (translated, edited and with an introduction by Gerth HH, Wright Mills C, Paul K)
Xiao G, Wang Z (2022) Digital transformation in higher education: key areas, content structures, and practice paths. Front Educ China 17(4):557–580. https://doi.org/10.3868/s110-007-022-0024-5
Yildirim KE (2023) Digital leadership in public administration. In: Management in the digital era: different perspectives, pp 25–45. https://www.scopus.com/inward/record.uri?eid=2-s2.0-85175502963&partnerID=40&md5=666f0608663b11f0c77d040c2cc6fa36
Zhang J, Chen Z (2024) Exploring human resource management digital transformation in the digital age. J Knowl Econ 15(1):1482–1498. https://doi.org/10.1007/s13132-023-01214-y
Živković S (2022) Inspiring digital transformation: an integrative leadership competency framework. Econ Thought Pract 31(1):237–254. https://doi.org/10.17818/EMIP/2022/1.11
Zuccarini M (2007) Transformation of government roles toward a digital state. In: Anttiroiko A-V, Mälkiä M (eds) Encyclopedia of digital government. Idea group reference, pp 1562–1565

The Digital Dimension as a Fundamental Variable for the Development of Human Capital in Public Administration: A Utilitarian Approach

15

João Rodrigues dos Santos, Andreia de Bem Machado, Maria José Sousa, Rui Brito Fonseca, and José Dias Lopes

Abstract

The article carries out a theoretical analysis of the concepts of Utilitarianism, the Principle of Diminishing Marginal Utility and the Laffer Curve, exploring their relevance in the context of investment in technology to increase the efficiency of Human Capital in public administration. Based on the "authorial intention" of the authors and the subsequent stages of bibliographical research and interpretation of concepts, the authors identify connections between these concepts

J. R. dos Santos (✉) · R. B. Fonseca · J. D. Lopes
Universidade Europeia/CETRAD, Lisbon, Portugal
e-mail: joao-rodrigues.santos@universidadeeuropeia.pt

R. B. Fonseca
e-mail: rui.fonseca@iseclisboa.pt

J. D. Lopes
e-mail: jose.diaslopes@iseclisboa.pt

J. R. dos Santos
Universidade Europeia, Lisbon, Portugal

Centre For Transdisciplinary Development Studies, Lisbon, Portugal

A. de Bem Machado
Universidade Federal de Santa Catarina, Florianopolis, Brazil
e-mail: andreia.bem@ufsc.br

M. J. Sousa
University of Lisbon, Lisboa, Portugal
e-mail: maria.jose.sousa@iscte-iul.pt

J. D. Lopes
Instituto Superior de Economia e Gestão, Lisboa, Portugal

M. J. Sousa et al. (eds.), *Economic Sustainability Potentiated by Artificial Intelligence*, Contributions to Environmental Sciences & Innovative Business Technology,
https://doi.org/10.1007/978-981-96-8089-4_15

and highlight how public investment in digital infrastructure and resources can maximize the public utility generated by each euro invested/spent in the public sphere. In the specific context of this article, the logicality of the conceptual relationships between the different elements studied emphasize the central role of Human Capital in public administration. Human Capital in the public sector, by benefiting from access to updated information and having the necessary skills to efficiently use available technology, assumes a crucial preponderance for optimizing investment and public spending decisions.

Keywords

Public administration • Human capital • Utilitarianism • Diminishing marginal utility • Laffer curve • Digital

15.1 Introduction

In a world in constant technological and economic evolution, efficiency in the management of human resources becomes crucial for the success and sustainability of public organizations.

Utilitarianism, an ethical theory developed by philosophers such as Jeremy Bentham and John Stuart Mill, postulates that actions should be evaluated based on their usefulness in maximizing general well-being. According to Mill (1861):

> The creed which accepts as the foundation of morals, Utility, or the Greatest Happiness Principle, holds that actions are right in proportion as they tend to promote happiness, wrong as they tend to produce the reverse of happiness.

To Bentham (1781, p. 14):

> By the principle of utility is meant that principle which approves or disapproves of every action whatsoever, according to the tendency it appears to have to augment or diminish the happiness of the party whose interest is in question.

Santos (2020) states that utilitarianism "*is a branch of moral philosophy applied to the area of economics. Within the scope of economic science, utilitarianism focuses on moral concerns that result from inequalities in the redistribution of wealth*" (p. 64).

In the context of utilitarianism, Sanchirico (2008) highlights that, in a generic way, literature "*views the labor income tax as (among other things) a means of equalizing individuals' economic well-being*" (p. 1599).

In Bentham and Mill's conception of utility (as in utilitarianism), the idea of "pleasure" (meaning the absence of "pain") groups, in a figurative sense, all the positive results that result from the actions that precede them.

In this context, investment in technology to improve the efficiency of Human Capital in Public Administration can be analyzed in light of this theory, considering its potential impact on well-being and the quality of services provided to citizens. Vellela et al. (2024) give us a perfect example in this sense. Regarding the crucial importance that technology has in water management - which is, in the twenty-first century, one of the most relevant responsibilities and one of the main challenges of the public sector in the twenty-first century, the authors highlight that "*water level sensors find applications in industries like agriculture, environmental monitoring, and wastewater management, providing crucial data for efficient water resource management*" (p. 294).

In theoretical terms, it is, in turn, the Principle of Diminishing Marginal Utility that allows this analysis. Todorova (2021) confirms that the "*dinishing marginal utility was first formulated by Herman Gossen (1854).*"According to the same author, Herman Gossen, regarding the dinishing marginal utility pattern, stated that *"the magnitude of one and the same satisfaction, when we continue to enjoy it without interruption continually decreases until satisfaction is reached."* Thus, according to its conceptual nature, the Principle of Diminishing Marginal Utility, which makes the concept of Utilitarianism operative, suggests that as the quantity of a given resource increases (such as investment in technology), the additional benefit obtained decreases progressively. For this reason, it is essential to understand the extent to which investment in technology can really contribute to increasing the efficiency of Human Capital in public administration, without reaching the point of diminished return.

In the context of technology applied to the fulfillment of public sector functions, Cuento et al. (2023) reinforce that "*e-government contributes to well-being by enabling more efficient management of public resources through paperless online applications and transactions, and by enhancing the quantity and quality of services offered to citizens*" (p. 13). The same authors also emphasize that this idea is supported by research.

Haynes & Haynes (2011) highlight that improvements in efficiency typically occur as a result of the emergence of new circumstances (for example, budgets that are unexpectedly reduced, technological developments, or learning from mistakes made). "*Efficiency savings, even before factoring out such improved technology or best practice improvements are marginal*" (p. 12).

The Laffer Curve, a concept widely discussed in economic theory, represents the relationship between tax rates and the State's tax revenues. The Laffer Curve is the main area of study of Artur Laffer, an American economist who first became recognized during the Reagan administration when he was a member of Reagan's Economic Policy Advisory Council, in the period 1981–1989. Slemrod & Slemrod (1994), regarding the crucial importance that the concept of the Laffer Curve had in the debates on fiscal policy in the 70 s and early 80 s, precisely defined Artur Laffer's conception in relation to the relationship between tax burden and productivity (which influences tax revenue):

> The "Laffer curve," the inverted-U-shaped plot of revenues against tax rate, encapsulated the idea that, because of the response of taxpayers, there was leakage in revenues when tax rates increased and that tax increases beyond some tax rate caused revenues to fall.

In the context of investment in technology in public administration, the Laffer Curve can be interpreted in order to examine the extent to which increased investment in technology results in an effective increase in the efficiency of Human Capital, without incurring excessive opportunity costs that could compromise the financial sustainability of the State. de Haan & Riener (2024), within the scope of their study entitled "Bargaining with Investment and Opportunity Costs," start from an absolute and empirically valid premise. They *"start from the notion that people may have a tendency to neglect opportunity costs"* (p. 23). This applies to all sectors of activity, including the public sector. This applies to all sectors of activity, including the public sector. In relation to the essence of the Laffer Curve, it is not so much about the tax revenue that is achieved, but rather what is no longer achieved when, after a certain level of taxation, tax revenue begins to shrink. This is the notion of "opportunity cost": what is not achieved when a decision is made.

In this way, this article (which has a very clear and objective authorial intention behind it) proposes to explore the articulation between these theoretical concepts, offering a critical and univocal analysis of the practical application of investment in technology to increase the efficiency of Capital Human in public administration. Regarding the concept of "authorial intention," Patterson (1989), in the context of the interpretation of literary and legal texts, making reference to Knapp & Michaels (1985), summarizes it protected from the point of view of hermeneutics as a fundamental science in the process court decision: *"when one asks, 'What does this text mean?' the question can only be answered by directing the questioner to the author's intentions"* (p. 79). As is well known, the decision-making process in a judicial case is based on the judges' own interpretation of the Law and the authorial intention of the parties who produced it. It is therefore a process that results from the very particular and ideosyncratic vision and framework of the "interpreter," who is the judge. His decisions are therefore based also (and, perhaps, above all) on the very specific framework that led him to the perception he had at each moment.

Thus, by addressing the intricately complex interactions between Utilitarianism, the Principle of Diminishing Marginal Utility, the Laffer Curve and technological investment that aims to increase the human capital of Human Resources in the public sector, we hope to contribute to an informed and reasoned debate on public investment policies and the efficient management of resources (human resources included) in government organizations.

15.2 Methodology

The methodological approach adopted for this research was interpretative qualitative. This methodology depends on the "authorial intention" of the researchers, allowing for an in-depth interpretative analysis and an original articulation of the concepts discussed.

In the qualitative-interpretive research process, the authors use their personal and intuitive interpretation, based on their expertise, to identify associations between apparently unconnected variables. As such, researchers' own everyday practice provides an opportunity to advance debates about rigor, allowing us as qualitative researchers to formulate a considered response that respects both the complexities and the unique contribution that qualitative research can make (Walther et al. 2017).

In this context, it is clear that this is the only methodological approach possible when the scientific production aims to substantiate a very specific "authorial intention" on the part of the authors, based precisely on their unique academic and professional backgrounds. This approach, resulting from a univocal vision of the articulation of concepts with no explored conceptual connection, can also be called "authorial reasoning," which corresponds to an analytical interpretation of the researchers who authored the article and their idiosyncrasies. This approach allows the researcher to recognize the basic data of the interpretative and cognitive processes inherent in social life (Bartoli, 2020).

In this type of approach, the authors rely fully on their expertise and personal experience to analyse and interpret the data or concepts. They apply their knowledge acquired throughout their academic and professional careers to develop a deep and original understanding of the subject in question, then put the result at the mercy of the scrutiny of their peers. The phenomenological basis of this approach relates to the idea that the "person" and the "world" are related through the experiences lived in that world (Zerguit, 2012).

This approach values the authors' critical reflection and creative thinking, allowing them to explore the topic in a unique and innovative way. However, it is important to emphasize that, although based on the authors' reasoning, this approach must be accompanied by a careful selection of relevant literature and a rigorous analysis of the available information, thus guaranteeing the logical coherence, validity, and credibility of the study.

In order to carry out the theoretical analysis of the concepts of Utilitarianism, the Principle of Decreasing Marginal Utility and the Laffer Curve, as well as their articulation with increasing the efficiency of Human Capital in the public administration through investment in technology, a methodological approach was adopted that consisted of several stages:

1. Bibliographic and Academic and Scientific Database Research: Initially, theoretical concepts were researched in various sources, including academic and scientific books and documents. Reliable scientific databases were used to ensure that the information gathered was comprehensive and up-to-date.

2. Interpretation of the Concepts: Each concept was carefully scrutinized and interpreted by the article's authors. This stage involved a critical analysis of the definitions and discussions found in the relevant literature, with the aim of deeply understanding the meaning of each concept and its theoretical implications. This interpretative qualitative methodological approach, reflective reasoning or authorial reasoning, depends on the "authorial intention" of the authors of this study. "Authorial intent" is a very specific theoretical concept.
3. Conceptual Association and the Creation of an Original Conception: After the individual interpretation of each concept, we proceeded to its conceptual association. This stage involved identifying the connections and interrelationships between the concepts of Utilitarianism, the Principle of Decreasing Marginal Utility, the Laffer Curve and investment in technology to increase the efficiency of Human Capital in public administration. Based on this analysis, the aim is to create an original and unique concept that allows for a deeper and more comprehensive understanding of the issue in question. Once again, it is reflective or authorial reasoning and the specific theoretical concept of "authorial intention" that guide this stage of the study.

It is important to emphasize that this qualitative methodological approach made it possible to explore in detail the nuances and complexities of the concepts covered, as well as to identify new perspectives and theoretical insights. The rigor and depth of this theoretical analysis served as a basis for consolidating the "authorial intention," which consisted of the subsequent articulation of the concepts with the central theme of the article, contributing to a critical and grounded analysis of the issues in focus.

For scientific research using the interpretative qualitative approach, which focuses on analyzing and integrating concepts, the steps can be adapted as follows:

1. Formulating the research question: Clearly defining the research question that will guide the study, identifying the key concepts to be analyzed and integrated.
2. Identification and selection of concepts: Select the concepts that will be analyzed and integrated based on their relevance to the research question and the availability of information in the literature.
3. Literature review: Carry out a comprehensive review of the literature relevant to the concepts under study, identifying existing perspectives and approaches.
4. Information gathering: In this case, information gathering may involve searching for and selecting materials, such as articles, books, documents, among others, that address the selected concepts (with special attention paid to authors who are consensually recognized within the scope of each concept).
5. Analyzing the data: Qualitatively interpreting the concepts found in the literature, seeking to identify patterns, relationships and insights relevant to the research question.
6. Interpretative integration of concepts: Integrate the different concepts analysed, exploring how they relate to each other and contribute to a broader understanding of the phenomenon under study.

7. Discussion and interpretation of results: Discuss the results of the analysis and integration of concepts, highlighting the conclusions reached and the implications for knowledge in the field of study.
8. Conclusions: Summarize the main conclusions of the study and highlight their contributions to understanding the concepts in question. Reflect on the research process, including the limitations encountered and possible directions for future research.

These steps emphasize the analysis and interpretation of the concepts found in the literature, as well as the integration of these concepts to generate deeper insights and understanding of the topic under study.

15.3 Public Administration, Public Sector Human Capital and Public Goods and Services: Conceptual Approach

In the context of this article, it is considered vital to explore the main elements of the interconnection between public administration, Public Sector Human Capital and Public Goods and Services. Public administration, as the entity responsible for managing and providing essential services to society, plays a central role in organizing and regulating government activities. According to Santos & Sousa (2022), "*Public Administration has an underlying structural objective: to provide goods and services of a public nature*" (p. 379). In turn, Public Sector Human Capital refers to the set of skills, knowledge and abilities of public servants, essential for the efficient performance of administrative tasks and the provision of quality public services. "*By human capital, we mean the stock of innate abilities and characteristics that people possess and the knowledge and skills they acquire over time*" (Christensen et. al. 2019, p. 127). Finally, Public Goods and Services are the products and benefits made available by the State to the population, aiming to satisfy collective needs and promote social well-being.

According White (2015, p. 28):

> A "public good" is one whose potential users are in the aggregate willing to pay enough to cover its production costs, yet its production cannot be financed on the usual user-pays basis because (with current technology) non-payers cannot be excluded from enjoying.

In the context of public administration, efficiency and effectiveness in the provision of Public Goods and Services are decisive for its performance and legitimacy. To achieve this objective, it is essential to invest in the development of Human Capital in the Public Sector. This investment includes the continuous training of public servants, the stimulation of innovation and the adoption of human resources management practices that promote the commitment, motivation and professional development of employees. Christensen et al. (2019), referindo-se a Becker (1993), a Grossman (1972), e a Laroche et al. (1999), destaca que o "*human capital can include knowledge, personality traits, health, experiences, education, and cognitive functioning"* (pp. 127–128). Qualified and motivated Human Capital is essential for the modernization and adaptation of public administration to contemporary challenges, including the increasing digitalization of society.

In this context, the Digital Dimension emerges as a fundamental variable for the development of Human Capital in public administration. The integration of digital technologies into administrative processes and the provision of public services offers numerous advantages, such as improving operational efficiency, optimizing resources and facilitating citizens' access to government services. However, to fully capitalize on the benefits of digitalization, it is necessary to ensure that public servants have the necessary skills and knowledge to effectively use digital tools. Precisely in this regard, Abdulraheem & Mohammed (2016) highlight that the "*enhancement of employees' ability to adapt to the changing and challenging business environment and technology for better performance*" (p. 62). Therefore, the development of Human Capital in the Public Sector must include training in relevant digital skills, as well as the promotion of an organizational culture oriented towards innovation and technological adaptation.

Furthermore, adopting a utilitarian approach in managing Human Capital and providing public services can provide a perspective oriented toward results and the common good. From this perspective, decisions regarding the development of Human Capital and the allocation of resources must be guided by maximizing public value and satisfying society's needs. "*In particular, welfare economics, and hence virtually every public policy recommendation formulated by economists, has for years been influenced by utilitarianism in some manner*" (Baujard, 2013, p. 45). The application of utilitarian criteria in the evaluation of government policies and programs can contribute to a more efficient and transparent management of public resources, ensuring equity in the distribution of benefits and promoting sustainable development. Mandler (2001), regarding the strong intuitive and qualitative dimensions of the principle of diminishing marginal utility, shows "*that diminishing marginal utility, rather than the standard stronger assumption of cardinality, also justifies utilitarian recommendations on redistribution*" (p. 2). And redistribution is, above all, carried out through public policies. Public goods are the best example of the redistribution process.

The development of Human Capital in the Public Sector and the integration of the Digital Dimension in public administration are interdependent elements that can enhance the efficiency, effectiveness, and legitimacy of the provision of Public Goods and Services. Abdulraheem & Mohammed (2016) emphasize that preventing skills obsolescence and dealing with new technological advances is a fundamental part in the "*standard of performance for the job*" (p. 45). A utilitarian approach can provide a useful conceptual framework to guide these initiatives, ensuring that they are aligned with the objectives of maximizing social well-being and promoting human development.

15.4 Utilitarianism and the Principle of Diminishing Marginal Utility

As an economic and social theory, or if we want, above all as a system of moral philosophy (which crosses law, economics and politics), ethical-normative oriented towards human development and the well-being of societies, Utilitarianism develops itself in a Europe still dominated by the principles of Enlightenment Reason.

Objectively, the central core of this theory is formed by a morality focused on human well-being, by the evaluation based on the consequences of the moral quality of actions or rules of action and by the conception that human well-being must be maximized and, consequently, suffering must be minimized. From the point of view of Utilitarianism as an ethical theory that assumes and defends that the best action is that which maximizes the happiness or general well-being of human societies, in other words that the main objective of any human action must be the production of the greatest quantity possible amount of happiness or pleasure and the least possible amount of pain or suffering (Bentham, 1907), in economic terms states that the economy can be an instrument capable of realizing human happiness and well-being. These characteristics constitute the so-called Principle of Utility.

Thus, Utilitarianism is based on the concept of utility, which can be understood as the ability of something to provide benefits or satisfaction. In this way, an action is considered good if it produces more benefits or satisfaction than any other alternative. The measure of this utility can be difficult to quantify, but the central idea is that the consequences of an action should be evaluated in terms of its contribution to the general well-being of society (Mill, 1863a, b).

Regarding the Principle of Diminishing Marginal Utility, it results from an economic idea connected with Utilitarian theory. Thus, the Principle of Diminishing Marginal Utility suggests that the utility or satisfaction a person derives from a good or service, decreases as he or she consumes more of that good or service. In other words, the more someone owns or consumes a good, the less additional value they attribute to each additional unit of that good. Note that this principle is fundamental to understanding consumer behavior and consumption decisions, especially in the consumer or hyper consumption societies in which we live today (Lipovetsky, 2010). Ultimately, we can take as an example the situation in which when a person is hungry, the value attributed to an additional unit of food is very high. However, as that person eats and becomes full, the value of each additional serving of food decreases. In fact, the satisfaction or utility obtained from each additional portion of food is less than the previous one (Marshall, 1982).

The Principle of Diminishing Marginal Utility has wide application and can be observed in different aspects of the economy and everyday life. For example, in economics, the Principle of Diminishing Marginal Utility is used to explain the law of supply and demand (Mankiw, 2014). Furthermore, in making investment decisions, the Principle of Diminishing Marginal Utility is used to evaluate the expected returns of different investment opportunities. Investing in a project that offers decreasing returns may not be as attractive as investing in projects with increasing or constant returns over time (Samuelson & Nordhaus, 2010).

Naturally, although Utilitarianism and the Principle of Diminishing Marginal Utility have their applications and clear advantages, like all science, they are also the target of criticism from other academics. One of the most common criticisms of Utilitarianism is that it can lead to decisions being made that sacrifice the rights or interests of minorities in favor of the majority, if this results in greater overall happiness. In this way, it can be interpreted as a violation of the principles of justice and individual rights (Rawls, 1971). Likewise, one of the criticisms of the Principle of Diminishing Marginal Utility is the fact that this power is not applicable to all socioeconomic or individual contexts. It should be noted that there are situations in which the usefulness of a good or service may increase with its consumption and not decrease. Furthermore, the measurement of usefulness can be subjective and can vary from person to person, making universal application of the principle difficult (Sen, 1977).

It should be noted that Utilitarianism and the Principle of Diminishing Marginal Utility are important concepts that have influenced both ethics and economics. These offer a rational framework for evaluating the consequences of our actions and decisions, whether in terms of general well-being or individual utility. However, as with all economic science, it is crucial to recognize its limitations, as ethics and economics are complex and multifaceted fields, and no theory or principle can fully capture the complexity and diversity of human experiences.

15.5 The Laffer Curve

The Laffer curve is an important contribution from supply-side economists, demonstrating the nonlinear effect of tax rates on total tax revenues (Tavor et. al. 2021). Changes in tax rates have two effects on tax revenues. The first effect is that due to the reduction in the tax rate there is a reduction in revenue from a given tax base. The second effect, possibly the opposite, should also be considered, since lower tax rates can encourage people to work and invest more. Thus, tax revenues may increase since a larger tax base is created due to the reduction in the tax rate. The overall effect of the tax rate on tax revenues is ambiguous. This can lead supply-side economists to the traditional view of Laffer and his followers that two different tax rates yield the same tax revenue (Laffer, 1979).

The question that needs to be asked is why only two points necessarily guarantee the same tax revenue that leads to only one maximum point on the Laffer curve. Other graphical interpretations of the Laffer curve start from the premise that it can have more than one inflection point. Henderson (1981) was the first to observe that the amount of tax revenue could vary depending on individual preferences in the case of lower taxes on labor income. The Laffer Curve provokes a reflection on the distortions and loss of incentives that tax rates cause in the economy, and these rates can, as they are increased, reduce tax revenue (Hájek, et al. 2021).

According to this theory, when the tax burden is low, the relationship between these two variables is direct, i.e., at first, a small increase in the tax burden leads to an increase in total tax collection by the public sector. The Laffer curve illustrates a theoretical relationship between tax rates and the resulting levels of government revenue (Lin & Jia, 2019).

However, after reaching a maximum revenue point, the curve suffers an inflection, and the relationship becomes inverse. If the government tries to raise this already very high tax burden even further, it begins to lose revenue instead of increasing it (Hájek, et al. 2021). Laffer (1979) points to the fact that when the tax burden is too high, it encourages tax evasion (tax avoidance) and encourages informal and illegal activities. This, in turn, discourages business in general in the formal economy, thus reducing the tax base.

This idea is corroborated by Hájek et al. (2020). The authors, referring to Moszer (1981), emphasize that, after a certain level of tax burden, "*any increase in the tax rate cannot lead to an increase in tax revenue, as with a higher tax burden, the taxpayer responds to this fact, and the work itself becomes insufficiently profitable due to the substitution effect*" (p. 4).

The results of the reduction in tax rates promoted in the United States by Ronald Reagan did not corroborate Laffer's (1979) idea, and there was a reduction in tax revenue. Although a general reduction in tax rates leads to a decrease in revenue, there is evidence that some taxpayers were on the wrong side of the Laffer curve (1979). The tax revenue collected in the 1980s from the richest Americans increased when these rates were reduced.

Laffer's (1979) arguments, in addition to provoking debate on the tax burden, can be more accepted when applied to countries where the tax burden is higher than in the United States. In Sweden, for example, at the beginning of the 1980s, workers were taxed 80% of their income, which was a disincentive to work. Studies have shown that if Sweden were to reduce the tax rate for workers, it could achieve an increase in revenue.

15.6 The Essence of the Principle of Diminishing Marginal Utility Applied to Public Sector Efficiency

In public management, the principle of decreasing marginal utility, also known as equimarginality, plays a crucial role in pursuing efficiency. The German economist Hermann Heinrich Gossen (1810–1858), alluded to by Greenwald et al. (2008), concluded that "*the amount of any pleasure is steadily decreasing as we continue until the last saturation is reached*" (p. 27). Regarding the central and conceptual importance that the principle of efficiency assumes in the sphere of public administration activity, Lai chimeng (2024), referring to Adegbami & Uche (2015) and Ali (2019), recalls that "*the theory of good governance encompasses various components, including the rule of law, accountability, transparency, efficiency, equity, responsiveness, participation, and consensus*" (p. 4958).

This principle of decreasing marginal utility, derived from economics, refers to the idea that as we increase the quantity of a resource applied in an activity, the additional benefit of that resource diminishes progressively. Precisely in this sense, Adner & Zemsky (2004) state that "*DMU [decreasing marginal utility] reduces the value of performance improvements over time*" (p. 5). In other words, the more resources allocated to a particular area, the less additional value is obtained with each additional unit of that resource.

When applying this principle to public administration, we can understand how resource allocation is crucial to maximizing results with minimal waste. Regarding the conceptualization of "governance," Srivastava (2009) highlights the notion of "resource management" by stating that "governance" is "*the process by which authority is exercised in the management of a country's economic and social resources*" (p. 3).

For example, consider the case of road construction in a city. Initially, building main roads brings huge benefits to mobility and economic development, increasing the overall efficiency of the transportation system. However, as more roads are built, the additional benefit of each new road decreases, since the main routes are already established and mobility needs are largely met. In this regard, the approach of Kleiven et al (2024) to what is called, in conceptual terms, the "profit-maximizing level of production" is considered relevant. In the field of planning investment policies in Hydropower Capacity, the authors recall that "*the optimal capacity installment is where the marginal value of installing capacity equals the marginal cost*" (p. 24).

Equimarginality in public administration means that resources should be allocated to maximize efficiency, considering the marginal costs and benefits of each additional unit of resource applied. This requires careful analysis of society's needs and priorities, as well as ongoing assessment of the impact of government policies and programs.

According to Santos & Sousa (2021, p. 383):

> when the logic of the equimarginal principle (rational expenditure rule) is partially applied it is easy to conclude that the investment in a public good or service, due to the specific characteristics of this type of good or service (see point 1), will allow it to generate more or less total utility, depending on the number of people who benefit from its consumption.

An illustrative quantitative example of applying this principle can be found in public health. In the health-economy interface space, the impact of adult mortality on economic growth has always been the subject of intense study by researchers. Laxminarayan et al. (2007) recall that "*greater adult mortality implies a lower rate of return to human capital investments, which in turn is a determinant of economic growth*" (p. 7). In this context, suppose a government is considering resource allocation for vaccination programs. Initially, investing in vaccination brings significant benefits in terms of reducing diseases and mortality, leading to a significant improvement in population health and economic productivity. However, as more resources are allocated to the vaccination program, the additional benefit of each new dose decreases, since most of the population is already immunized against major diseases.

The equimarginal rule (or rational expenditure rule) allows us to clearly understand the diminishing marginal utility concept (see Tables 15.1 and 15.2). Greenwald et al. (2008, p. 28), once again referring to the German economist Hermann Heinrich Gossen (1810–1858), pose the following example as an explanation of the Equimarginal Principle:

> If a man is free to choose among several pleasures but has not time to afford them all to their full extent, then in order to maximize the sum of his pleasures he must engage in them all to at least some extent before enjoying the largest one fully, so that the amount of each pleasure is the same at the moment when it is stopped; and this however different the absolute magnitude of the various pleasures may be.

Thus, considering the logic of the diminishing marginal utility, when the individual decides to replace the consumption of the fifth unit of good A with the consumption of the fourth unit of good B, he achieves a greater total utility (400) and,

Table 15.1 Diminishing marginal utility (consumption of goods A and B)

Consumption level	Good A expense (€)	Good A utility (Marginal ut)	Good B expense (€)	Good B utility (Marginal ut)	UT per 1€ invested (total ut / total € invested)
1	2,00	100	2,00	100	
2	2,00	60	2,00	60	
3	2,00	30	2,00	30	
4	2,00	10	0,00	0	
5	2,00	5	0,00	0	
6	0,00	0	0,00	0	
Total	1,00	205	6,00	190	395 total ut/ 16.00€ = 24.7

Source Authors' elaboration (2024)

Table 15.2 Diminishing marginal utility (consumption of goods A and B): maximizing utility

Consumption Level	Good A Expense (€)	Utility (*ut*) Good A (marginal *ut*)	Good B Expense (€)	Good B (marginal *ut*)	*ut* per €1 invested (total ut / total € invested)
1	2,00 €	100	2,00 €	100	
2	2,00 €	60	2,00 €	60	
3	2,00 €	30	2,00 €	30	
4	2,00 €	10 →	(2,00 €)	→ (10)	
5	~~2,00 €~~	~~5~~	0,00 €	0	
6	0,00 €	0	0,00 €	0	
Total	8,00 €	200	8,00 €	200	400 total *ut* / €16,00 = 25,0

Source Authors' elaboration (2024)

also, a higher utility per euro invested. If we consider that resources are scarce and that, therefore, they should generate the greatest possible utility (*ut*), this is, according to the principle of scarcity in Economics, a rational choice. Precisely in this sense, Santos (2020) highlights *that "economics is the study of the way people make choices in conditions of scarcity and the results of these choices for society"* (p. 21).

Thus, within the scope of expenditure/investment choices made in the sphere of public administration, economic rationality (in the context of the principle of diminishing utility and the rule of equimarginality) assumes decisive preponderance.

For example, suppose a public investment of € 5000 in a municipal territory. For purely academic purposes, we will consider that each individual resident or present who is covered by the investment achieves 1000 utility units (*ut*) each time they use it. So, it is possible to conclude that each € 1 of the € 5000 triggers 0.2*ut* (0.2 *ut*/€1), that is, 0.2 *ut*/€1 = 1000*ut*/€5000 for each individual. According to the same logic, if the investment covers two people, 0.4 *ut*/€1 is reached. If the investment covers 10 people, the efficiency level will be equal to 2.0 *ut*/€1; if it covers 100 people, the efficiency level will be equal to 20.0 *ut*/€1 and so on. The example may seem strange, but it is not at all. Let's imagine installing four benches and a public garden table. This urban furniture corresponded to an investment of 5000 euros. Now, just transpose the previous example to this case. If the street furniture has been used by one person in a certain period of time, then each €1 of the €5000 will have triggered 0.2 *ut* (0.2 *ut*/€1). If the street furniture has been used by 100 people in the same period of time, then each €1 of the €5000 will have triggered 20.0 *ut* (20.0 *ut*/€1).

Once again, according to Santos & Sousa (2022), *"it is about efficiency in the management of public resources that we are dealing with"* (p. 383).

Therefore, considering the logic exposed, when going back to the example of vaccination and the point at which its presentation remained, equimarginality suggests that the government should balance the marginal costs and benefits of expanding the vaccination program. This may involve considering alternatives, such as investing in other public health programs or allocating resources to areas where the need is more urgent. Considering that it is using the tax burden applied to taxpayers that a vaccination program is carried out by Governments, attention is drawn to the Marginal Revenue Rule by Weisbach et al. (2018), which *"instructs that the marginal increase in revenue resulting from the change in taxpayer behavior be counted as a benefit in the cost–benefit calculation"* (p. 1511). In this way, b

By applying the principle of decreasing marginal utility, the government can maximize efficiency in resource allocation and ensure that social benefits are optimized. It is the economic rationality (in the economic sense of utilitarianism) of choices that this logic focuses on. Obrimah (2022), in his study entitled "The marginal utility for wealth: Effects of either of heterogeneity of risk preferences or irrationality of risk seeking agents," highlights that *"if risk seeking agents act rationally, they only are parameterized by decreasing marginal utility for wealth"* (p. 1).

According to this logic, then, the utility quotient (*ut*) for each euro (€1) invested in the public sphere will increase if the investment is carried out in order to meet the greatest amount of needs in a territory. The dimension of human needs at each moment can, theoretically and subjectively, be evaluated by the total utility that their supply will trigger. Therefore, the more calibrated and adjusted public decisions are, the greater the efficiency in the management of each monetary unit available to the public administration and applied in the form of public investment or even public current expenditure. And this calibration is absolutely essential, because the greater the tax burden, the greater the drain on liquidity from the economy carried out by civil society. In effect, *"increases in capital taxes raise private and public capital as long as the increase in investment in public capital (the second term) outweighs the negative effect of capital taxes on the return to private capital"* (p. 9).

And it is here, precisely, that Human Resources in the Public sphere, having access to updated and permanent information—only possible, today, with the use of increasingly sophisticated Information and Communication Technologies—increase and improve their human capital so that in each public expenditure decision all potentially achievable dimensions of information are considered. Lai chimeng (2024), regarding the importance of accessing and using information from the various platforms available today, states that *"data entry, processing, analysis, and management are all essential professional competencies, meticulous planning is necessary with regard to training, recruitment, and knowledge management"* (p. 4959).

In this way, with the use of the immense technological platforms available, by ensuring, at each moment, the consideration of all available information relevant to the decision-making process, decisions will invariably result in reaching the total potential utility (*ut*).

15.7 Digital and Technological Resources and Increased Utility (in the Economic Sense of Utilitarianism) Triggered by Public Sector Expenditure

In the digital Era we live in, investing in technology and digital resources plays a pivotal role in optimizing the management of public resources. Srivastava (2009), comparing efficiency in resource management between private companies and government entities, highlights that *"the growing consensus among development specialists in this regard is that, with the limited access of governments to information, markets generally allocate resources more efficiently"* (p. 7). By allowing Human Resources in public administration to use real-time updated information for every decision made, not only does decision-making efficiency increase, but the utility of available resources is also maximized. In this context, the principle of decreasing marginal utility emerges as an essential tool to guide investment and allocation of technological resources effectively and efficiently.

Regardless of the personal preferences of decision makers, utilitarian rationality must dictate the rules. With a conception that fully fits the essentials of this article, Friedman-Sokuler & Senik (2023) show "*that in order to maximize one's total "utility" or happiness, it is not enough to spend more time on pleasant activities*" (p 30).

The principle of decreasing marginal utility suggests that as we increase the quantity of a specific resource applied in an activity, the additional benefit of that resource decreases progressively. Adner & Zemsky (2004), in this regard, give an example that is both simple and very clear: "*the utility from increasing the maximum speed of an automobile is much greater when the maximum speed is 40 miles per hour than when it is 100 miles per hour*" (p. 7). Applied to the management of public resources, this means that continuous increase in investment in digital technology may lead to diminishing returns in terms of efficiency if not done strategically. Instead, it is crucial that investment in technology is articulated to maximize the utility of available resources and to achieve the objectives set by public administration. Azzutti et al. (2023) in the context of Artificial Intelligence, highlights precisely this logic, stating that "*the use of AI tools to support and potentially replace human judgment in administrative decision-making is proposed to deliver efficiency gains and reduce human biases*" (p 14).

By integrating digital technology into public administration, it is possible to provide human resources with access to real-time updated information, enabling more informed and agile decision-making. Also Mikesell (2014), cited by Kim (2015), in the context of fiscal administration and public sector, highlights that the "*use of new technology to build up the competence and efficacy of financial and budgetary function*" (p. 15). For example, imagine a government agency responsible for managing traffic in a city and, therefore, for the public expenditure inherent to this responsibility. By investing in real-time traffic monitoring systems and data analysis algorithms, managers can make more effective decisions to optimize traffic flow, reduce congestion, and improve road safety. Still Lai chimeng (2024), in the field of the functioning of justice and courts (one of the most important competencies in the sphere of public administration), emphasizes the importance of technology in "establishing information exchange channels between courts, and judicial departments. This will enable the efficient utilization of data [...], leading to improved performance and value" (p. 4959).

However, it is important to recognize that the added value of each new technology or digital resource decreases as more resources are allocated to that specific area. Kögel (2009), in the context of environmental sustainability, making reference to Vilfredo Pareto's conception of efficiency ("Pareto Efficiency"), highlights that the "*marginal utility of goods consumption to be decreasing in environmental consumption*" (p. 3). For example, in another domain within the sphere of public administration, after implementing advanced traffic monitoring systems, the additional benefits of investing in additional technologies may be lower, since the main traffic management issues have already been addressed. At this point, the principle of decreasing marginal utility guides public managers to consider other areas where investment may bring greater marginal benefits.

Furthermore, investment in digital technology in public administration can also enhance transparency, accountability, and civic participation. In the context of innovation in public services, Carrubbo & Sarno (2011), highlight that Information Communication Technologies são ferramentas fundamentais "*to support change in public administration fostering an increase of the service exchange improving both internal relationships and the interactions with all stakeholders*" (p. 18). By providing real-time information about government decisions and actions, citizens can closely monitor government performance and contribute to the decision-making process. This not only increases efficiency in managing public resources but also strengthens democracy and trust in public administration.

Investment in digital technology in public administration has the potential to significantly increase utility in public resource management by providing Human Resources with access to real-time updated information for every decision made. However, it is crucial that this investment is articulated according to the principle of decreasing marginal utility, ensuring that resources are allocated effectively and efficiently to maximize social benefits.

15.8 Conclusive Summary

The theoretical analysis of the concepts of Utilitarianism, the Principle of Diminishing Marginal Utility and the Laffer Curve reveals the importance of investing in technology to improve Human Capital in public administration. This research highlights the conceptual relationship between public administration, Human Capital, Utilitarianism, Diminishing Marginal Utility, Laffer Curve, and Digitalization. The evidence in Tables 15.1 and 15.2 elucidates how economic rationality, combined with the principle of diminishing marginal utility, can maximize efficiency in the allocation of public resources. For example, when considering the substitution of consumption between goods, the logic of diminishing marginal utility shows that each euro invested can generate greater public utility when allocated strategically. In this context, the crucial role of Human Capital in public administration is highlighted, where access to updated information and the efficient use of available technologies are fundamental to optimizing investment decisions and public spending.

Future developments may include approaching the implementation of policies that promote continuous digitalization in public administration and the systematic technological training of Human Capital in the public sector, guaranteeing public management processes potential access and continued and effective use of updated information. The materialization of this process will allow optimizing investment and public spending decisions, ultimately aiming to maximize the potential utility achieved through the application of public administration budgets.

References

Abdulraheem S, Mohammed R (2016) The impact of training and development on employees performance and productivity. Int J Manag Sci Bus Res 5(7). Available at: https://ssrn.com/abstract=2849769

Adegbami A, Uche CIN (2015) Good governance and prison congestion in Nigeria: the case of maximum security Prison, Ilesa. Public Policy Adm Rev 3(2). Available at: https://doi.org/10.15640/ppar.v3n2a3

Adner R, Zemsky PB (2004) A demand based view of sustainable competitive advantage. Available at: https://doi.org/10.2139/ssrn.651184

Ali M (2019) Governance and good governance—a conceptual perspective. Dialogue 10(1):65–77

Azzutti A, Batista PM, Ringe W-G (2023) Navigating the legal landscape of AI-enhanced banking supervision: protecting eu fundamental rights and ensuring good administration. European Banking Institute Working Paper Series 2023–no. 140. Available at: https://doi.org/10.2139/ssrn.4430642

Bartoli A (2020) Every picture tells a story: combining interpretative phenomenological analysis with visual research. Qual Soc Work 19(5–6):1007–1021

Baujard A (2013) Utilitarianism and anti-utilitarianism. GATE Working Paper No. 1332. Available at: https://doi.org/10.2139/ssrn.2357441

Becker Gary S (1993) Human capital: a theoretical and empirical analysis, with special reference to education (3rd ed.). University of Chicago Press

Bentham J (1781) An introduction to the principles of morals and legislation. In: Batoche books kitchener. Available at: https://historyofeconomicthought.mcmaster.ca/bentham/morals.pdf.=

Carrubbo L, Sarno S (2011) Service innovation within new public governance theories E-platforms as smart service systems for public utilities. In: Proceedings of 1st cooperlink workshop, vol 3(3), pp 1972–2036. Available at: https://ssrn.com/abstract=1903973

Christensen J, Aarøe L, Bækgaard M, Herd P, Moynihan DP (2019) Human capital and administrative burden: the role of cognitive resources in citizen-state interactions. Public Adm Rev 80:127–136. Available at: https://doi.org/10.1111/puar.13134

Cuento M, Timpug J, Calabio G, Vigonte F, Abante MV (2023) Public value and electronic governance development in the social security system (SSS). Available at: https://doi.org/10.2139/ssrn.4621317

de Haan T, Riener G (2024) Bargaining with investment and opportunity costs. Available at: https://doi.org/10.2139/ssrn.4696273

Friedman-Sokuler N, Senik C (2023) Time-use and subjective well-being: is diversity really the spice of life? IZA Discussion Paper No. 16090. Available at: https://doi.org/10.2139/ssrn.4431363

Greenwald A, Naroditskiy V, Odean T, Ramirez M, Sodomka E, Zimmerman J, Cutler C (2008) Marginal bidding: an application of the equimarginal principle to bidding in TAC SCM. In: Collins J et al. (eds) Agent-mediated electronic commerce and trading agent design and analysis, Springer, pp 44–57. Available at: https://doi.org/10.1007/978-3-540-88713-3_15

Grossman M (1972) On the concept of health capital and the demand for health. J Polit Econ 80(2):223–55. Available at: https://mgrossman.ws.gc.cuny.edu/files/2017/06/conceptofhealthcap.pdf

Hájek J, Šafr K, Rotschedl J, Cadil J (2020) The Laffer curve decomposed. Ekonomický časopis. Available at: https://ssrn.com/abstract=3749646

Henderson D (1981) Limitations of the Laffer curve as a justification for tax cuts. Cato J 1:45. Available at: https://doi.org/10.5296/jpag.v9i3.15417

Kim K (2015) Fiscal administration and public sector. Available at: https://doi.org/10.2139/ssrn.2586039

Kleiven A, Nadarajah S, Fleten S-E (2024) Hierarchical planning for hydropower capacity upgrade: exploiting structure in reoptimization and investment policies. Available at: https://doi.org/10.2139/ssrn.4728674

Kögel T (2009) On the relation between dual-rate discounting and substitutability. Economics Discussion Paper No. 2009–10. Available at: https://doi.org/10.2139/ssrn.1726709
Laffer AB (1979) An equilibrium rational macroeconomic framework. Econ Issues Eighties 44:44–57
Lai Chimeng P (2024) A perspective of good governance on the reform of correctional governance system and capabilities in the digital era. Int J Multidiscip Res Anal. Available at: https://ssrn.com/abstract=4734147
Laroche M, Marcel M, Ruggeri GC (1999) On the concept and dimensions of human capital in a knowledge-based economy context. Can Public Policy 25(1):87–100. Available at: https://doi.org/10.2307/3551403
Laxminarayan R, Klein EY, Dye C, Floyd K, Darley S, Adeyi O (2007) Economic benefit of tuberculosis control. World Bank Policy Research Working Paper No. 4295. Available at: https://doi.org/10.1596/1813-9450-4295
Lin B, Jia Z (2019) Tax rate, government revenue and economic performance: a perspective of Laffer curve. China Econ Rev 56:101307. https://doi.org/10.1016/j.chieco.2019.101307
Lipovetsky G (2010) A Felicidade Paradoxal—Ensaio sobre a sociedade do hiperconsumo. Edições 70
Mandler M (2001) Compromises between cardinality and ordinality in preference theory and social choice. Available at: http://fmwww.bc.edu/repec/es2000/1360.pdf
Mankiw NG (2014) Principles of Economics. Cengage Learning
Marshall A (1982) Princípios de economia. São Paulo: editora Abril cultural
Mill JS (1863) Utilitarianism (Capítulo 2). Hackett Publishing Company. Available at: https://spot.colorado.edu/~heathwoo/readings/mill.pdf
Mill JS (1863) Utilitarianism. London: Parker, Son, and Bourn. Available at: https://www.loc.gov/resource/gdcmassbookdig.utilitarianism01mill/?st=pdf
Obrimah OA (2022) The marginal utility for wealth: effects of either of heterogeneity of risk preferences or irrationality of risk seeking agents. Available at: https://doi.org/10.2139/ssrn.3985501
Rawls J (1971) A theory of justice. Harvard University Press, Harvard
Samuelson PA, Nordhaus WD (2010) Economics. McGraw-Hill, New York
Sanchirico CW (2008) Progressivity and potential income: measuring the effect of changing work patterns on income tax progressivity. C Law Rev 108:1551. Disponível em SSRN: https://ssrn.com/abstract=961310
Santos JR (2020) Economia indispensável. Lisbon International Press, Portugal
Santos JR, Sousa MJ (2022) Public administration and new information: increasing utility. In: Ullah A, Anwar S, Rocha Á, Gill S (eds) Proceedings of international conference on information technology and applications, Springer, pp 306–326. https://doi.org/10.1007/978-981-16-7618-5_33
Sen A (1977) Rational fools: a critique of the behavioral foundations of economic theory. Philos Public Aff 6(4):317–344. Available at: http://fs2.american.edu/dfagel/www/Class%20Readings/GeneralSocialScience/Rational%20Fools_%20A%20Critique%20of%20the%20Behavioral%20Foundations%20of%20Economic%20Theory.pdf.
Slemrod J, Slemrod J (1994) On the high-income Laffer curve. In: Tax progressivity and income inequality, Cambridge University Press, pp 177–210. https://doi.org/10.1017/CBO9780511571824.011
Stinespring JR (2006) Can tax cuts pay for themselves? an examination of dynamic scoring with public capital. Colorado College Working Paper No. 2006–05. Available at: https://doi.org/10.2139/ssrn.944068
Tavor T, Gonen LD, Spiegel U (2021) Reservations on the classical Laffer curve. Rev Austrian Econ 34:1–15. https://doi.org/10.1007/s11138-019-00483-8
Todorova T (2021) Diminishing marginal utility and the teaching of economics: a note. J Res Educ Sci 12(14):25–31. Available at: https://doi.org/10.14505/jres.v12.14.02.

Vellela SS, Babu BV, Mahendra YB (2024) IoT-based tank water monitoring systems: enhancing efficiency and sustainability. Int J Mod Trends Sci Technol 10(02):291–298. https://doi.org/10.1016/j.gltp.2021.08.062

Walther J, Sochacka NW, Benson LC, Bumbaco AE, Kellam N, Pawley AL, Phillips CM (2017) Qualitative research quality: a collaborative inquiry across multiple methodological perspectives. J Eng Educ 106(3):398–430. https://doi.org/10.1002/jee.20170

Weisbach D, Hemel DJ, Nou J (2018) The marginal revenue rule in cost-benefit analysis. Tax Notes 160(11):1507–1528. Available at: https://chicagounbound.uchicago.edu/cgi/viewcontent.cgi?article=2258&context=public_law_and_legal_theory.

White LH (2015) What economics can and cannot say about egalitarian redistribution. GMU working paper in economics, pp 15–48. Available at: https://doi.org/10.2139/ssrn.2686423.

Zerguit L (2012) The study of strategy-making process in UK Universities: an interpretative approach. In: European conference on research methodology for business and management studies, Academic Conferences International Limited, pp 527–535

The Role and Challenges of IT Project Management in the Portuguese Public Administration: An Analysis of the Adoption of PM² Methodology

16

Márcia Maria Pereira Pinheiro

Abstract

Administrative modernization and digital transformation have become central to improving operational efficiency, transparency, and accountability in public administration (PA). In the Portuguese context, the management of information technology (IT) projects faces significant challenges, including administrative complexity, strict legal requirements, the involvement of multiple stakeholders, and limited resources. These challenges often lead to delays, cost overruns, and unmet objectives. This study investigated IT project management practices and vulnerabilities within the Portuguese Public Administration, with a particular focus on the adoption of the PM² methodology developed by the European Commission. This methodology is designed to promote transparency, efficiency and alignment with European policies. The research adopted a quantitative methodology, using a structured questionnaire distributed across the Portuguese PA. The sample consisted of 200 participants, including managers and team members involved in IT projects, representing different academic backgrounds and different levels of experience. Descriptive and inferential statistical analyses revealed critical weaknesses in key areas such as time management, risk management and quality management, particularly during the planning, monitoring and control, and closing phases of projects. Recurring issues included the limited use of technological tools and insufficient specialized training. Notably, only 7.5% of respondents reported using the PM² methodology, highlighting cultural barriers and the urgent need for capacity building initiatives. On the other hand, agile methodologies, such as Agile and Scrum, were more widely adopted, albeit inconsistently and often without full integration. The

M. M. P. Pinheiro (✉)
Digitalization in Public Administration, State University Institute of Lisbon (ISCTE), Lisbon, Portugal
e-mail: marciapereirapinheiro@hotmail.com

M. J. Sousa et al. (eds.), *Economic Sustainability Potentiated by Artificial Intelligence*, Contributions to Environmental Sciences & Innovative Business Technology,
https://doi.org/10.1007/978-981-96-8089-4_16

findings highlight the need for organizational and cultural change to improve IT project management within the Portuguese PA. Recommendations include the promotion of the PM^2 methodology, the establishment of dedicated Project Management Offices (PMOs), the strengthening of project management training, and the adoption of advanced technological tools. The study provides a detailed diagnosis of current practices and challenges and presents actionable recommendations for public bodies such as the Shared Services Entity of Public Administration, I.P. (eSPap), which oversees shared services in Project Management Offices (PMOs). Ensuring alignment with international and European standards is highlighted as essential to improve the effectiveness of project management and support the ongoing digital transition in Portuguese public administration. Future research could explore qualitative analysis, international comparisons, and the impact of emerging technologies on project management practices.

Keywords

Information technology project management • European commission PM^2 methodology • Public administration • Administrative modernization • Digital transformation

16.1 IT Project Management in Portuguese Public Administration

Digital transformation has emerged as one of the main drivers of modernization in public administration (PA), playing a crucial role in improving operational efficiency, process transparency, accountability, and service delivery to citizens. This transformation process, while creating opportunities, also poses significant challenges to administrative modernization (Araújo and Azevedo 2018). In the Portuguese context, the management of information technology (IT) projects is of strategic importance, as it aligns organizational goals with technological requirements and the growing expectations of citizens (Project Management Institute [PMI] 2021). However, the implementation of such projects faces significant challenges, including administrative complexity, regulatory compliance, multiple stakeholders, and limited resources (PMI 2021). These obstacles often result in delays, cost overruns, and failure to meet initial objectives (Kerzner 2017; Flyvbjerg 2007).

The importance of IT project management is reinforced by the rapid pace of technological development and its interdisciplinary nature. Studies show that structured methodologies can reduce risk, improve cost and time control, and ensure that results are better aligned with organizational objectives (PMI 2021). Among these methodologies, PM^2, developed and recommended by the European Commission, stands out for its simplicity, adaptability, and adaptation to the specificities of the public sector and transnational projects (European Commission [EC] 2023).

In Portugal, the modernization of the public administration is a strategic priority, in line with European objectives for efficient government and digital transformation. However, the literature shows a limited adoption of structured methodologies in IT project management, such as PM^2, exacerbated by a lack of specific training and the absence of support structures, such as project management offices (PMOs), which are essential to coordinate and ensure the success of initiatives (Costa and Sousa 2021).

In this context, identifying existing weaknesses and proposing practical solutions is imperative for the development of Portuguese public administration, which is the guiding objective of this research. The need for greater efficiency and sustainability is accompanied by the urgency to meet citizens' expectations while optimizing public resources. Although the PM^2 methodology has been widely promoted by the European Commission as an efficient tool for managing public projects, there is a clear lack of studies evaluating its application in the Portuguese context. The current literature focuses mainly on the adoption of methodologies such as PMBOK®, PRINCE2, and SCRUM, neglecting specific analyses of the adoption and impact of PM^2 in public sector IT projects.

This research aims to fill this gap by addressing the following research question *What are the main weaknesses in IT Project Management in the Portuguese Public Administration and how can the PM^2 methodology contribute to mitigate them?* To this end, three specific objectives were outlined: (1) to assess the weaknesses in IT project management in the Portuguese public administration; (2) to identify the most common project management practices in the public sector; and (3) to analyze the level of use and knowledge of the PM^2 methodology among IT project managers.

The results of this research provide a diagnosis of the weaknesses and practices in project management within the Portuguese PA, accompanied by a set of practical recommendations that can be implemented by eSPap and other public entities. The choice of the PM^2 methodology is justified by its adoption by the European Commission and its suitability to the needs of the public sector, providing processes, templates, and guidelines that promote transparency, control, and efficiency in project execution (Vicente et al. 2018; EC 2020).

The findings are relevant to other European public administrations facing similar challenges and contribute to the debate on the importance of professional training and strengthening organizational structures, such as PMOs, to ensure the success of public sector projects.

16.2 Project Management in Public Administration

Project management is a discipline that combines the principles of planning, execution, and control with the goal of achieving specific results efficiently and effectively. In the public sector, project management takes on unique characteristics, marked by complex regulations, limited budgets, and the need to meet the demands of multiple stakeholders (Kerzner 2017). In contrast to the private sector, where financial returns are the priority, projects in Public Administration (PA) often aim to maximize social impact and promote transparency (PMI 2021). In this context, project success should be understood not only in terms of cost, time and quality but also in multidimensional terms that incorporate organizational learning and public value (Bannerman 2008).

Flyvbjerg (2007) identifies cost and schedule underestimation as one of the most common problems in public projects, resulting from inadequate planning and scope management. These challenges are exacerbated by an often risk-averse organizational culture and the lack of standardized tools to monitor and control project progress (Schwalbe 2020). In the case of the Portuguese public administration, these difficulties are exacerbated by the predominance of rigid hierarchical structures and the absence of a generalized project management culture, which often leads to failures in the execution of IT projects.

In addition, the diversity of stakeholders in PA—including citizens, local governments, suppliers, and regulatory entities—imposes significant challenges. Each group has different expectations, making communication and expectation management essential components for the success of public projects (Silva and Ribeiro 2020). In this regard, the literature on project management in the public sector highlights unique demands compared to the private sector, particularly the need for greater alignment with governance and accountability frameworks (Bottini 2012).

Project management methodologies are recognized as essential to project success, providing frameworks that help structure processes and mitigate risks. Among the methodologies most used in the public and private sectors are PMBOK®, PRINCE2, and SCRUM. PMBOK® (Project Management Body of Knowledge), published by the Project Management Institute (PMI), is a guide that standardizes best practices in project management, organizing them into ten knowledge areas such as scope, time, cost, quality, and risk management. Despite its widespread use, the PMBOK® is often criticized for its complexity, which can hinder its application in organizations with a low level of project management maturity (PMI 2021).

Developed in the UK, PRINCE2 (Projects in Controlled Environments) is widely used in the public sector because of its focus on structured processes and governance. This methodology emphasizes stakeholder roles and risk management and is particularly effective in public contexts where accountability is a priority. On the other hand, SCRUM, an agile approach, is characterized by its flexibility and iterative methodology and is widely used in software development projects. However, it has limitations in public contexts where more rigorous processes and robust governance structures are required (Schwaber and Sutherland 2020).

Although effective, these methodologies face challenges in adapting to the public context, where the training of the professionals involved is a crucial factor for their implementation. In this scenario, the PM^2 methodology, developed by the European Commission, emerges as a solution adapted to the specificities of the public sector, offering a simplified and adaptable approach that facilitates effective project management in compliance with European requirements. PM^2 integrates the practices of several recognized methodologies, promoting greater transparency, efficiency, and alignment with public policies, making it a relevant tool to address the challenges of project management in public administration.

16.3 European Commission PM^2 Methodology

The PM^2 methodology, developed by the European Commission, was created to address recurring weaknesses in public project management. This methodology combines best practices from different project management approaches and presents itself as a solution tailored to the specific needs of the public sector. It is characterized by simplicity, flexibility and ease of implementation, making it particularly suitable for European public institutions. PM^2 promotes harmonized communication and efficient project management, ensuring efficiency, transparency, and compliance with Community regulations, especially in projects funded by the European Union (European Commission 2023; Vicente et al. 2018).

The management of uncertainty, often present in public projects, reinforces the relevance of methodologies such as PM^2, which help mitigate risks in complex and volatile contexts (Geraldi 2019). The structure of PM^2 is based on four fundamental pillars. First, the governance model establishes clear roles and responsibilities between stakeholders, including the project owner, the project manager and the project team, promoting a transparent and effective organization. Second, the project lifecycle is divided into four main phases: initiation, planning, execution and closure, each with specific activities and artifacts. Third, management processes organize and control project progress and ensure alignment with strategic objectives. Fourth, artifacts and tools, such as the project handbook and progress reports, standardize, and facilitate effective communication throughout all project phases (European Commission 2023; Takagi and Varajão 2019). Figure 16.1 illustrates the PM^2 model.

The PM^2 project lifecycle consists of four main phases. Each phase plays a critical role in the success of the project. In the initiation phase, objectives are defined, the team is assembled, stakeholders are identified, and the project is formalized through the Project Initiation Request (PIR). During the planning phase, resources, schedules, and project-related risks are detailed in the project work plan (PWP). The execution phase focuses on carrying out the planned activities, ensuring continuous monitoring and producing progress reports. The Closure phase evaluates results, captures lessons learned, and formally closes the project. These phases are complemented by continuous monitoring and control activities to ensure risk identification and corrective actions (European Commission 2023; PM^2 Alliance 2020). Figure 16.2 shows the project life cycle, activities and artifacts.

PM^2 offers many benefits, including its simplicity, which makes it easy to implement in organizations with low levels of project management maturity, and its flexibility, which allows it to be adapted to different contexts and project types. It also promotes consistency and transparency by providing a common language and standardized practices, facilitating communication, and collaboration between

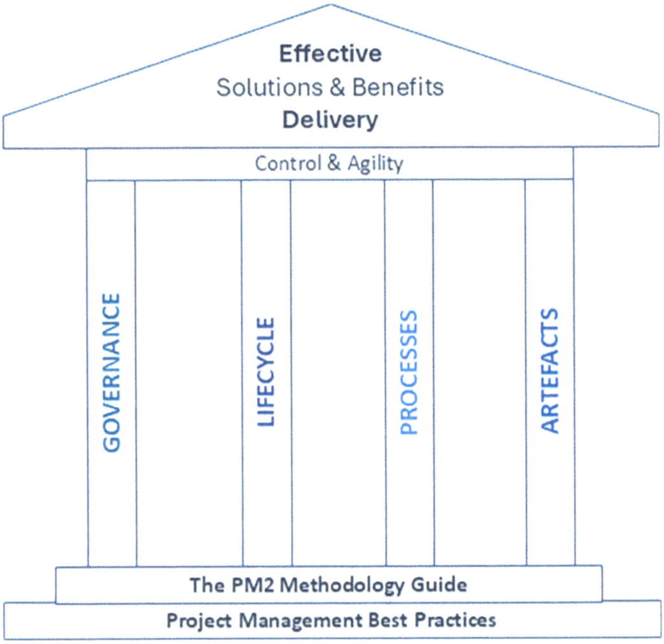

Fig. 16.1 The House of PM^2

stakeholders. The methodology also ensures clarity in defining roles and responsibilities, improving accountability and efficiency through recommended practices and appropriate tools (European Commission 2023; Vicente et al. 2018).

However, there are challenges in implementing PM^2. The need for training and capacity building may be an initial barrier, especially in organizations that are less familiar with structured methodologies (Silva and Ribeiro 2020). In addition, the effectiveness of the methodology may be limited if the organizational culture is not aligned with its principles, which may hinder full adoption. In smaller projects, PM^2 may be perceived as overly bureaucratic, requiring contextual adaptations to ensure its applicability (Costa and Sousa 2021; Martínez 2019).

Studies have highlighted that methodologies such as PM^2 have a positive impact on mitigating risks and increasing efficiency throughout the life cycle of projects (Xue et al. 2009; Krpan et al. 2023). PM^2 is particularly effective in projects funded by the European Union, where compliance with EU regulations is essential (European Commission 2023). Its structured approach covers the entire project lifecycle, promotes better outcomes and identifies areas for continuous improvement. Despite challenges related to training and cultural adaptation, the implementation of PM^2 provides a solid foundation for modernizing and improving public project management, positioning it as an indispensable solution in the European context (European Commission 2023; Takagi and Varajão 2019).

Fig. 16.2 Project lifecycle, activities, and artifacts

16.4 Knowledge Areas in Project Management

Project management is a multifaceted discipline that encompasses different areas of knowledge, each of which plays a crucial role in the planning, execution and effective control of projects in different organizational contexts. In the context of the Portuguese Public Administration, particularly in the management of information technology (IT) projects, the identification of weaknesses within each knowledge area is fundamental to the success of projects. The PM^2 methodology, although relatively new, integrates several knowledge areas that are also present in widely recognized methodologies such as PMBOK®, PRINCE2, and Agile, allowing its application in different organizational contexts (European Commission 2023; Takagi and Varajão 2019).

Different project management methodologies present specific approaches but share common areas that aim to ensure effective management. For example, PMBOK® is widely recognized as a global reference and provides a comprehensive structure that includes ten knowledge areas: integration, scope, time, cost, quality, human resources, communication, risk, procurement and stakeholder management (PMI 2021). In contrast, PRINCE2, developed in the UK and widely used in the public sector, focuses on areas such as business case management, project organization, progress control, quality management, risk management, and benefits realization. This approach is particularly effective in contexts where accountability and governance are central (OGC 2009). On the other hand, agile methodologies such as Scrum and Kanban prioritize iterative deliverables, continuous improvement and team management, making them particularly suited to dynamic contexts and software development environments (Schwaber and Sutherland 2017). In the Portuguese context, studies on project management maturity also point to structural limitations that condition the adoption of these methodologies (Carvalho 2019).

The PM^2 methodology, developed by the European Commission, takes an integrated approach, combining elements from several established methodologies. This integration allows PM^2 to cover not only traditional knowledge areas such as scope, time, cost, quality, resource and communication management, but also specific areas such as transition and configuration management. These areas reflect the specificities of projects funded by the European Union, where compliance and transparency are critical (European Commission 2023). Each knowledge area associated with PM^2 has defined processes, such as scope definition, schedule creation, risk planning, and continuous stakeholder monitoring, to ensure a structured and effective approach (European Commission 2023).

Each of these knowledge areas plays a critical role in project management, especially for complex projects such as IT initiatives. Identifying and mitigating weaknesses in specific areas, such as risk management or time management, is essential for developing strategies to ensure project success. This detailed analysis, coupled with the implementation of methodologies such as PM^2, strengthens the ability of public organizations to achieve their strategic goals and promotes efficiency and effectiveness in project delivery (Takagi and Varajão 2019; European Commission 2023).

16.5 Challenges and Critical Success Factors

The management of information technology (IT) projects in the Portuguese public administration poses significant challenges, affecting the efficient and effective implementation of initiatives essential for the modernization of the sector. Among the main obstacles is resistance to organizational change, often rooted in a culture that is averse to new methodologies and technologies, which limits the capacity for innovation and continuous improvement (Martins and Cunha 2015). This cultural resistance is often exacerbated by a lack of adequate training in project management, making it difficult to adopt structured methodologies such as PM^2 (Costa and Sousa 2021). Furthermore, the inherent complexity of bureaucratic processes and the need to align projects with public policies make planning and execution even more challenging (Ribeiro and Gomes 2019). Additionally, early studies on eGovernment already highlighted the difficulties of implementing technological projects in the public sector, associated with institutional and organizational factors (Heeks 2006).

Another factor that complicates IT project management in public administration is the multiplicity of stakeholders involved, each with different interests and expectations. Effective stakeholder management is crucial to avoid conflicts and misalignments that could put at risk the success of the project (Brookes and Clark 2009). In addition, compliance with legal and regulatory requirements adds another layer of complexity, requiring teams to be prepared to navigate strict regulatory environments (European Commission 2020).

The lack of specialized skills also emerges as a critical challenge. Ongoing training is essential to enable managers and teams to apply modern methodologies and drive execution in line with the strategic objectives of public organizations (Vicente et al. 2018). Without this training, project efficiency can be severely compromised.

The success of IT projects in both the public and private sectors depends on several critical factors. The selection and adaptation of project management methodologies appropriate to the type of project are fundamental elements. In the private sector, agile methodologies such as Scrum and Kanban have proven effective in enabling rapid adaptation to change. In public administration, the PM^2 methodology has been widely adopted to ensure compliance with European standards and align projects with public policy (Kloppenborg and Tesch 2017; European Commission 2020). Empirical evidence also demonstrates that professionalized project management has a direct positive impact on organizational performance (Mir and Pinnington 2014).

Other success factors include clear requirements definition, careful planning, adequate team training, and strategic alignment of projects with organizational goals. Unclear requirements or frequent changes during development can have a negative impact on project outcomes (Levinson 2015). In addition, failures in activity, resource and schedule planning often lead to delays and additional costs (Kerzner 2017). Training and skills development in IT project management are therefore essential to mitigate these challenges and address the technical and operational specificities of projects in public administration and the private sector (Ribeiro and Gomes 2019).

The lack of project alignment with the strategic goals of the organization is another factor that increases the likelihood of failure, as it reduces the impact and value delivered by projects (Liu et al. 2016). In addition, continuous support from senior management and effective stakeholder management plays an essential role in ensuring the success of IT projects (Martins and Cunha 2015). Finally, the early identification of risks and the implementation of proactive management strategies are essential to minimize adverse impacts and ensure that projects are delivered on time and within budget (Brookes and Clark 2009). This set of factors highlights the need for integrated approaches and robust strategies to overcome the challenges and ensure the success of IT projects in the Portuguese public administration.

16.6 Project Management Practices: Results of Quantitative Analysis

The research carried out on project management practices in the Portuguese public administration was designed with methodological rigor, ensuring the robustness and validity of the results. The study was applied in nature, aiming to provide practical solutions to specific problems in the management of information technology (IT) projects. Classified as descriptive and exploratory, it used quantitative methods to characterize practices, challenges and opportunities in the public sector and to better understand the adoption of the PM^2 methodology (Sousa and Baptista 2011; Gil, 2019). The choice of the questionnaire survey as a data collection instrument is justified by its ability to obtain standardized and comparable data in large samples (Hill and Hill 2008).

The quantitative approach proved appropriate due to its ability to collect and analyze numerical data, allowing the identification of patterns and hypothesis testing (Hair et al. 2010). Data were collected using a structured questionnaire, an efficient tool for obtaining standardized and comparable responses (Creswell 2014). The target population included professionals involved in IT project management in the Portuguese public administration, using the database of the State Organizational Information System (SIOE), managed by the Directorate General of Administration and Public Employment (DGAEP). After exclusions, a final sample of 1964 entities was obtained, representing different ministries and departments (Hair et. al. 2010).

The questionnaire was divided into four sections. The first collected demographic data such as age, gender and educational background. The second explored respondents' familiarity with project management methodologies, including PM^2. The third examined management practices throughout the project lifecycle, using Likert scales to assess frequency and challenges. Finally, the fourth section focused on weaknesses encountered, also allowing for open-ended responses for greater analytical depth (Sousa and Baptista 2011). The clarity of the questions was previously evaluated by 13 professionals, and suggested changes were implemented to ensure the quality of the data collection instrument.

Data analysis was carried out using SPSS and Excel software, using descriptive and inferential analysis to identify significant patterns and relationships (Field 2020; Pallant 2020). The research adopted the Generic Roadmap Framework model, adapted from Phaal et. al. (2004) and Albright (2009), which structured the work around identifying weaknesses, improving transparency and promoting efficiency in public services.

The demographic analysis of the respondents showed a significant distribution across sectors, with the highest representation in the Municipal Administration (38.19%) and the Ministry of Finance (24.12%). In terms of professional roles, most respondents were "project team members" (34.5%) and "project managers" (28%), reflecting a focus on project delivery and direct coordination, which influences perceptions of management practices. The age distribution showed a predominance of professionals aged 50–59 (39.5%), followed by the 40–49 age group (36.5%), indicating a sample of experienced professionals at an advanced stage of their careers. Figure 16.3 shows the distribution of respondents by role and age.

In terms of gender, most respondents were male (56%), with a similar age distribution for both sexes, concentrated mainly in the 40–49 and 50–59 age groups. Figure 16.4 shows the distribution of respondents by gender.

The academic background of the respondents also deserves attention, with 88% holding at least a bachelor's degree. The most common degrees were Bachelor's (33.0%), Master's (29.5%), and Postgraduate Diplomas (25.5%). This qualification profile shows that professionals involved in the management of IT projects in the PA have a solid base of technical and theoretical knowledge, which is essential to meet the challenges associated with these projects. Figure 16.5 shows the educational background of the respondents.

In addition, 50.5% of respondents have more than 10 years of experience in the field, indicating a correlation between experience and roles with greater responsibility, such as project manager and director. Table 16.1 shows the years of experience in project management and the roles held.

In terms of project management practices, 79.5% of respondents reported that their organization undertook IT projects. However, only 31.5% had a dedicated project management department and 23.5% had a project management office (PMO). Figure 16.6 shows the respondents' project management practices.

The analysis of management practices revealed differences between organizations with dedicated departments or PMOs and those without such structures. The presence of these units is associated with a higher adoption of formal methodologies and more consistent practices. For example, 35.22% of organizations undertaking IT projects have a dedicated department, compared to only 17% of organizations not undertaking IT projects. Figure 16.7 illustrates the project management practices of organizations that undertake IT projects compared to those that do not.

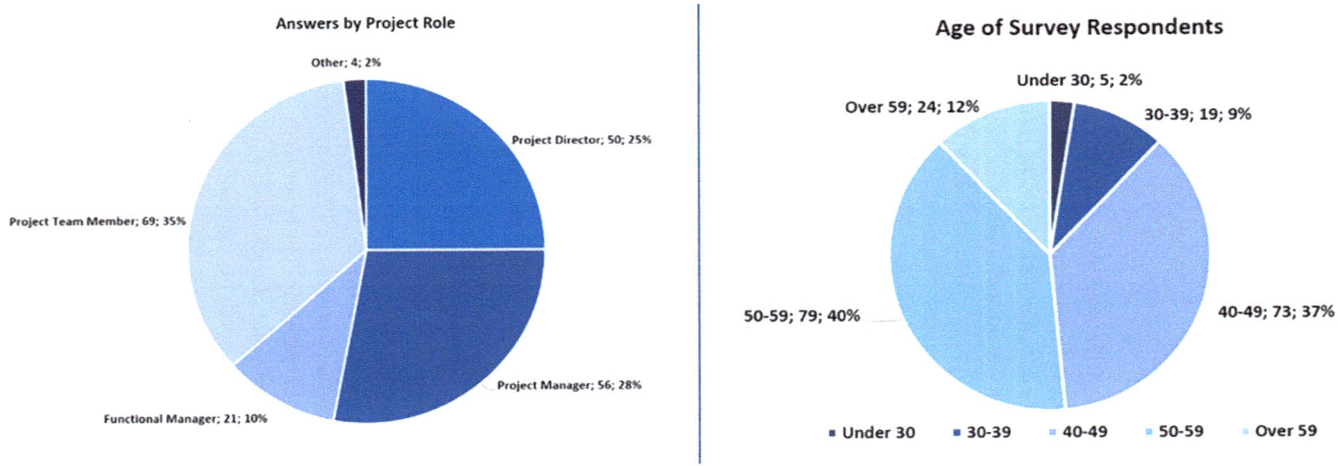

Fig. 16.3 Distribution of respondents by roles and ages

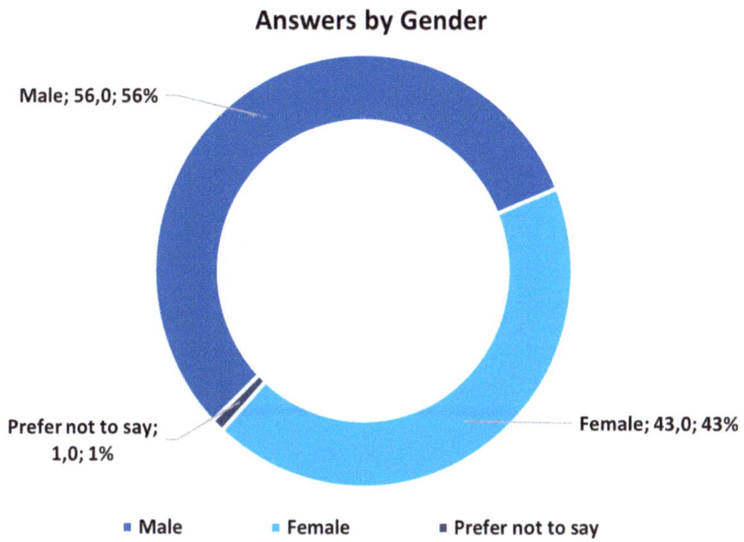

Fig. 16.4 Distribution of respondents by gender

Entities that implement IT projects are more likely to use external managers (57.23%) than those not undertaking IT projects (41.46%).

Adoption of formal project management methodologies was relatively low, with 53% of respondents reporting sporadic or no use. Among the most used methodologies, agile approaches such as Agile (25%) and Scrum (18.5%) stood out, while traditional methodologies such as PRINCE2 (0.5%) and Six Sigma (2.5%) showed marginal adoption. The PM^2 methodology, adopted by 7.5% of respondents, is still not widely used compared to others such as the PMBOK (16.5%). Its adoption may be linked to organizations seeking alignment with specific European standards. Figure 16.8 illustrates the adoption of methodologies.

Statistical analysis revealed that specific training is a determining factor in the adoption of methodologies, with significant positive correlations between training and methodologies such as PMBOK ($r = 0.645$, $p < 0.001$), PM^2 ($r = 0.645$, $p < 0.001$), PRINCE2 ($r = 0.476$, $p < 0.001$), Waterfall ($r = 0.523$, $p < 0.001$), Agile ($r = 0.430$, $p < 0.001$), Scrum ($r = 0.424$, $p < 0.001$), and Kanban ($r = 0.483$, $p < 0.001$). In addition, a correlation was observed between complementary methodologies such as Agile and Scrum, suggesting the combined use of these approaches in certain contexts.

The motivations for adopting the PM^2 methodology were analyzed and revealed that its use is strongly driven by external requirements, such as European Commission recommendations, with less emphasis on factors such as the effectiveness or ease of use of the methodology. This pattern of adoption suggests that regulatory compliance plays a critical role in the decision to use PM^2, while operational and technical aspects are secondary. For those adopting PM^2, compatibility with other

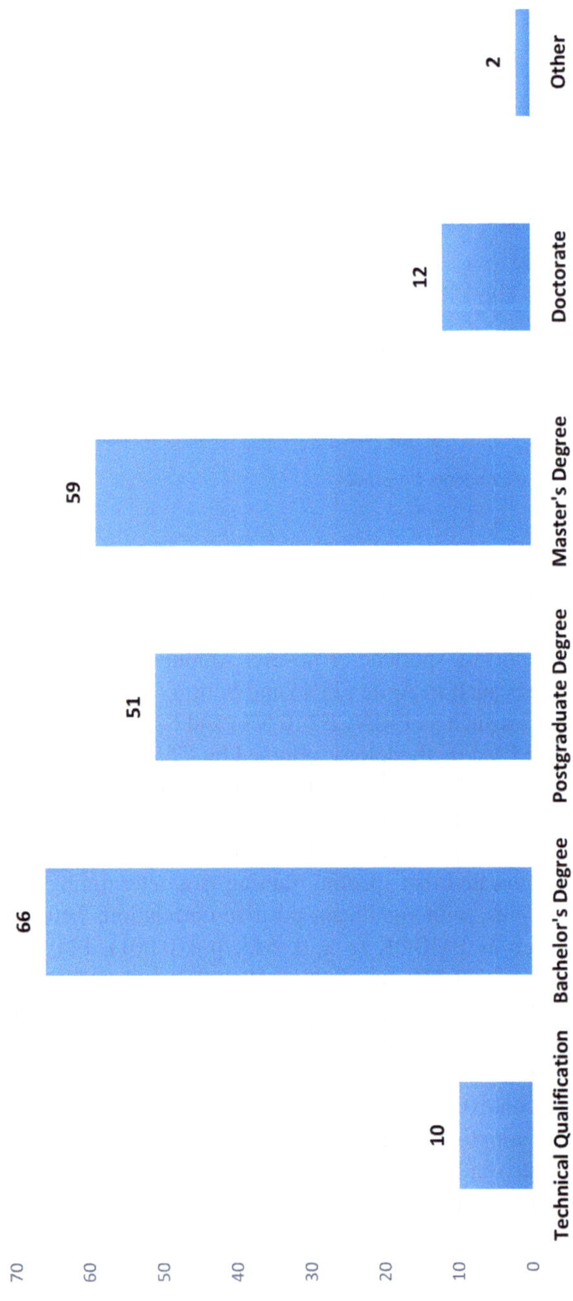

Fig. 16.5 Academic qualifications of respondents

Table 16.1 Roles by years of experience in project management

Roles	None	Less than 1 year	1–3 years	4–6 years	7–10 years	More than 10 years	Total
Project director	5	0	3	4	4	34	50
Project manager	2	3	6	11	4	30	56
Functional manager	3	2	1	6	1	8	21
Project team member	20	2	11	4	7	25	69
Other	0	0	0	0	0	4	4
TOTAL	30	7	21	25	16	101	200

methods and employer recommendations are also important factors. Table 16.2 shows the reasons that led respondents to adopt the PM^2 methodology.

The data show that the PM^2 methodology is little used by the respondents, with the majority stating that they do not use it, reinforcing the idea that this methodology is still in the growth phase of adoption.

The analysis of weaknesses in the knowledge areas revealed significant issues that threaten the efficiency and success of projects. Among the most critical areas were time management (4.42) and risk management (4.29), which respondents rated on a scale of 1–7.

Time management proved problematic due to the lack of realistic schedules and limited use of technology tools, often resulting in delays that negatively affected all subsequent phases of the project lifecycle. Risk management was found to be deficient due to the lack of formal management plans and proactive mitigation strategies, exposing projects to uncertainties that could have been avoided.

Weaknesses were also found in other areas. Procurement and contract management, for example, showed significant variation, with a variance of 3.7, reflecting the lack of standardization in contracting processes. Human resources management, on the other hand, showed difficulties in properly allocating competencies and coordinating teams, contributing to work overload and operational inefficiencies. Although less critical, areas such as integration management and communication management also showed gaps, particularly in documentation and information. The analysis of the project lifecycle identified weaknesses in all phases.

In the initiation phase, there was a lack of clarity in scope definition and insufficient stakeholder involvement, with average scores of 3.45–3.44, respectively. These shortcomings threaten the foundation of the project, leading to scope deviations, conflicts, and difficulties in subsequent phases.

The planning stage was the most problematic, with low use of project management tools (average score 2.58) and the absence of formal quality (2.44) and

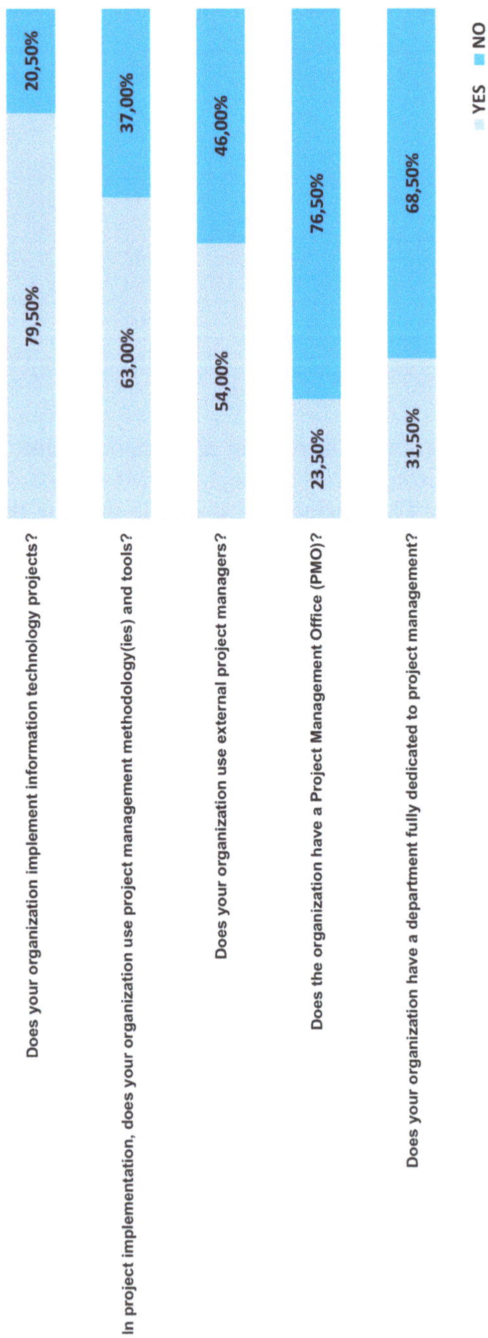

Fig. 16.6 Project management practices

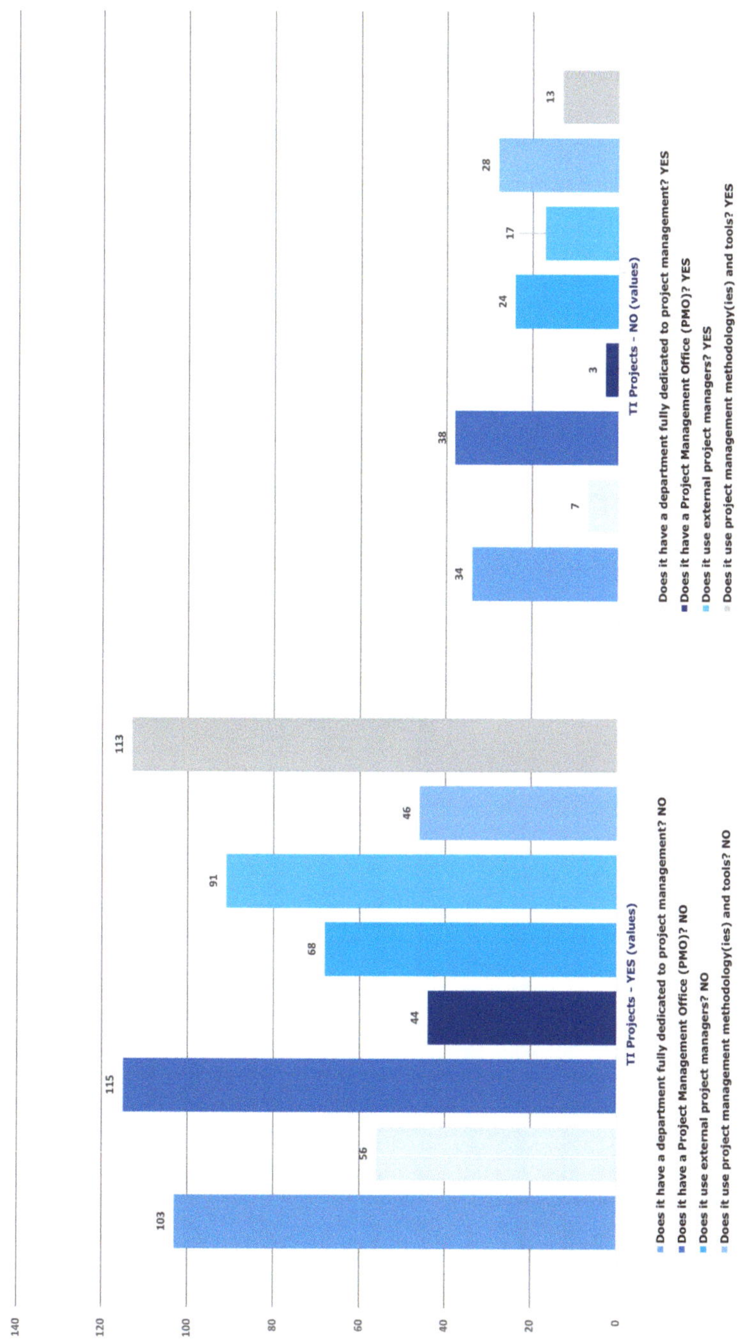

Fig. 16.7 Project management practices

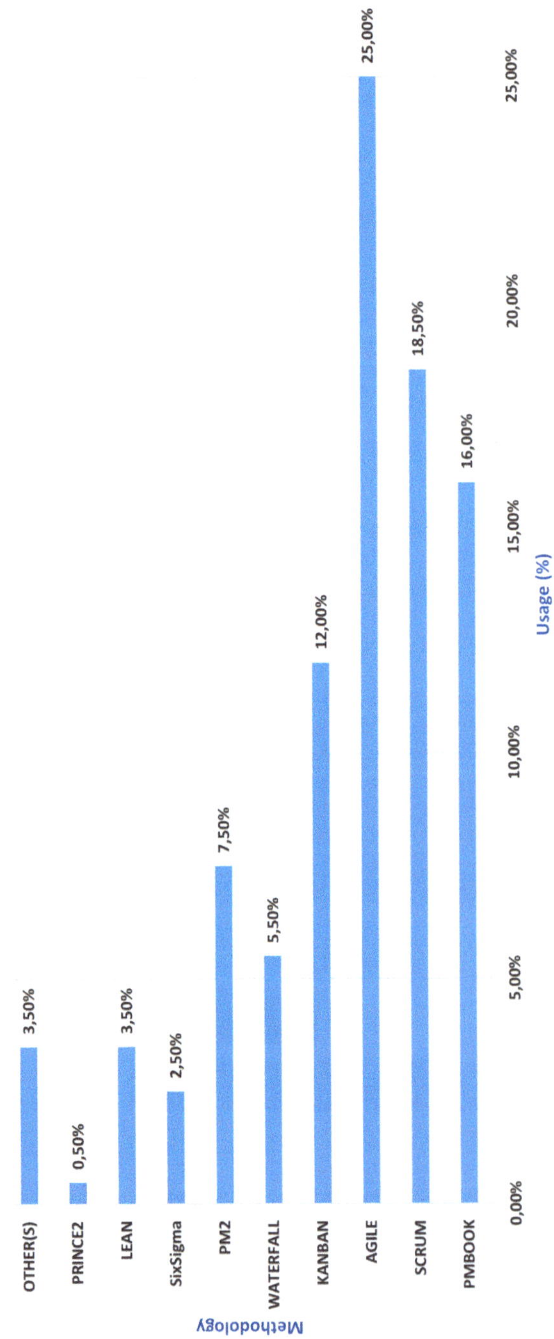

Fig. 16.8 Adoption of methodologies

Table 16.2 Reasons for adopting the PM^2 methodology

	I don't use PM^2	EU recommendation	Recommendation from my entity	Proven effectiveness	Ease of use	Compatibility with other methodologies
Valid	200	200	200	200	200	200
Omitted	0	0	0	0	0	0
Mean	0.88	0.1	0.05	0.02	0.03	0.03
Median	1	0	0	0	0	0
Mode	1	0	0	0	0	0
Error deviation	0.332	0.301	0.208	0.122	0.157	0.171
Variance	0.11	0.09	0.043	0.015	0.024	0.029
Sum	175	20	9	3	5	6

risk (2.62) management plans, jeopardizing the ability to anticipate and manage problems.

At the implementation stage, weaknesses were identified in relation to ineffective management of human resources, with an average score of 3.18, and inadequate quality reviews, with an average score of 2.50, resulting in outputs that often did not meet expected standards.

The monitoring and control phase revealed inconsistencies in risk management (2.80) and communication (2.63), resulting in difficulties in maintaining effective control over project progress. Finally, the closure phase revealed significant gaps in the documentation of lessons learned (2.76) and in the validation of deliverables, limiting continuous improvement and the use of institutional knowledge for future projects.

The analysis of the correlations between the knowledge areas and the life cycle phases revealed significant interdependencies. Specific weaknesses in knowledge areas tend to have a negative impact on the corresponding project phases, reinforcing the need for integrated approaches.

In the initiation phase, difficulties in scope management ($r = 0.471$, $p < 0.01$) and stakeholder management ($r = 0.443$, $p < 0.01$) correlate with problems in the initiation phase. The lack of clear objectives and inadequate stakeholder involvement undermines the foundation of the project, leading to scope deviations and conflicts during development.

At the planning stage, weaknesses in time management ($r = 0.430$, $p < 0.01$), risk management ($r = 0.462$, $p < 0.01$), and quality management ($r = 0.415$, $p < 0.05$) are directly related to planning problems. The lack of management tools and formal plans prevents effective planning and exposes projects to delays and uncertainties.

During execution, weaknesses in human resource management ($r = 0.398$, $p < 0.05$) and quality management ($r = 0.410$, $p < 0.05$) emerge as critical factors. Inadequate management of human resources leads to work overload and low efficiency, while the lack of regular audits jeopardizes compliance with deliverables.

Monitoring and control ensure that the project stays on track, but data on risk management ($r = 0.463$, $p < 0.01$) and communication management ($r = 0.417$, $p < 0.05$) show a significant relationship with difficulties at this stage. Effective monitoring depends on integrated practices and standardized processes, which are still inconsistent in many public administrations.

In the final phase, weaknesses in integration management ($r = 0.389$, $p < 0.05$) and transition management ($r = 0.418$, $p < 0.05$) jeopardize the successful completion of projects. The lack of documentation of lessons learned limits continuous improvement, while the lack of formal procedures affects knowledge transfer and business continuity. Weaknesses in this phase reflect a lost opportunity to consolidate good practices and to use lessons learned in future projects.

The analysis shows that specific weaknesses in knowledge areas have a transversal impact on all phases of the project life cycle. Specific efforts in these areas, combined with the adoption of standardized practices and improved professional training, are essential to strengthen the project management capacity of the Portuguese public administration. In addition, the systematic documentation of lessons learned, and the adoption of robust technological tools can reduce risks and increase project effectiveness.

The open comments reinforce the statistical conclusions, pointing to the need for greater clarity in defining objectives and early involvement of stakeholders. Resistance to adopting new methodologies such as PM^2 was frequently mentioned, highlighting the importance of training and cultural change.

16.7 Discussion and Implications

This research provides a comprehensive and detailed overview of the practices and vulnerabilities associated with the management of Information Technology (IT) projects in the Portuguese Public Administration (PA), aligning the findings with the specific objectives proposed and the existing literature. The discussion focuses on the practical and theoretical implications of the findings, exploring the structural and organizational factors that influence the adoption of project management methodologies and performance throughout the project lifecycle.

About the first specific objective, which aimed to identify weaknesses in IT project management, the results highlighted time management and risk management as the most critical areas, with mean scores of 4.42–4.29, respectively. These findings are in line with studies by Flyvbjerg (2007) and Kerzner (2017), which highlight the relevance of these dimensions to project effectiveness. Respondents indicated that the lack of appropriate technological tools and inadequate training in risk management are significant gaps that undermine effectiveness. These challenges, which are widely discussed in the literature (Schwalbe 2020), result in delays, cost overruns, and substandard deliverables. Furthermore, the analysis revealed that these weaknesses particularly affect the planning and monitoring phases, underlining the need for targeted interventions at these stages.

Regarding the second specific objective, which analyzed the most used project management practices, the results showed that the implementation of IT projects is a common practice in 79.5% of the PA units. However, only 53% of professionals reported frequent use of formal methodologies, highlighting the low frequency of structured application. These findings are consistent with the literature, which identifies cultural resistance and lack of training as significant barriers to the adoption of methodologies (Martins and Cunha 2015; Vicente et al. 2018). Agile methodologies such as Agile (25%) and Scrum (18.5%) stand out as the most widely used, while traditional approaches such as PRINCE2 (0.5%) and PMBOK (16.5%) have a lower prevalence. These results reflect a growing trend toward flexible approaches in the public sector, but also reveal gaps in training and awareness of the benefits of structured methodologies (Schwaber and Sutherland 2020).

Regarding the third specific objective, which aimed to assess the level of use and knowledge of the PM^2 methodology, the results indicate that this methodology is still underutilized (7.5%), despite being promoted as a solution tailored to the needs of the European public sector (European Commission 2023). Its low uptake suggests the existence of cultural and operational barriers, in line with the challenges highlighted by Silva and Ribeiro (2020). The analysis also showed that training is a determining factor for the adoption of PM^2, supporting studies that highlight the importance of equipping managers to apply structured methodologies (Takagi and Varajão 2019). Furthermore, significant correlations were found between specific training and the use of methodologies such as PMBOK ($r = 0.645$, $p < 0.001$) and Agile ($r = 0.430$, $p < 0.001$), reinforcing the role of continuous training in promoting effective practices (Kloppenborg and Tesch 2017). Beyond the methodological perspective, the literature on programme management also emphasizes the importance of linking projects with strategic programmes, ensuring organizational alignment and sustainability (PM^2 Alliance 2021).

The analysis of the project lifecycle revealed weaknesses in all phases. In the Initiation phase, issues related to inadequate scope definition and insufficient stakeholder involvement were prominent, in line with literature highlighting these challenges in public administration (Martins and Cunha 2015). In the Planning phase, the low use of technological tools and the absence of formal quality and risk management plans undermined the ability to anticipate and address problems, as highlighted by Brookes and Clark (2009). In the Execution phase, ineffective human resource management and the lack of quality audits were identified as structural weaknesses (Levinson 2015). Finally, in the Closure phase, inadequate documentation of lessons learned, and insufficient validation of deliverables reflected a missed opportunity to consolidate best practice and promote continuous improvement (Krpan et al. 2023).

The findings highlight several practical implications. First, the need to promote cultural and organizational change to support the adoption of structured methodologies such as PM^2 is evident. Furthermore, the low frequency of use of the methodology suggests that public policies and training programs need to be restructured to address gaps in critical areas such as risk management, time management, and quality. The establishment of PMOs could provide an effective solution to many of these challenges by promoting consistency and efficiency in processes.

The qualitative analysis of open-ended comments reinforces these conclusions, highlighting the importance of clarifying objectives from the outset and involving all stakeholders. These observations, in line with the quantitative results, underline the need for organizational and cultural changes that support more effective management practices.

Finally, the practical implications of these findings are clear: the adoption of standardized practices, such as PM^2, and investment in continuous training are essential to mitigate the identified weaknesses and promote more efficient and sustainable project management. In addition, the establishment of PMOs and the use of robust technology tools can help to improve the transparency, efficiency and consistency of project management, thereby meeting the growing demands for efficiency in the public sector.

16.8 Conclusion

This study offers a significant contribution to understanding the practices, vulnerabilities, and potential of information technology (IT) project management in Portuguese Public Administration (PA). The findings indicate that the effectiveness of project management in PA is often compromised by structural and operational weaknesses, particularly in critical areas such as time management and risk management. These dimensions were identified as major challenges resulting from factors such as the lack of adequate technological tools and insufficient specialized training. In addition, certain phases of the project lifecycle, notably planning and monitoring and control, were found to be particularly vulnerable to inconsistent practices and lack of standardized approaches, undermining the overall success of projects.

While the use of project management methodologies is present in some PA entities, the results indicate that their application remains sporadic and limited. Agile methodologies such as Agile and Scrum emerged as the most widely used, while traditional approaches such as PRINCE2 and PMBOK showed significantly lower adoption rates. The PM^2 methodology, promoted by the European Commission as a solution tailored to the public sector, showed limited adoption, with only 7.5% of respondents reporting its use. This low level of adoption reflects challenges related to awareness, training and cultural adaptation required for effective implementation, in line with previous studies (European Commission 2023; Silva and Ribeiro 2020).

The limitations of this study must be acknowledged. The moderate response rate (10.31%), although statistically valid, limits the generalizability of the findings to the Portuguese PA population. In addition, the reliance on self-reported data may have introduced response bias, influencing the perceptions reported by participants. Another limiting factor was the focus on specific methodologies, such as PM^2, Agile and PMBOK, which excluded the analysis of other potentially relevant approaches in different organizational contexts. Finally, the specific context of the Portuguese PA limits the direct applicability of the conclusions to other countries or sectors, which requires caution in the interpretation of the results.

Based on the findings and identified limitations, this study suggests promising directions for future research. Complementary qualitative research, such as interviews with project managers and teams, could provide deeper insights into the challenges and practices observed. Expanding the sample with strategies to increase the participation of a larger number of organizations would strengthen the conclusions. Longitudinal studies could track the implementation of the proposed recommendations, allowing an assessment of their impact over time. In addition, international comparative analysis would be valuable in identifying best practice in other European countries, particularly those that have adopted the PM^2 methodology. Future studies could also explore the impact of emerging technologies, such as artificial intelligence and Big Data, in improving IT project management, opening new avenues for innovation in the field.

The findings of this study highlight the urgent need for cultural and organizational change in project management within the Portuguese public administration. The adoption of standardized practices, such as the PM^2 methodology, together with improved professional training and the implementation of robust technological tools, is an essential step in addressing the identified weaknesses. These changes are essential to ensure more efficient and sustainable project management. Implementing the recommendations proposed in this study has the potential to make a significant contribution to a more efficient, transparent and digitally enabled public administration, promoting modernization and effectiveness in the delivery of public services.

References

Albright RE (2009) Innovation roadmapping: how visual planning tools can help align business and technology for improved performance. Technology Futures Inc.

Araújo J, Azevedo R (2018) Digital transformation in public administration: challenges and opportunities. Public Adm Rev 78(6):1234–1246

Bannerman PL (2008) Defining project success: a multilevel framework. In: Proceedings of the PMI research conference. Project Management Institute

Bottini R (2012) Gerenciamento de projetos no setor público. Encontro Nacional de Gerenciamento de Projetos, Cuiabá, MT

Brookes NJ, Clark R (2009) Using maturity models to improve project management practice. Proj Perspect 31(1):34–39

Carvalho MM (2019) Metodologia de gestão de projetos: Aplicações no setor público e privado. Atlas

Costa A, Sousa J (2021) Project management practices in the Portuguese public sector: a maturity perspective. Int J Proj Manag 39(5):451–463
Creswell JW (2014) Research design: qualitative, quantitative, and mixed methods approaches, 4th edn. SAGE Publications
European Commission (2020) PM^2 project management methodology guide—Open edition v3.0. Publications Office of the European Union
European Commission (2023) PM^2 project management methodology guide—Open edition v3.1. Publications Office of the European Union
Field A (2020) Discovering statistics using IBM SPSS statistics, 5th edn. SAGE Publications
Flyvbjerg B (2007) Cost overruns and demand shortfalls in urban rail and other infrastructure. Transp Plann Technol 30(1):9–30
Geraldi J (2019) Project management and uncertainty: managing the unknown in projects. Routledge
Hair JF, Black WC, Babin BJ, Anderson RE (2010) Multivariate data analysis, 7th edn. Pearson
Heeks R (2006) Implementing and managing eGovernment: an international text. SAGE Publications
Hill MM, Hill A (2008) Investigação por questionário, 2.ª edn. Edições Sílabo
Kerzner H (2017) Project management: a systems approach to planning, scheduling, and controlling, 12th edn. Wiley
Kloppenborg TJ, Tesch D (2017) Project success and executive sponsor behaviors: empirical life cycle stage investigations. Project Management Institute
Krpan D, Milinković D, Šimunović I (2023) PM^2 methodology as a driver of project success in EU-funded projects. J Contemp Manag Issues 28(1):45–63
Levinson M (2015) Project management for the real world: case studies in IT and business. Harvard Business Review Press
Liu L, Wang Y, Wilkinson S (2016) Identifying critical factors affecting the effectiveness and efficiency of tendering processes in public projects. Int J Proj Manag 34(1):1–11. https://doi.org/10.1016/j.ijproman.2015.08.004
Martínez J (2019) Implementation challenges of PM^2 in small and medium organizations. Eur J Proj Manag 7(2):65–78
Martins LL, Cunha MP (2015) Organizational change and innovation in public administration: the role of project management. Revista de Administração Pública 49(3):709–732
Mir FA, Pinnington AH (2014) Exploring the value of project management: linking project management performance and project success. Int J Proj Manag 32(2):202–217
Office of Government Commerce (OGC) (2009) Managing successful projects with PRINCE2, 5th edn. The Stationery Office
Pallant J (2020) SPSS survival manual: a step by step guide to data analysis using IBM SPSS, 7th edn. Open University Press
Phaal R, Farrukh C, Probert D (2004) Technology roadmapping—a planning framework for evolution and revolution. Technol Forecast Soc Change 71(1–2):5–26
PM^2 Alliance (2021) PM^2 programme management guide. PM^2 Alliance
Project Management Institute (PMI) (2021) A guide to the project management body of knowledge (PMBOK® guide, 7th edn. Project Management Institute
Ribeiro F, Gomes J (2019) Critical success factors for IT projects in public administration: insights from Portugal. Int J Inf Syst Proj Manag 7(3):65–81
Schwalbe K (2020) Information technology project management, 9th edn. Cengage Learning
Schwaber K, Sutherland J (2017) The Scrum guide: the definitive guide to Scrum: the rules of the game. Scrum.org
Silva A, Ribeiro P (2020) Stakeholder management in public sector projects: a Portuguese case study. J Public Adm Policy Res 12(4):101–113
Sousa M, Baptista C (2011) Metodologia de investigação científica em ciências sociais e humanas. Gradiva
Takagi H, Varajão J (2019) PM^2 methodology in practice: case studies from European Union projects. Procedia Compute Sci 164:523–532. https://doi.org/10.1016/j.procs.2019.12.209

Vicente A, Varajão J, Amaral L (2018) PM^2 methodology in practice: a case study from a Portuguese public institution. Procedia Comput Sci 138:654–662. https://doi.org/10.1016/j.procs.2018.10.089

Xue X, Shen Q, Ren Z (2009) Critical review of collaborative working in construction projects: business environment and human behaviors. J Manag Eng 26(4):196–208. https://doi.org/10.1061/(ASCE)ME.1943-5479.0000025

Index

M. J. Sousa et al. (eds.), *Economic Sustainability Potentiated by Artificial Intelligence*, Contributions to Environmental Sciences & Innovative Business Technology,
https://doi.org/10.1007/978-981-96-8089-4